# Lecture Notes in Mobility

*Series Editor*

Gereon Meyer

For further volumes:
http://www.springer.com/series/11573

Jan Fischer-Wolfarth · Gereon Meyer
Editors

# Advanced Microsystems for Automotive Applications 2013

## Smart Systems for Safe and Green Vehicles

Springer

*Editors*
Jan Fischer-Wolfarth
Gereon Meyer
VDI/VDE Innovation + Technik GmbH
Berlin
Germany

ISSN 2196-5544 ISSN 2196-5552 (electronic)
ISBN 978-3-319-00475-4 ISBN 978-3-319-00476-1 (eBook)
DOI 10.1007/978-3-319-00476-1
Springer Cham Heidelberg New York Dordrecht London

Library of Congress Control Number: 2013936959



Printed on acid-free paper

Springer is part of Springer Science+Business Media (www.springer.com)

# Preface

Strategies for the sustainable development of road transport like those described in the European white paper are built on two major long-term visions: zero road fatalities and a shift away from conventionally fuelled vehicles. Such goals will significantly influence the priorities of public funding programs for research and innovation as well as the regulatory framework related to road transport, and thus have an impact on the activities of the automotive industry. From the perspective of policy makers, electric vehicles appear to be the most promising alternative power train option, and it can be argued that autonomous vehicles would seriously increase road safety. The introduction of these technologies, however, faces hurdles in terms of liability and customer acceptance. Hence, from the point of view of the automotive industry, a broader portfolio of technologies has to be developed and deployed at the right time, supported by appropriate incentives, legal frameworks and infrastructures. A well-balanced match of policy goals and industrial priorities is reflected, e.g. in the focus of the European Green Vehicles Initiative Public–Private Partnership on energy efficiency of vehicles and alternative powertrains.

The automobile of the future will particularly embrace innovations from three major automotive technology fields: driver assistance systems, vehicle networking and alternative propulsion. Smart systems such as adaptive ICT components and MEMS devices, novel network architectures, integrated sensor systems, intelligent interfaces and functional materials form the basis of these features and permit their successful and synergetic integration. They appear to be the key enabling technologies for safe and green vehicles including autonomous and electric vehicles, and they may play a key role where the two worlds meet, e.g. in safety systems like electronic stability control adapted to hybrid or electric power trains. Representing a domain of particular strength of the European industry, they may significantly contribute to the unique selling proposition of the automobile made in Europe in the future.

For more than 15 years the International Forum on Advanced Microsystems for Automotive Applications (AMAA) has been successful in detecting novel trends and in discussing the technological implications from early on. The topic of the AMAA 2013, held in Berlin on June 17–18, 2013, is “Smart Systems for Safe and Green Vehicles”. The organisers are VDI/VDE Innovation + Technik GmbH and the European Technology Platform on Smart Systems Integration (EPoSS), supported by the EU-funded Coordination Actions CAPIRE and Smart EV-VC.

This book contains peer-reviewed papers written by leading engineers and researchers who presented their ongoing research and novel developments at the AMAA 2013 conference. It is published worldwide as part of the new book series Lecture Notes in Mobility that reports on the latest advances in research, development and innovations for the intelligent, connected and sustainable transportation systems of the future.

In our roles as co-chairman and chairman of the AMAA 2013 we would like to thank all the authors for preparing most excellent contributions to the conference and to the book. We gratefully acknowledge the important help by the members of the AMAA Steering Committee in selecting the best papers for this publication. And, last but not least we would like to point out that the conference and this book are built on the tremendous support we received from our colleagues at VDI/VDE-IT, in particular Beate Müller, Frauke Bierau and Christian Martin.

Berlin, June 2013

Jan Fischer-Wolfarth
Gereon Meyer

# Advanced Microsystems for Automotive Applications 2013

## Funding Authority

European Commission

## Supporting Organisations

European Council for Automotive R&D (EUCAR)

European Association of Automotive Suppliers (CLEPA)

Strategy Board on Electric Mobility (eNOVA)

Zentralverband Elektrotechnik- und Elektronikindustrie e.V. (ZVEI)

Mikrosystemtechnik Baden-Württemberg e.V.

Hanser Automotive

## Organisers

European Technology Platform on Smart Systems Integration (EPoSS)

Coordination Action "Smart Electric Vehicle Value Chains" (SMART EV-VC)

Coordination Action "PPP Implementation for Road Transport Electrification" (CAPIRE)

VDI|VDE Innovation + Technik GmbH

## Steering Committee

| | |
|---|---|
| Mike Babala | TRW Automotive, Livonia MI, USA |
| Serge Boverie | Continental AG, Toulouse, France |
| Geoff Callow | Technical & Engineering Consulting, London, UK |
| Bernhard Fuchsbauer | Audi AG, Ingolstadt, Germany |
| Kay Fürstenberg | Sick AG, Hamburg, Germany |
| Wolfgang Gessner | VDI\|VDE-IT, Berlin, Germany |
| Roger Grace | Roger Grace Associates, Naples FL, USA |
| Klaus Gresser | BMW Forschung und Technik GmbH, Munich, Germany |
| Riccardo Groppo | Ideas & Motion, Cavaller Maggiore, Italy |
| Hannu Laatikainen | VTI Technologies Oy, Vantaa, Finland |
| Jochen Langheim | ST Microelectronics, Paris, France |
| Günter Lugert | Siemens AG, Munich, Germany |
| Steffen Müller | NXP Semiconductors, Hamburg, Germany |
| Roland Müller-Fiedler | Robert Bosch GmbH, Stuttgart, Germany |
| Andy Noble | Ricardo Consulting Engineers Ltd. Shoreham-by-Sea, UK |
| Pietro Perlo | IFEVS, Sommariva del Bosco, Italy |
| Detlef E. Ricken | Delphi Delco Electronics Europe GmbH, Rüsselsheim, Germany |
| Christian Rousseau | Renault SA, Guyancourt, France |
| Jürgen Valldorf | VDI\|VDE-IT, Berlin, Germany |
| Egon Vetter | Ceramet Technologies Ltd. Melbourne, Australia |
| David Ward | MIRA Ltd. Nuneaton, UK |

## Conference Chairs

| | |
|---|---|
| Gereon Meyer | VDI\|VDE-IT, Berlin, Germany |
| Jan Fischer-Wolfarth | VDI\|VDE-IT, Berlin, Germany |

# Contents

## Part IV Energy Efficiency



## Part V Components and Systems

# Contributors


**Felix Acker** Centre for Technology Infusion, La Trobe University, Melbourne, VIC 3086, Australia, e-mail: f.acker@latrobe.edu.au

**Heinrich Acker** Continental, Division Chassis and Safety, Business Unit Passive Safety and Sensorics, Guerickestr. 7, 60488 Frankfurt, Germany, e-mail: Heinrich.Acker@continental-corporation.com

**Christian Adam** Chemnitz University of Technology, Chemnitz, Germany, e-mail: christian.adam@baselabs.de

**Sagar Akhegaonkar** INTEDIS GmbH and Co. KG, Max-Mengeringhausen-Strae 5, 97084 Wuerzburg, Germany, e-mail: Sagar.Akhegaonkar@intedis.com

**Luciano Altomare** Department of Info-Telematic Systems, Centro Ricerche FIAT—Trento Branch, Via della Stazione, 27, 38123 Trento, Italy, e-mail: luciano.altomare@crf.it

**Blanca Araujo Pérez** Cidaut Foundation, Parque tecnologica de Boecillo, P 209, Boecillo, Spain, e-mail: blaraa@cidaut.es

**Bert Auerbach** IAV GmbH, Carnotstraße 1, 10587 Berlin, Germany, e-mail: bert.auerbach@iav.de

**Sven Bauer** Chemnitz University of Technology, Chemnitz, Germany, e-mail: sven.bauer@etit.tu-chemnitz.de

**Stefan Bauer** Robert Bosch GmbH, Wernerstraße 51, 70469 Stuttgart, Germany, e-mail: stefan.bauer5@de.bosch.com

**Thilo Bein** Structural Durability and System Reliability, Fraunhofer Institute, Darmstadt, Germany, e-mail: thilo.bein@lbf.fraunhofer.de

**Olivier Brunel** Electricfil Automotive, 77, allée des Grandes Combes, Z.I. Ouest Beynost 01708, Miribel cedex, France, e-mail: Olivier.brunel@electricfil.com

**Ulrich Büker** HELLA KGaA Hueck & Co, Beckumer Straße 130, 59552 Lippstadt, Germany, e-mail: ulrich.bueker@hella.com

**Robert Büthorn** IAV GmbH, Carnotstraße 1, 10587 Berlin, Germany, e-mail: robert.buethorn@iav.de

**Esteban Cañibano Álvarez** Cidaut Foundation, Parque tecnologica de Boecillo, P 209, Boecillo, Spain, e-mail: estcan@cidaut.es

**Roberto Cavaliere** TIS Innovation PARK, Free Software and Open Technologies, Via Siemens 19, Bolzano 39100, Italy, e-mail: Roberto.Cavaliere@tis.bz.it

**Yann Creff** IFP Énergies nouvelles, Rond point de l'échangeur de solaize, BP3, 69360 Solaize, France, e-mail: Yann.Creff@ifpen.fr

**Leandro D'Orazio** Department of Info-Telematic Systems, Centro Ricerche FIAT—Trento Branch, Via della Stazione, 27, 38123 Trento, Italy, e-mail: leandro.dorazio@crf.it

**Andrea Dardanelli** Dipartimento di Elettronica e Informazione, Politecnico di Milano, Piazza L. da Vinci, 32, 20133 Milan, Italy, e-mail: dardanelli@elet.polimi.it

**Marco Darin** Department of Info-Telematic Systems, Centro Ricerche FIAT—Trento Branch, Via della Stazione, 27, 38123 Trento, Italy, e-mail: marco.darin@crf.it

**Klaus David** Chair of Communication Technology, University of Kassel, Wilhelmshöher Allee 73, 34121 Kassel, Germany, e-mail: david@uni-kassel.de

**Nikolaus Decius** HELLA KGaA Hueck & Co, Beckumer Straße 130, 59552 Lippstadt, Germany, e-mail: nikolaus.decius@hella.com

**Arie P. van den Beukel** University of Twente, Laboratory of Design, Production and Management, Postbus 217, 7500 AE ENSCHEDE, The Netherlands, e-mail: a.p.vandenbeukel@utwente.nl

**Mascha C. van der Voort** University of Twente, Laboratory of Design, Production and Management, Postbus 217, 7500 AE ENSCHEDE, The Netherlands, e-mail: m.c.vandervoort@utwente.nl

**Anirudha Desai** Centre for Technology Infusion, La Trobe University, Melbourne, VIC 3086, Australia, e-mail: a.desai@latrobe.edu.au

**Klaus Dietmayer** Institute of Measurement, Control, and Microtechnology, Ulm University, Albert-Einstein-Allee 41, 89081 Ulm, Germany, e-mail: klaus.dietmayer@uni-ulm.de

**Stanley Ding** Centre for Technology Infusion, La Trobe University, Melbourne, VIC 3086, Australia, e-mail: s.ding@latrobe.edu.au

**Rik W. De Doncker** Institute for Power Electronics and Electrical Driver (ISEA), RWTH Aachen University, Jaegerstr. 17-19, 52066 Aachen, Germany, e-mail: post@isea.rwth-aachen.de

**Frederic Duault** Electricfil Automotive, 77, allée des Grandes Combes, Z.I. Ouest Beynost, F-01708 Miribel cedex, France, e-mail: Frederic.duault@electricfil.com

**Andreas von Eichhorn** BMW Group Research and Technology, BMW Forschung und Technik GmbH, Hanauerstraße 46, 80992 Munich, Germany, e-mail: andreas.an.eichhorn@bmw.de

**Sebastian Engel** Audi AG, Department of Vehicle Safety, 85045 Ingolstadt, Germany, e-mail: sebastian.engel@audi.de

**Alberto Fraile Del Pozo** Department of Mechanical Engineering, University of Zaragoza, C/Maria de Luna s/n, Zaragoza 50018, Spain, e-mail: afrailep@unizar.es

**Marco Ferraresi** STMicroelectronics, Via Tolomeo, 1, 20010 Cornaredo, MI, Italy, e-mail: marco.ferraresi@st.com

**Alberto Fraile** Campus Río Ebro, Edificio Betancourt, Universidad de Zaragoza, C/María de Luna s/n, Zaragoza 50018, Spain, e-mail: afrailep@unizar.es

**Franco Fresolone** Safety and Security Department, AIT Austrian Institute of Technology GmbH, Donau-City-Straße 1, Vienna 1220, Austria, e-mail: Franco.Fresolone@ait.ac.at

**Martin Fritzsche** Group Research and Advanced Engineering, Daimler AG, Wilhelm-Runge-Str. 11, 89081 Ulm, Germany, e-mail: martin.fritzsche@daimler.com

**Michael Gabb** Institute of Measurement, Control, and Microtechnology, Ulm University, Albert-Einstein-Allee 41, 89081 Ulm, Germany, e-mail: michael.gabb@uni-ulm.de

**Gianvito Giuffrida** STMicroelectronics, Strada Primosole, 50, 95121 Catania, CT, Italy, e-mail: gianvito.giuffrida@st.com

**Sebastien Glaser** LIVIC (Laboratory on Interactions Vehicles-Infrastructure-Drivers), A research unit of IFSTTAR, Batiment 824, 14 Route de la Minière, 78000 Versailles, France, e-mail: Sebastien.glaser@ifsttar.fr

**Farid Golnaraghi** Mechatronic Systems Engineering, Simon Fraser University, Room 4240, Surrey BC V3T 0A3, Canada, e-mail: mfgolnar@sfu.ca

**Manuel Ignacio González Hernández** Cidaut Foundation, Parque tecnologica de Boecillo, P 209, Boecillo, Spain, e-mail: mangon@cidaut.es

**Helge Graßhoff** Continental, Chassis and Safety Division, Passive Safety and Sensorics Business Unit, Guerickestraße 7, 60489 Frankfurt, Germany, e-mail: helge.grasshoff@continental-corporation.com

**Christian Michael Gruber** BMW Group, Knorrstrasse 147, 80788 Munich, Germany, e-mail: christian-michael.gruber@bmw.de

**Frank Grunwald** Continental, Division Chassis and Safety, Business Unit Passive Safety and Sensorics, Guerickestr. 7, 60488 Frankfurt, Germany, e-mail: Frank.Grunwald@continental-corporation.com

**Stefan Günthner** Continental, Chassis and Safety Division, Passive Safety and Sensorics Business Unit, Guerickestraße 7, 60489 Frankfurt, Germany, e-mail: stefan.guenthner@continental-corporation.com

**Alexander Hanzlik** AIT Austrian Institute of Technology GmbH, Donau-City-Straße 1, Vienna, Austria, e-mail: alexander.hanzlik.fl@ait.ac.at; ahanzlik@gmx.at

**Eva Hirtz** Fraunhofer-Institute for Structural Durability and System Reliability LBF, Bartningstraße 47, 64287 Darmstadt, Germany, e-mail: eva.maria.hirtz@lbf.fraunhofer.de

**Hauke van Hoek** Institute for Power Electronics and Electrical Driver (ISEA), RWTH Aachen University, Jaegerstr. 17-19, 52066 Aachen, Germany, e-mail: hauke.vanhoek@isea.rwth-aachen.de

**Klaus Höhne** Fraunhofer-Institute for Structural Durability and System Reliability LBF, Bartningstraße 47, 64287 Darmstadt, Germany, e-mail: hoehne.klaus@web.de

**Frederic Holzmann** INTEDIS GmbH and Co. KG, Max-Mengeringhausen-Strae 5, 97084 Wuerzburg, Germany, e-mail: Frederic.Holzmann@intedis.com

**Chen-Yu Hsieh** Mechatronic Systems Engineering, Simon Fraser University, Room 4240, Surrey BC V3T 0A3, Canada, e-mail: chenyuh@sfu.ca

**Bo Huang** Mechatronic Systems Engineering, Simon Fraser University, Room 4240, Surrey BC V3T 0A3, Canada, e-mail: bha23@sfu.ca

**Stephen Jones** AVL LIST GMBH, Hans List Platz 1, A-8020 Graz, Austria, e-mail: Stephen.Jones@avl.com

**Michael Karner** Virtual Vehicle Research and Test Center, Inffeldgasse 21a/1, 8010 Graz, Austria, e-mail: michael.karner@v2c2.at

**Reinhard Kloibhofer** Safety and Security Department, AIT Austrian Institute of Technology GmbH, Donau-City-Straße 1, Vienna 1220, Austria, e-mail: Reinhard.Kloibhofer@ait.ac.at

**Kosmas Knödler** ROBERT BOSCH GMBH, Robert-Bosch-Allee 1, D-74232 Abstatt, Germany, e-mail: Kosmas.Knoedler@de.bosch.com

**Jürgen Kogler** AIT—Austrian Institute of Technology GmbH, Donau-City-Strasse 1, 1220 Vienna, Austria, e-mail: Juergen.kogler@ait.ac.at

**Ulrich Köhler** HELLA KGaA Hueck & Co, Beckumer Straße 130, 59552 Lippstadt, Germany, e-mail: ulrich.koehler@hella.com

**Uwe Konzelmann** Robert Bosch GmbH, Wernerstraße 51, 70469 Stuttgart, Germany, e-mail: uwe.konzelmann@de.bosch.com

**Martin Krammer** Virtual Vehicle Research and Test Center, Inffeldgasse 21a/1, 8010 Graz, Austria, e-mail: martin.krammer@v2c2.at

**Claudia Kratzsch** Audi AG, Department of Vehicle Safety, 85045 Ingolstadt, Germany, e-mail: claudia.kratzsch@audi.de

**Erwin Kristen** AIT Austrian Institute of Technology GmbH, Donau-City-Straße 1, Vienna, Austria, e-mail: erwin.kristen@ait.ac.at

**Klaus Krumbiegel** IAV GmbH Ingenieurgesellschaft Auto und Verkehr, Carnotstraße 1, 10587 Berlin, Germany, e-mail: klaus.krumbiegel@iav.de

**Emre Kural** AVL LIST GMBH, Hans List Platz 1, A-8020 Graz, Austria, e-mail: Emre.Kural@avl.com

**Tobias Lange** Institute for Power Electronics and Electrical Driver (ISEA), RWTH Aachen University, Jaegerstr. 17-19, 52066 Aachen, Germany, e-mail: tobias.lange@isea.rwth-aachen.de

**Jacques Lavy** IFP Énergies nouvelles, Rond point de l'échangeur de solaize, BP3, 69360 Solaize, France, e-mail: Jacques.Lavy@ifpen.fr

**Sören Lehmann** Continental, Division Chassis and Safety, Business Unit Passive Safety and Sensorics, Guerickestr. 7, 60488 Frankfurt, Germany, e-mail: Soeren.Lehmann@continental-corporation.com

**Juan Sabas Martín Sánchez** Cidaut Foundation, Parque tecnologica de Boecillo, P 209, Boecillo, Spain, e-mail: juasan@cidaut.es

**Christopher Masjosthusmann** HELLA KGaA Hueck & Co, Beckumer Straße 130, 59552 Lippstadt, Germany, e-mail: c.masjosthusmann@hella.com

**Dirk Mayer** Structural Durability and System Reliability, Fraunhofer Institute, Darmstadt, Germany, e-mail: dirk.mayer@lbf.fraunhofer.de

**Mehrdad Moallem** Mechatronic Systems Engineering, Simon Fraser University, Room 4240, Surrey BC V3T 0A3, Canada

**Rainer Moller** Electricfil Automotive, 77, allée des Grandes Combes, Z.I. Ouest Beynost 01708, Miribel cedex, France, e-mail: rainer.moller@electricfil.com

**Sara Sánchez Monclus** Department of Mechanical Engineering, University of Zaragoza, C/Maria de Luna s/n, Zaragoza 50018, Spain, e-mail: ssanchez@unizar.es

**Rainer Moritz** Robert Bosch GmbH, Wernerstraße 51, 70469 Stuttgart, Germany, e-mail: rainer.moritz@de.bosch.com

**Steffen Müller** NXP Semiconductors Germany GmbH, Stresemannallee 101, 22529 Hamburg, Germany, e-mail: st.mueller@nxp.com

**Marc Muntzinger** Group Research and Advanced Engineering, Daimler AG, Wilhelm-Runge-Str. 11, 89081 Ulm, Germany, e-mail: marc.muntzinger@daimler.com

**Peter Nelson-Furnell** Public Transport Victoria, State Government of Victoria, Melbourne 3086, VIC, Australia, e-mail: peter.nelson-furnell@ptv.vic.gov.au

**Lydie Nouveliere** LIVIC (Laboratory on Interactions Vehicles-Infrastructure-Drivers), A research unit of IFSTTAR, Batiment 824, 14 Route de la Minière, 78000 Versailles, France, e-mail: lydie.nouveliere@ifsttar.fr

**Marcus Obst** Chemnitz University of Technology, Chemnitz, Germany, e-mail: marcus.obst@etit.tu-chemnitz.de

**Olivier Orfila** LIVIC (Laboratory on Interactions Vehicles-Infrastructure-Drivers), A research unit of IFSTTAR, Batiment 824, 14 Route de la Minière, 78000 Versailles, France, e-mail: olivier.orfila@ifsttar.fr

**Nicola Palella** STMicroelectronics, Via C. Olivetti, 2, 20864 Agrate Brianza, MB, Italy, e-mail: nicola.palella@st.com

**Timo Pech** Chemnitz University of Technology, Chemnitz, Germany, e-mail: timo.pech@etit.tu-chemnitz.de

**Emilio Larrodé Pellicer** Department of Mechanical Engineering, University of Zaragoza, C/Maria de Luna s/n, Zaragoza 50018, Spain, e-mail: elarrode@unizar.es

**Aashath Abdoul Rachide** Centre for Technology Infusion, La Trobe University, Melbourne, VIC 3086, Australia, e-mail: a.abdoulrachide@latrobe.edu.au

**Pierre Reisdorf** Chemnitz University of Technology, Chemnitz, Germany, e-mail: pierre.reisdorf@etit.tu-chemnitz.de

**Michael Rittmann** Robert Bosch GmbH, Wernerstraße 51, 70469 Stuttgart, Germany, e-mail: michael.rittmann@de.bosch.com

**Sara Sánchez** Campus Río Ebro, Edificio Betancourt, Universidad de Zaragoza, C/María de Luna s/n, Zaragoza 50018, Spain, e-mail: ssanchez@unizar.es

**Mario Santucci** Piaggio Spa, Viale Rinaldo Piaggio, 25, 56025 Pontedera, Pisa, Italy, e-mail: mario.santucci@piaggio.com

**Sergio M. Savaresi** Dipartimento di Elettronica e Informazione, Politecnico di Milano, Piazza L. da Vinci, 32, 20133 Milan, Italy, e-mail: savaresi@elet.polimi.it

**Frank Schade** NXP Semiconductors Germany GmbH, Stresemannallee 101, 22529 Hamburg, Germany, e-mail: frank.schade@nxp.com

**Christoph Schäper** Institute for Power Electronics and Electrical Driver (ISEA), RWTH Aachen University, Jaegerstr. 17-19, 52066 Aachen, Germany, e-mail: christoph.schaeper@isea.rwth-aachen.de

**Volker Scheuch** INTEDIS GmbH and Co. KG, Max-Mengeringhausen-Strae 5, 97084 Wuerzburg, Germany, e-mail: volker.scheuch@intedis.com

**Bernhard Schmid** Continental, Division Chassis and Safety, Business Unit Passive Safety and Sensorics, Guerickestr. 7, 60488 Frankfurt, Germany, e-mail: Bernhard.Schmid@continental-corporation.com

**Erwin Schoitsch** AIT—Austrian Institute of Technology GmbH, Donau-City-Strasse 1, 1220 Vienna, Austria, e-mail: Erwin.schoitsch@ait.ac.at

**Dieter Schramm** Institute of Mechatronics, University Duisburg-Essen, Lotharstraße 1, 47057 Duisburg, Germany, e-mail: dieter.schramm@uni-due.de

**Markus Schratter** Virtual Vehicle Research and Test Center, Inffeldgasse 21a/1, 8010 Graz, Austria, e-mail: markus.schratter@v2c2.at

**Robin Schubert** BASELABS GmbH, Chemnitz, Germany, e-mail: robin.schubert@baselabs.de

**Karsten Schulze** IAV GmbH, Carnotstraße 1, 10587 Berlin, Germany, e-mail: Karsten.schulze@iav.de

**Robert Schwaiger** IAV GmbH Ingenieurgesellschaft Auto und Verkehr, Carnotstraße 1, 10587 Berlin, Germany, e-mail: robert.schwaiger@iav.de

**Jugdutt Singh** Centre for Technology Infusion, La Trobe University, Melbourne, VIC 3086, Australia, e-mail: Jack.Singh@latrobe.edu.au

**Jochen Steinmann** ROBERT BOSCH GMBH, Robert-Bosch-Allee 1, D-74232 Abstatt, Germany, e-mail: Jochen.Steinmann@de.bosch.com

**Robin Streiter** Chemnitz University of Technology, Chemnitz, Germany, e-mail: robin.streiter@etit.tu-chemnitz.de

**Christoph Sulzbachner** AIT—Austrian Institute of Technology GmbH, Donau-City-Strasse 1, 1220 Vienna, Austria, e-mail: Christoph.sulzbachner@ait.ac.at

**Hadj Hamma Tadjine** IAV GmbH Ingenieurgesellschaft Auto und Verkehr, Carnotstraße 1, 10587 Berlin, Germany, e-mail: hadj.hamma.tadjine@iav.de

**Mara Tanelli** Dipartimento di Elettronica e Informazione, Politecnico di Milano, Piazza L. da Vinci, 32, 20133 Milan, Italy, e-mail: tanelli@elet.polimi.it

**Onorino di Tanna** Piaggio Spa, Viale Rinaldo Piaggio, 25, 56025 Pontedera, Pisa, Italy, e-mail: onorino.ditanna@piaggio.com

**Jan Thomanek** IAV GmbH, Chemnitz, Germany, e-mail: jan.thomanek@iav.de

**Anthony Thomson** Qualcomm Europe, Chiswick Park, 566 Chiswick High Rd, Chiswick W4 5YE, London, e-mail: athomson@qti.qualcomm.com

**Filippo Visintainer** Department of Info-Telematic Systems, Centro Ricerche FIAT—Trento Branch, Via della Stazione, 27, 38123 Trento, Italy, e-mail: filippo.visintainer@crf.it

**Gerd Wanielik** Chemnitz University of Technology, Chemnitz, Germany, e-mail: gerd.wanielik@etit.tu-chemnitz.de

**Daniel Watzenig** Virtual Vehicle Research and Test Center, Inffeldgasse 21a/1, 8010 Graz, Austria, e-mail: daniel.watzenig@v2c2.at

**Bernd X. Weis** BlackForestLightning, Haeberlinstr. 29b, Stuttgart 70563, Germany, e-mail: Bernd.Weis@BlackForestLightning.de

**André Welzel** Chemnitz University of Technology, Chemnitz, Germany, e-mail: andre.welzel@etit.tu-chemnitz.de

**Antje Westenberger** Group Research and Advanced Engineering, Daimler AG, Wilhelm-Runge-Str. 11, 89081 Ulm, Germany, e-mail: antje.westenberger@daimler.com

**Peter Wimmer** Virtual Vehicle Research and Test Center, Inffeldgasse 21a/1, 8010 Graz, Austria, e-mail: peter.wimmer@v2c2.at

**Bilal Youssef** IFP Énergies nouvelles, Rond point de l'échangeur de solaize, BP3, 69360 Solaize, France, e-mail: Bilal.Youssef@ifpen.fr

**Peter Zahn** BMW Group Research and Technology, BMW Forschung und Technik GmbH, Hanauerstraße 46, 80992 Munich, Germany, e-mail: peter.zahn@bmw.de

# Part I
# Driver Assistance and Road Safety

# A Warning Algorithm for Intersection Collision Avoidance

**Andreas von Eichhorn, Peter Zahn and Dieter Schramm**

**Abstract** This paper addresses two challenges for successful implementations of driver assistance systems for providing intersection collision avoidance based on Vehicle-to-Vehicle communication. Firstly, satellite based positioning typically comes along with varying precision, which has to be considered by the warning function. Secondly, a distinction has to be made between critical and safe driver behavior. An ideal warning function for the use in critical driver behavior is one which warns the accident perpetrator early enough so that the driver can defuse the situation on his own, and which remains silent during normal safe driving behavior. The following article develops such a warning algorithm and verifies it using data obtained in a field test. The validation considers to what extent parameterization of the algorithm can resolve the conflict mentioned above between the two goals of a low false-positive warning rate and high capability for accidence avoidance.

**Keywords** Advanced driver assistance systems · Intersection collision avoidance · V2V · Sim$^{TD}$

A. von Eichhorn (✉) · P. Zahn
BMW Group Research and Technology, BMW Forschung und Technik GmbH, Hanauerstraße 46 80992 Munich, Germany
e-mail: andreas.an.eichhorn@bmw.de

P. Zahn
e-mail: peter.zahn@bmw.de

D. Schramm
Institute of Mechatronics, University Duisburg-Essen, Lotharstraße 1 47057 Duisburg, Germany
e-mail: dieter.schramm@uni-due.de

J. Fischer-Wolfarth and G. Meyer (eds.), *Advanced Microsystems for Automotive Applications 2013*, Lecture Notes in Mobility, DOI: 10.1007/978-3-319-00476-1_1,

## 1 Problem Statement and Concept of Evaluation

The analysis of car accidents reveals three significant types of accidents, namely accidents while normal driving, rear end collisions and accidents at intersections. For the first two types mentioned, there already exist active safety systems to prevent an accident or mitigate its impact, e.g. rear end collision or lane departure warning systems. Due to the lack of driver assistance systems for collision avoidance at intersections as well as for demographic reasons, it is estimated that in the foreseeable future the significance of intersection related accidents will increase when compared to other types of accidents. One reason for the absence of intersection collision avoidance systems today is that on-board sensors like cameras or radar sensors have a field of view that is limited to the area in front of the car. As a consequence, object-crossing cannot be recognized early enough in order to prevent an accident. V2V (Vehicle-to-Vehicle) communication is a promising technology to overcome this limitation. Vehicles equipped with that technology determine their position within a global coordinate system with a GNSS-receiver and transmit it to other traffic participants via dedicated short-range communication units for automotive safety applications. Thereby, lateral object-crossing can be detected earlier, in many cases even before a direct line of sight between two vehicles is available.

As mentioned above, safety-oriented ADAS[1] typically suffer from the well-known warning dilemma: An early warning increases the potential for accident mitigation and avoidance while increasing the false-warning rate of safe drivers at the same time. When developing a warning algorithm there usually is much measurement data available of safe drivers from field operational tests (FOT), which enables the parameterization for a low false-positive warning rate of a given system. Unfortunately, it is more challenging to verify the capability for accident avoidance because accidents are unlikely recorded. One approach to deal with this restriction is to reconstruct the trajectories of vehicles involved in real accidents, as is contained e.g. in the GIDAS data base [1]. In this case, no measurement data of the target-system is available. To overcome this restriction a concept for the evaluation of a CICAS[2] is introduced in the following, that is based on data of intersection crossing maneuvers without any accident.

The concept is based on the idea that possible driver errors at intersection can be classified in two main categories. On the one hand, a driver may not be aware of the fact that he has to yield his right of way. In this case he is likely to show a similar velocity profile as a driver who has the actual priority. On the other hand, a driver may notice the traffic regulation correctly but overlook other traffic participants that have right of way. In this case the velocity profile will be similar to that of a driver who has to yield his right of way and only then crosses the intersection when no priority traffic is present. Previous investigations showed that

[1] ADAS: Advanced Driver Assistance Systems.

[2] CICAS: Cooperative Intersection Collision Avoidance System.

**Table 1** Setting of parameters used within this work

| Symbol | Name | Value | Unit |
|---|---|---|---|
| $a_{max}$ | Braking deceleration | 7 | $m/s^2$ |
| $d_{la}$ | Width of lane | 3 | m |
| $n_s$ | Samples used for evaluation | 200 | – |
| $t_R$ | Driver reaction time | 1 | s |
| $v_{obj}$ | Speed of crossing vehicle | 50 | km/h |
| $\sigma_d$ | Standard deviation of position error | 2.3 | m |

the erroneous behavior of the first case mentioned can be discriminated from correct behavior with a low false-positive warning rate [2, 3]. The second case offers more potential for optimization; hence it is focused by the following investigations.

Given these assumptions, all approaches of a given data set of intersection crossing maneuvers are classified as safe, whenever a driver actually stops within a certain area ahead of the stop bar. This subset is referred to as "stop approaches" (SA) in the following. As few as possible warnings should be triggered during these approaches.

The remaining maneuvers, referred to as drive-through maneuvers (DM), are regarded as accident perpetrators by assuming that virtual priority traffic is present, which reaches the intersection at just about the same time as the subject vehicle. Based on the warning algorithm and a hypothetical driver reaction, it is assumed in this paper that the potential for collision mitigation can be evaluated based on this data subset. The goal is to find a warning algorithm which balances the conflicting goals of a low false-warning rate of the safe drivers and a high capability for collision avoidance.

The advantage of this method is that its evaluation can be based on a large set of data obtained with the target system e.g. from a FOT and can hence account for the variances in measurement precision and driver behavior.

## 2 Data Collection and Preprocessing

The data used for the subsequent investigations were obtained from the FOT of the project sim$^{TD}$. The project aims at researching and testing V2V communication and its applications. Therefore, 120 experimental vehicles were equipped with dedicated hardware for wireless communication and GNSS-localization. Several controlled experiments were performed by unbiased test drivers. The tests for the intersection collision avoidance system were carried out repeatedly with different sets of drivers at three intersections on a cordoned off test-site in Friedberg, Germany. The drivers were instructed to drive freely on the terrain and decide on their own if they turn or cross every time they reach any of the intersections.

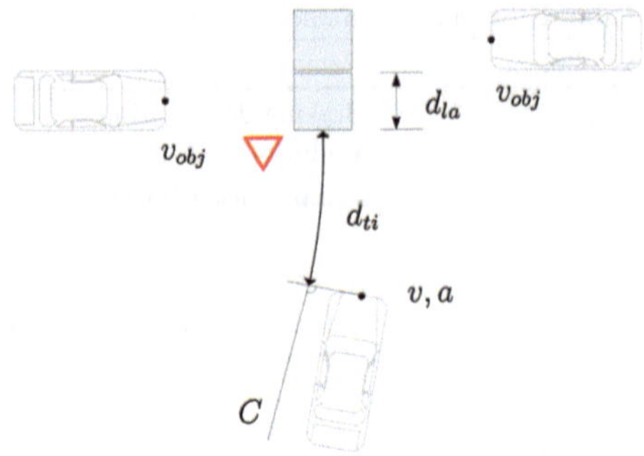


**Fig. 1** Schematic illustration of the examined scenario

Each vehicle uses the same hardware setup and determines its position with an Extended Kalman filter for GPS/INS[3]-fusion. Thus short GPS outages can be bridged and the position update frequency is increased from 1 to 10 Hz. The positioning precision is improved using correction data for GPS provided via mobile connection (Differential GPS). On the test site good sky view is ensured due to the absence of large buildings. The system used was evaluated with a high-precision reference system under comparable conditions in previous investigations and achieved a precision of typically $\sigma_d = 2{,}3$ m.[4] The longitudinal acceleration of the INS is processed offline with a zero-phase low pass filter, before it is used in the warning algorithm.

Based on the obtained data, a map is generated by b-spline curve approximation as described in [4]. This yields curves which approximate the actually driven paths and average out the GPS measurement errors to a large extent. For each curve representing one lane, the latest possible stop point is located where the driver can stop without disturbing priority traffic. The distance to intersection $d_{ti}$ of a given position is obtained by determining the length along the curve $C$ between the closest foot point on the curve to the latest possible stop point (see Fig. 1).

For further evaluation those maneuvers are selected which fulfill the following criteria: The driver approaches on a lane where he has to yield his right of way. No other vehicle is ahead during the approach which might influence the velocity behavior. This results in a total number of 619 approaches that are suitable for further processing.

## 3 Warning Algorithm

For the assessment of the criticality of traffic situations time measures have been established in the literature, such as Time-To-Steer or Time-To-Brake [5]. In order to compute these measures different emergency maneuvers are considered to avoid an imminent collision, like maximum steering, braking or accelerating. The

[3] INS: Inertial Navigation System.

[4] The setting of the constant variables used within this work are summarized in Table 1.

maximum time is calculated by which the driver or an actuator should execute any one of these maneuvers to avoid a collision [6].

For the subsequent considerations it is assumed that in the case of a critical situation the driver is warned and will have to defuse the situation on his own. Moreover, for the purpose of evaluating longitudinal driver behavior at intersections only braking is considered in the following. The warning system calculates the latest possible warning point, which is the driver's reaction time $t_R$ earlier than the latest Time-To-Maneuver. In the following, the Time-To-Warn $t_{tw}$ is used, which estimates the reaction time available to the driver in addition to the time $t_R$.

During the reaction time span $t_R$, it is assumed that the driver moves with constant acceleration. Based on the current velocity $v$ and acceleration $a$, the velocity at the onset of the driver reaction $v_R$ is

$$v_R = v + at_R. \tag{1}$$

For further investigations only $v_R > 0$ is treated, because otherwise it is assumed that the driver is already about to stop and no warning is necessary. The distance needed to stop $d_{St}$ comprises the distance passed during the driver reaction $d_R$ and the distance passed while braking $d_{Br}$ with maximal deceleration $a_{max}$:

$$d_{St} = \underbrace{vt_R + \frac{1}{2}at_R^2}_{d_R} + \underbrace{\frac{v_R^2}{2a_{max}}}_{d_{Br}} \tag{2}$$

The Time-To-Warn $t_{tw}$ is obtained by the difference of the distance to the object $d_{obj}$ and $d_{St}$, passed with constant velocity $v_R$

$$t_{tw} = \frac{d_{obj} - d_{St}}{v_R}. \tag{3}$$

The common approach is to trigger a warning when $t_{tw}$ falls below a predefined threshold $t_{tw,th}$. As a major drawback, this approach cannot take into account varying uncertainties induced by measurements or the prediction of other traffic participants. Therefore, it is proposed to take the decision at a minimal required confidence level $P_{conf}$ [7]:

$$p\left(t_{tw} < t_{tw,th}\right) > P_{conf}. \tag{4}$$

For the subsequent considerations it is assumed that the measurement uncertainty of $d_{obj}$ is the dominant influence on the variance of $t_{tw}$. The measurement error is modeled as normally distributed with zero mean, as it is common when dealing with GNSS-based localization:

$$d_{obj} = \mathcal{N}\left(\mu_d, \sigma_d^2\right) \tag{5}$$

The distribution of the error can consider the total error of the localization as well as uncertainties caused by imprecise knowledge of the stop line. For the further calculations a substitution is applied, which separates the $d_{obj}$-dependent part of Eq. (3):

$$t_{tw} = f_1(v,a)d_{obj} - f_2(v,a) \tag{6}$$

with

$$f_1(v,a) = \frac{1}{v_R}, f_2(v,a) = \frac{d_{St}}{v_R}. \tag{7}$$

Under these assumptions, $t_{tw}$ is also normally distributed.

$$X_{\text{ttw}} = \mathcal{N}(\mu_{ttw}, \sigma_{ttw}) \tag{8}$$

Standard deviation and variance can be obtained from $d_{obj}$,$f_1$ and $f_2$:

$$\mu_{ttw} = f_1\mu_d - f_2 \tag{9}$$

$$\sigma_{ttw} = f_1\sigma_d \tag{10}$$

Substituting results in the standard deviation of $X_{ttw}$:

$$\sigma_{ttw} = \frac{\sigma_d}{v + at_R} \tag{11}$$

It is expected that with an increasing variance of $X_{ttw}$ the rate of false positive warnings increases as well.

# 4 Evaluation

In the following the approach of “A Flexible Environment Perception Framework for Advanced Driver Assistance Systems” is applied to the measured data described in “Time-to-Collision Estimation in Automotive Multi-Sensor Fusion with Delayed Measurements” for the evaluation and parameterization. For this purpose two different cases of virtual vehicles approaching from any of two sides are considered, as illustrated in Fig. 1. This results in a different calculation of $t_{tw}$ in Eq. (3), assuming right-hand driving: For the crossing vehicle approaching from the left $d_{\text{obj}} = \text{d}_{\text{ti}}$, whereas for the vehicle from the right additionally the width of one lane is considered: $d_{obj} = \text{d}_{\text{ti}} + \text{d}_{\text{la}}$.

## *4.1 Dynamic States at the Time of the Warning*

In Fig. 2 velocity and acceleration are illustrated when reaching $t_{tw} \leq 0\text{s}$ for both directions and both SA and DM. This is the last instance at which the driver can

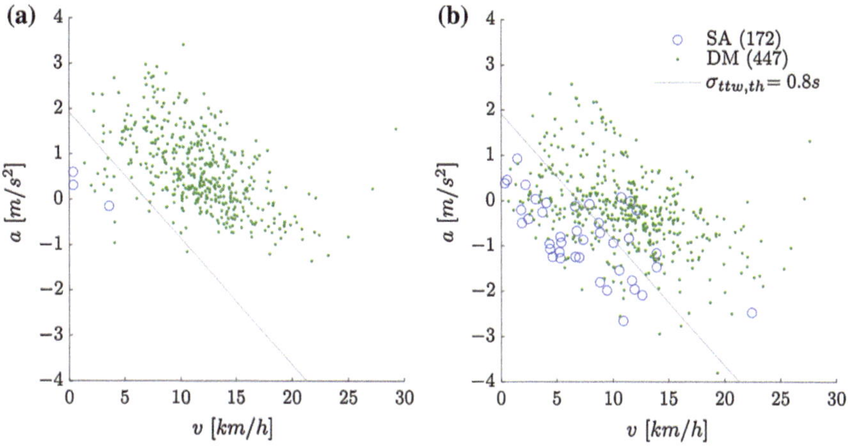


**Fig. 2** Longitudinal dynamic when reaching $t_{tw} = 0s$. **a** Crossing vehicle from the *right*. **b** Crossing vehicle from the *left*

prevent the accident with the assumed reaction to the warning as described above. Ideally as few as possible SA should appear.

In Fig. 2a it can be seen that in the case of a crossing vehicle from the right only three out of 172 (1.74 %) SA got a warning. This is due to the fact that drivers typically stop at the stop bar, regardless of the direction of the crossing traffic. When considering the vehicle approaching from the left in Fig. 2b, a higher false positive rate can be observed.

To distinguish SA and DM, the probabilistic approach described in Ch. 3 is utilized. For illustrative purposes the straight line resulting from Eq. (11) for a particular threshold of $\sigma_{\mathrm{ttw,th}}$ is additionally depicted in Fig. 2. When demanding $\sigma_{\mathrm{ttw}} < \sigma_{\mathrm{ttw,th}}$, all maneuvers under that line do not get a warning. SA typically have a smaller velocity and acceleration when reaching $t_{tw} = 0s$ and therefore, are more likely to be excluded by this condition. When decreasing $\sigma_{\mathrm{ttw,th}}$ it can be pondered to which amount the SA are excluded while also including an increasing subset of the DM. The diagram demonstrates that the approach is capable of differentiating between the conflicting goals described in the introduction. Moreover, it becomes obvious that the quality of the distinction of SA and DM benefits from the combined evaluation of both longitudinal motion states $v$ and $a$.

## 4.2 Variation of the Warning Parameters

For further investigations only the case of a crossing vehicle from the left is considered because it is the more demanding use case, as shown above. The false positive warning rate $p_{fp}$ is calculated by the ratio of warned to all SA.

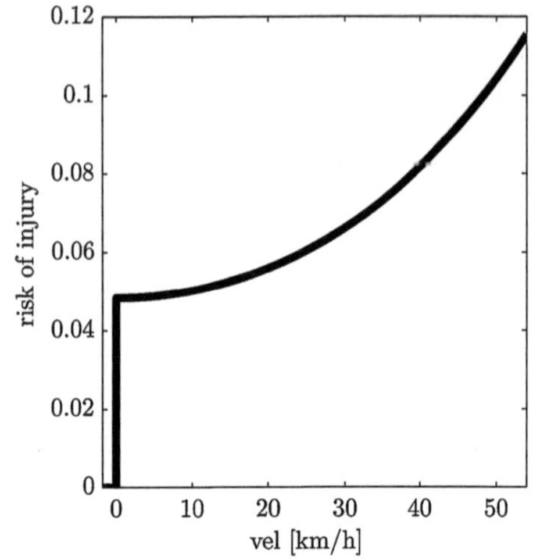


**Fig. 3** Weighting function for the approximation of the severity of the accident consequences when colliding with a crossing vehicle in dependence of the subject velocity

Concerning the capability of accident mitigation and avoidance the DM are regarded. To assess the severity of the accident consequences a weighting function is assumed, which estimates the risk of an injury $r$. It depends on the velocity at which the ego vehicle collides as illustrated in Fig. 3. For every DM the risk of injury $r_w$ respectively $r_{\underline{w}}$ is calculated, with and without the triggering of a warning. After the warning is released, it is assumed that the driver brakes with $a_{max}$ when the reaction time $t_R$ has passed. Since the position measurement in the used data is noisy, the evaluation is repeated $\mathrm{n_s}$ times with an additional position error sampled from a zero mean normal distribution with standard deviation $\sigma_\mathrm{d}$, which yields $r_{w,i}$. The potential for accident avoidance $r_{avo}$ is obtained for every maneuver by

$$r_{avo} = 1 - \frac{\sum_i r_{w,i}}{n_s r_{\underline{w}}} \tag{11}$$

Two different warning algorithms are considered:

- **A**: Warning, when $(t_{tw} < t_{tw,th}) \wedge (\mathrm{v} > \mathrm{v}_{min})$
- **B**: Warning, when $p(t_{tw} < t_{tw,th}) > P_{conf}$.

For every warning algorithm, two parameters are varied (**A**: $t_{tw,th}$ and $v_{min}$, **B**: $t_{tw,th}$ and $P_{conf}$) and the false positive warning rate $p_{fp}$ as well as the potential for accident avoidance $r_{avo}$ is calculated from all intersection crossing maneuvers. Every configuration of the parameter set results in one point in the diagram of Fig. 4. From this set of points only those points are depicted, which are Pareto-optimal for each of the two warning algorithms.

The figure shows that the probabilistic algorithm B is superior to the common approach A in a Pareto-optimal sense. Based on this analysis the optimal parameterization for the desired warning properties can be obtained.

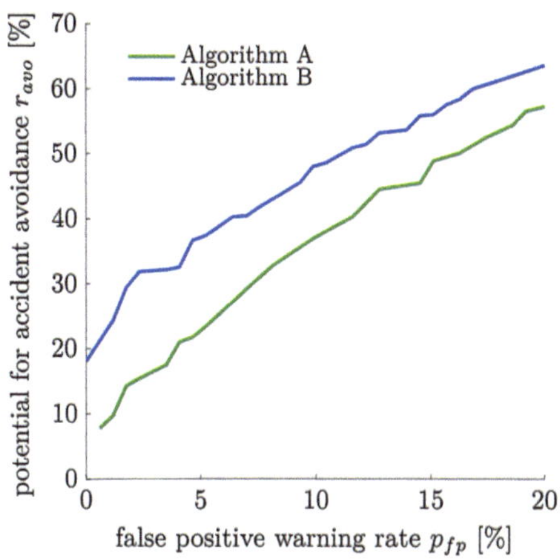


**Fig. 4** Comparison of the two warning algorithms

### *4.3 Discussion*

The precision $\sigma_{\mathrm{d}}$ is assumed to be constant throughout the examined measurement data in this work. The only influence considered for the uncertainty of $t_{\mathrm{tw}}$ is the current motion state. When dealing with data from urban areas with a partially blocked view of sky, the conditions may deteriorate considerably, though in many cases this can be detected, by the number of visible satellites and their geometric constellation for example. In contrast, the precision of the estimation of $d_{\mathrm{obj}}$ may be increased in some scenarios if on-board sensors are available and sensor data fusion is applied. Further sources for the variance in quality of precision can be thought of. Given the knowledge of the precision of the current position $\sigma_{\mathrm{d}}$, the approach of Eq. (4) allows to assess the system boundaries online in a continuous way without the necessity for pre-defined thresholds on the velocity or measurement precision.

For the evaluation several worst case assumptions were applied, namely the exclusion of a preceding vehicle, the scenario with a vehicle approaching from the left and the more challenging of the two mentioned driver errors. Therefore, it is supposed that in practice the system performs better than illustrated in Fig. 4.

## 5 Conclusion

In the present work two different warning algorithms for an intersection collision avoidance system are described and applied to measurement data obtained in a FOT. For evaluation the two conflicting goals of warning functions are described and it is shown how these can be determined based on the measurement data. The results show, that for the performance of a CICAS the direction of the crossing vehicle is important. The probabilistic approach shows better performance when compared to the common approach not considering uncertainties.

**Acknowledgments** This work was funded within the project sim$^{TD}$ by the German Federal Ministries of Economics and Technology as well as Education and Research, and supported by the Federal Ministry of Transport, Building and Urban Development.

## References

1. GIDAS (German In-Depth Accident Study) http://www.gidas.org/
2. Mages M (2008) Top-Down-Funktionsentwicklung eines Einbiege- und Kreuzenassistenten. Dissertation, Technischen Universität Darmstadt
3. Meitinger K (2008) Top-Down-Entwicklung von Aktiven Sicherheitssystemen für Kreuzungen. Dissertation, Technischen Universität München
4. Eichhorn A, Zahn P, Schramm D (2013) Automatic generation of intersection topologies using numerous gps traces. In: Proceedings of 5th IEEE international symposium on wireless vehicular communications
5. Hillenbrand J (2007) Fahrerassistenz zur Kollisionsvermeidung. Dissertation, Universität Karlsruhe
6. Tamke A, Dang T, Breuel G (2011) A flexible method for criticality assessment in driver assistance systems. In: Proceedings of IEEE intelligent vehicles symposium (IV), pp 697–702
7. Berthelot A, Tamke A, Dang T (2011) Handling uncertainties in criticality assessment. In: Proceedings of IEEE intelligent vehicles symposium (IV), pp 571–576

# Time-to-Collision Estimation in Automotive Multi-Sensor Fusion with Delayed Measurements

**Antje Westenberger, Marc Muntzinger, Michael Gabb, Martin Fritzsche and Klaus Dietmayer**

**Abstract** This paper presents a time-to-collision estimation in the context of multi-sensor fusion. Several asynchronous sensors are fused where the measurements arrive at the fusion unit out-of-sequence, i.e., some measurements are temporally more delayed than others. The adequate out-of-sequence handling is crucial for time-critical applications such as pre-crash systems. Several methods are discussed and compared with respect to accuracy and computational costs. In addition, a reduced out-of-sequence algorithm for practical application is derived. The performance of the pre-crash system is evaluated using real-world data from crash tests. To this end, a soft crash target is used with a position ground truth accurate to the centimeter and a contact sensor as temporal ground truth.

**Keywords** Driver assistance systems · Pre-crash · Multi-sensor fusion · Out-of-sequence · Time-to-collision

A. Westenberger (✉) · M. Muntzinger · M. Fritzsche
Group Research and Advanced Engineering, Daimler AG, Wilhelm-Runge-Str. 11, 89081 Ulm, Germany
e-mail: antje.westenberger@daimler.com

M. Muntzinger
e-mail: marc.muntzinger@daimler.com

M. Fritzsche
e-mail: martin.fritzsche@daimler.com

M. Gabb · K. Dietmayer
Institute of Measurement, Control, and Microtechnology, Ulm University, Albert-Einstein-Allee 41, 89081 Ulm, Germany
e-mail: michael.gabb@uni-ulm.de

K. Dietmayer
e-mail: klaus.dietmayer@uni-ulm.de

J. Fischer-Wolfarth and G. Meyer (eds.), *Advanced Microsystems for Automotive Applications 2013*, Lecture Notes in Mobility, DOI: 10.1007/978-3-319-00476-1_2,

# 1 Introduction

Future driver assistance and safety systems have to face more and more challenging tasks, like autonomous driving or active collision avoidance. In order to guarantee the required ASIL level, more than one sensor has to be used to compensate a possible failure of one sensor. Advanced sensor fusion techniques are therefore, crucial.

This paper deals with sensor fusion algorithms in time-critical contexts, specifically in a pre-crash system that calculates a time-to-collision. In order to quickly react to an imminent collision, it is important to avoid all additional delays in the filtering and to integrate all measurements immediately upon arrival. This is not always fulfilled in state-of-the-art fusion systems, as will be discussed in the following.

In sensor fusion, a frequent problem is the so-called out-of-sequence problem, which means that one sensor is slower than another. I.e., the pre-processing time between the raw measurement and the arrival at the fusion system differs for each sensor. This means that the original order of the measurements is not guaranteed in the fusion system.

Figure 1 shows an example of an out-of-sequence problem with two sensors. Here the measurement from time $t_{k_0}$ is the oldest, but arrives third at the fusion system, which makes it a delayed measurement, a so-called out-of-sequence measurement (OOSM). Methods to deal with these out-of-sequence measurements are required, as will be discussed in this chapter.

# 2 Filtering with Out-of-Sequence Measurements

## *2.1 Problem Description*

The well-known Kalman filter [1] estimates the state at time $t_k$ by doing a prediction from time $t_{k-1}$ to time $t_k$, yielding an *a priori* state estimate $x_{k|k-1}$ with corresponding estimation error covariance $P_{k|k-1}$, followed by an innovation with the latest measurement $z_k$, which gives an *a posteriori* state estimate $x_{k|k}$ with covariance $P_{k|k}$. It is assumed that all measurements

$$Z^k := \{z_1, \ldots, z_k\}$$

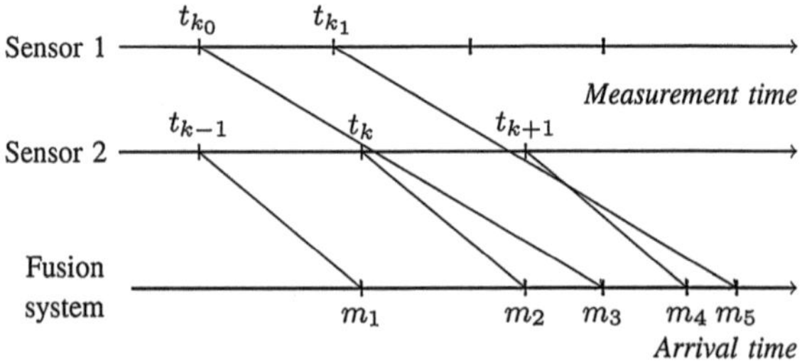


**Fig. 1** Out-of-sequence problem with two different asynchronous sensors

arrive at the fusion unit in the correct temporal order. However, when more than one sensor is used, measurements in practical applications often are out-of-sequence: After the innovation with $z_k$, an older measurement $z_{k_0}$ arrives. This out-of-sequence measurement cannot be directly integrated using the standard Kalman filter. Therefore, advanced OOSM algorithms have to be used, which will be described in the following.

## 2.2 Out-of-Sequence Algorithms

The simplest method to deal with out-of-sequence measurements is the so-called *Buffering*, which simply stores all measurements first without using them. Only when all sensors have provided at least one measurement, the oldest measurement can be integrated, since no OOSM can be missing by then any more [2]. However, Buffering is clearly an undesirable algorithm in time-critical applications, since additional delays make the system slow and not all information is used at most times. Quick reactions like emergency braking are not possible.

Another straight-forward method is *Reprocessing* (see for example [3, 4]). Here all measurements are processed immediately upon arrival. In case an OOSM arrives, all later measurements are integrated again, thus correcting the error due to the previously missing OOSM. Clearly this is the most accurate algorithm, but also the most expensive one, therefore in practical applications cheaper algorithms are needed that perform comparably.

One advanced OOSM algorithm is the so-called *Retrodiction* [5]. Upon OOSM arrival, a backward prediction in the past to the OOSM timestamp is calculated, followed by an innovation with the OOSM. A practical modification called *Reduced Retrodiction* is given in the following chapter.

Another method to deal with OOSM is the *Forward-Prediction Fusion and Decorrelation (FPFD)*. Other that the Retrodiction, it does not use a backward prediction, but a decorrelation step in information space. See [6, 7] for details.

Both Retrodiction and FPFD have much lower computational costs than Reprocessing at comparable performance [3, 4]. Therefore, Retrodiction or FPFD should be used instead of Buffering or Reprocessing. A practical comparison of these algorithms will be given in "Flexible environment perception for advanced driver assistance systems".

## 2.3 The Reduced Retrodiction Algorithm

The Retrodiction algorithm [5] was derived under the assumption that at the latest timestamp $t_k$, an innovation with a measurement $z_k$ is done. Based on this estimate $x_{k|k}$, the Retrodiction to an out-of-sequence timestamp $t_{k_0}$ is calculated. However, in practical applications, there may not be an innovation at every timestamp, even if a measurement arrives. For example, a gating procedure may exclude all measurements if the Mahalanobis distance is too big. In this case, the Retrodiction must be calculated based on the predicted *a priori* estimate $x_{k|k-1}$ instead of the *a posteriori* estimate $x_{k|k}$. Therefore, the derivation of a modified Retrodiction is needed, called Reduced Retrodiction in the following.

### 2.3.1 Reduced State and Covariance Retrodiction

In the usual Retrodiction algorithm, the retrodicted state is derived from the system dynamics

$$x_k = F_{k,k0} x_{k0} + v_{k,k0}$$

by inverting the system matrix $F_{k,k_0}$, where $v_{k,k_0}$ denotes the process noise between $t_{k_0}$ and $t_k$. The state at the OOSM timestamp $t_{k_0}$ can therefore be calculated as [5]

$$x_{k_0|k} = F_{k,k_0}^{-1}\left[x_{k|k} - v_{k,k_0|k}\right],$$

with the expected *a posteriori* process noise

$$v_{k,k_0|k} = E\left[v_{k,k_0}|Z^k\right] - Q_{k,k_0} H_k^T S_k^{-1} \gamma_k.$$

Here $Q_{k,k_0}$ denotes the process noise covariance, $H_k$ is the measurement matrix, and $\gamma_k = z_k - H_k x_{k|k-1}$ is the residual with corresponding covariance $S_k$. However, if at time $t_k$ no innovation with $z_k$ has been done, the state Retrodiction reduces to

$$x_{k_0|k-1} = F_{k,k_0}^{-1}\left[x_{k|k-1} - v_{k,k_0|k-1}\right] = F_{k,k_0}^{-1} x_{k|k-1},$$

since the *a priori* expectation value for the process noise is zero:

$$v_{k,k_0|k-1} = E\left[v_{k,k_0}|Z^{k-1}\right] = 0.$$

The corresponding reduced covariance Retrodiction can be calculated as

$$P_{k_0|k-1} = F_{k,k_0}^{-1}\left[P_{k|k-1} - Q_{k,k_0}\right]\left(F_{k,k_0}^{-1}\right)^T.$$

This completes the reduced state and covariance Retrodiction.

### 2.3.2 Reduced OOSM Innovation

After the Retrodiction to the past timestamp $t_{k_0}$, the innovation with the OOSM $z_{k0}$ and simultaneously the prediction to the latest timestamp $t_k$ are done. Again, this OOSM innovation reduces in case no innovation has been done at time $t_k$. The cross covariance between state and measurement as in [5] is needed, where some terms now vanish in the reduced form:

$$P^{xz}_{k,k_0|k-1} = \left[P_{k,k-1} - Q_{k,k_0}\right] F^T_{k,k_0} H^T_{k_0}.$$

The modified Kalman gain then is

$$W_{k_0} = P^{xz}_{k,k_0|k-1} S^{-1}_{k_0}.$$

With that, the reduced OOSM state innovation is as usual

$$x_{k|k_0} = x_{k|k-1} + W_{k_0}\left[z_{k_0} - H_{k_0} x_{k_0|k-1}\right].$$

The corresponding reduced OOSM covariance innovation is

$$P_{k|k_0} = P_{k|k-1} - P^{xz}_{k,k_0|k-1} S^{-1}_{k_0} \left(P^{xz}_{k,k_0|k-1}\right)^T.$$

With this step, the reduced OOSM innovation is completed.

# 3 Practical Evaluation

## *3.1 System Description and Sensor Setup*

The proposed algorithms, including the Reduced Retrodiction derived in the previous section, are tested using data from crash tests with a soft crash target (see Fig. 2b). In this paper, a frontal collision scenario with constant velocity is evaluated. The sensor vehicle is equipped with a monocamera, two short range radars and a long range radar, as shown in Fig. 2a. Whereas the monocamera is used for classification only, the radars result in an out-of-sequence problem that has to be accounted for. The ground truth is provided by an inertial measurement unit for accurate positioning as well as a contact sensor on the front bumper for the exact timestamp of the collision.

In order to apply OOSM algorithms, the temporal characteristics of each sensor (such as cycle time and measurement delay) have to be known. Methods to temporally calibrate the sensors can be found in, e.g., [8].

## *3.2 Results*

This chapter shows the results from the different OOSM algorithms. The proposed Reduced Retrodiction is combined with the usual Retrodiction and applied depending on whether an innovation has been done or not.

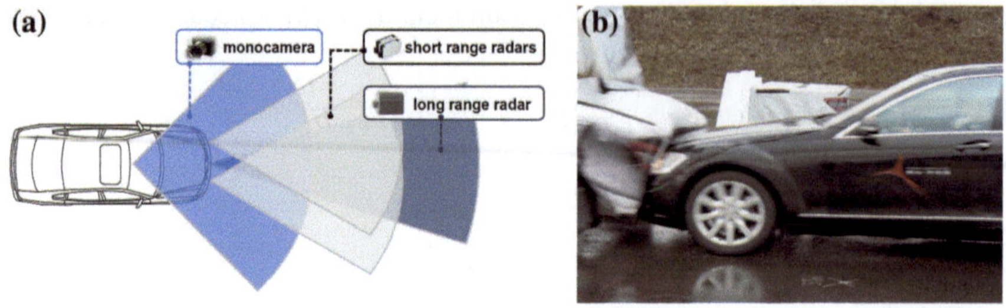


**Fig. 2** **a** Sensor setup with monocamera as well as short and long range radars. **b** Crash test with a soft crash target. The temporal ground truth is provided by a contact sensor on the front bumper

First, the one-sigma-volume of the estimation error covariance of all trackers is shown in Fig. 3. This volume can be interpreted as the uncertainty of the corresponding tracker. Clearly Buffering yields a very uncertain state estimate, since the measurements used are mainly outdated. The covariance volume is much bigger than with all of the other trackers. Due to this uncertainty, Buffering should not be used in case a high reliability of the fusion system is needed. Therefore, Buffering is not considered in the following evaluation any more.

The next plot, Fig. 4, shows the position error of the different OOSM algorithms. Clearly all methods perform nearly identical. This is due to the fact that in the one-step lag case, i.e., the case when the OOSM lags behind only one measurement, the algorithms are equivalent [5–7]. In the multi-step lag case, like in this test setup, the algorithms differ slightly, but the results are still very similar.

This shows that due to their lower computational costs, FPFD or Retrodiction should be preferred to Reprocessing. Clearly the derived Reduced Retrodiction (included in the usual Retrodiction) performs satisfactory as well.

Finally the time-to-collision estimation is evaluated. The estimation error can be seen in Fig. 5. Again, the differences in the OOSM algorithms are neglectible, all algorithms yield sufficiently precise time-to-collision estimates. Again, this shows the advantages of the cheaper OOSM algorithms Retrodiction and FPFD compared to Reprocessing, as well as the accuracy of the derived Reduced Retrodiction algorithm.

**Fig. 3** One-sigma-volume of the estimation error covariance of the different trackers. The crash time is depicted by the *vertical black line*

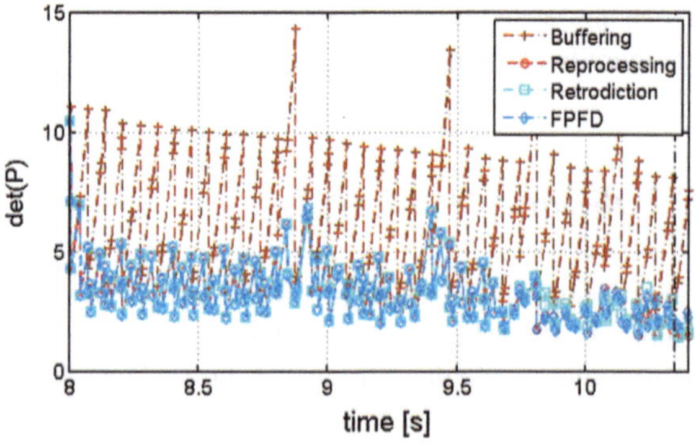

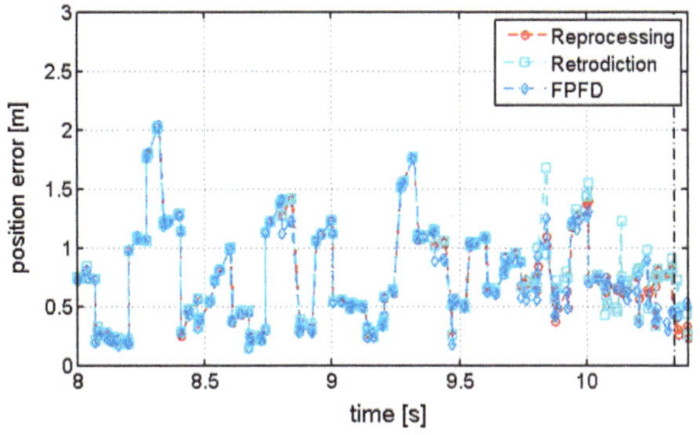


**Fig. 4** Position error of the different trackers. The crash time is depicted by the *vertical black line*

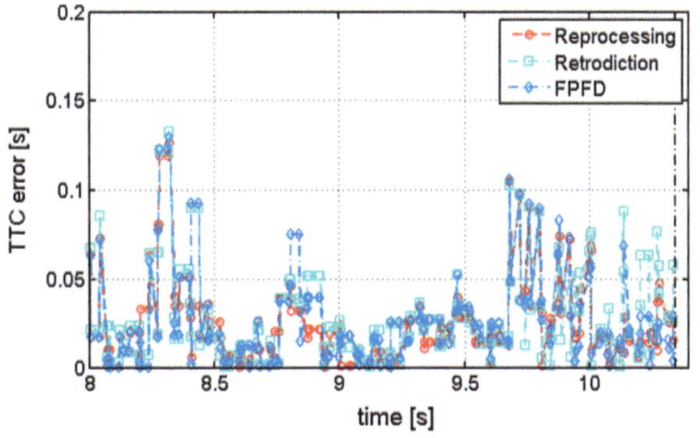


**Fig. 5** Error in the time-to-collision estimation of the different trackers. The crash time is depicted by the *vertical black line*

# 4 Conclusion and Outlook

This paper has demonstrated the effects of different out-of-sequence algorithms in practical applications. A time-to-collision based on the different algorithms was evaluated using real-world data from crash tests. A modification of the main out-of-sequence algorithm for practical applications, the so-called Reduced Retrodiction, was derived. This modification was validated and its applicability was shown.

Future work will concentrate on the application of OOSM algorithms not only in state estimation, but also in existence estimation. This topic of current research is discussed in [9, 10] and will be further investigated in the future.

# References

1. Kalman R (1960) A new approach to linear filtering and prediction problems. Trans ASME J Basic Eng 82:35–45
2. Kämpchen N, Dietmayer K (2003) Data synchronization strategies for multi-sensor fusion. In: Proceedings of world congress on intelligent transport systems, vol 10

3. Muntzinger M, Aeberhard M, Schröder F, Sarholz F, Dietmayer K (2009) Tracking in a cluttered environment with out-of-sequence measurements. In: Proceedings of IEEE international conference on vehicular electronics and safety. pp 56–61
4. Muntzinger M, Aeberhard M, Zuther S, Schmid M, Dickmann J, Dietmayer K (2010) Reliable automotive pre-crash system with out-of-sequence measurement processing. In: Proceedings of IEEE conference on intelligent vehicles symposium. pp 1022–1027
5. Bar-Shalom Y (2000) Update with out-of-sequence measurements in tracking: exact solution. Proc SPIE, Signal Data Process Small Targets 4048:541–556
6. Lanzkron P, Bar-Shalom Y (2007) A two-step method for out-of-sequence measurements. In: Proceedings of the IEEE aerospace conference, vol 3
7. Rheaume F, Benaskeur A (2008) Forward prediction-based approach to target-tracking with out-of-sequence measurements. In: Proceedings of IEEE conference on decision and control. pp 1326–1333
8. Westenberger A, Huck T, Fritzsche M, Schwarz T, Dietmayer K (2011) Temporal synchronization in multi-sensor fusion for future driver assistance systems. In: Proceedings of international IEEE symposium on precision clock synchronization for measurement, control and communication
9. Westenberger A, Duraisamy B, Munz M, Muntzinger M, Fritzsche M, Dietmayer K (2012) Impact of out-of-sequence measurements on the joint integrated probabilistic data association filter for vehicle safety systems. In: Proceedings of IEEE intelligent vehicles symposium (IV)
10. Westenberger A, Wäldele S, Dora B, Duraisamy B, Muntzinger M, Dietmayer K (2013) Multi-sensor fusion with out-of-sequence measurements for vehicle environment perception. In: Proceedings of IEEE international conference on robotics and automation, to appear

# A Flexible Environment Perception Framework for Advanced Driver Assistance Systems

**Markus Schütz and Klaus Dietmayer**

**Abstract** The advanced driver assistance systems (ADAS) of the future will rely heavily on an accurate and extensive description of the ego vehicle's environment based on multiple sensors. This paper gives an overview of a framework tackling this task by using Simultaneous Localization And Mapping (SLAM) and Detection And Tracking of Moving Objects (DATMO) techniques in real world scenarios using multiple 4-layer lidar sensors and a radar sensor. After a short literature overview, a modular framework is proposed which allows testing and evaluating different algorithms in outdoor scenarios. An advanced particle filter based tracking algorithm is presented to estimate the shape of extended objects with arbitrary forms. Results are evaluated based on sensor data.

**Keywords** SLAM · DATMO · Tracking · Shape estimation · Pre-crash

## 1 Introduction

Advanced driver assistance systems (ADAS) of the future will rely heavily on an accurate and extensive description of the ego vehicle's environment based on multiple sensors. The proposed frontal perception system is based on two 4-layer laser scanners and a long- and mid-range radar.

Although the "Simultaneous Localization and Mapping" (SLAM) with the "Detection and Tracking of Moving Objects" (DATMO) has gained a lot of

M. Schütz (✉)
Daimler AG, Wilhelm-Runge-Str. 11 89081 Ulm, Germany
e-mail: markus.m.schuetz@daimler.com

K. Dietmayer
Institute of Measurement, Control, and Microtechnology, University of Ulm, Albert-Einstein-Allee 41 89081 Ulm, Germany
e-mail: klaus.dietmayer@uni-ulm.de

J. Fischer-Wolfarth and G. Meyer (eds.), *Advanced Microsystems for Automotive Applications 2013*, Lecture Notes in Mobility, DOI: 10.1007/978-3-319-00476-1_3,

attention in recent years, the problems remain challenging when applied to real world outdoor scenarios.

Many algorithms that perform well in indoor areas are not applicable when it comes to outdoor environments. For example, it is hard to robustly find suitable landmarks in different scenarios with range sensors and therefore to localize the ego vehicle properly with many algorithms. This is mainly because of areas where there are many unstructured roadside objects such as bushes.

The system's localization approach is occupancy grid map based. In particular, it assumes that there is no external information about the environment available. This assumption is adequate, since GPS often fails in urban canyons (and if not its accuracy is not appropriate for driver assistance systems) and it is hard to create and update an accurate a priori digital map, or even a laser-generated map, for large areas. The goal is a pre-crash system that can robustly handle highly dynamic urban scenarios. For a system requiring high accuracy for the prediction of a crash situation, it is a good benchmark to evaluate the quality of the environment perception. A newly developed tracking algorithm is presented. For any in-depth explanation, however, another publication is referenced.

The system is intended to deliver the environmental description of the area in front of the host vehicle. The main goal is to report all relevant obstacles that can be detected by the sensor platform. To realize comfort and safety applications on the basis of the resulting information, the detections of single sensors have to be fused and filtered in time to stabilize the environment description and derive stable additional information describing e.g. dynamic parameters of each object.

The perception framework is intended to work only on the basis of the measurements delivered by the range sensors and the ego vehicle dynamics.

## 2 Related Literature

SLAM and DATMO provide concepts to perceive the environment of the vehicle. The goal is to obtain a consistent description of the environment. The algorithms are extensively studied in the field of robotics and in the field of autonomous cars.

The DATMO process deals with dynamic objects. A challenging task is to detect dynamic objects in laser scans. Sensor readings, classified as "dynamic", are used to estimate the dynamic state of dynamic objects. These are described by their dynamic state (position, orientation, velocity and yaw rate) and often by their size.

The SLAM process deals with the static environment. The map is created only by integrating previously received static measurements. New static sensor readings update the map and the location of the ego-vehicle simultaneously, while dynamic sensor readings are regarded as noise and left out [1]. Also, vehicle odometry provides proprioceptive data to improve the location of the vehicle. The results of the SLAM process are a map and the location of the ego-vehicle in that map. In a succeeding step, static objects are extracted from the raw occupation information in the map.

Appropriate approximations must be made to reduce the computational complexity of solving both processes simultaneously.

Wang [1] integrated SLAM and DATMO together first. He estimated SLAM and DATMO posteriors in a Bayesian formulation separately. Object movements are tracked by an IMM filter. For SLAM he used a scan-matching technique, the Iterative Closest Point (ICP) algorithm. The system was tested in urban scenarios in a vehicle equipped with laser scanners and sensors to improve the localization (e.g. GPS, compass, IMU).

Vu [2] introduced a new method to improve the localization obtained by odometry. In a grid-based SLAM approach he was able to detect moving object by inconsistencies between free-space and occupation. Tracking was achieved by an IMM filter.

Most approaches base on Bayesian statistics. Other mathematical theories used for SLAM and DATMO can also be found in literature, e.g. based on fuzzy logic and Dempster-Shafer theory [3].

## 3 Demonstrator

As a demonstrator car we use a Mercedes-Benz E 500 (W212). It is equipped with a long-/mid-range radar sensor and two laser scanners. The sensors are mounted at the front of the vehicle (see Fig. 1).

The laser scanners have a maximum range of 100 m and a field of view of 100° with an angular resolution of 0.25°. The scan rate is set to 25 Hz. To increase the common field of view, the laser scanners are turned outwards on each side. For this reason, the common field of view is 140°.

**Fig. 1** A Mercedes E-Class is used as a demonstrator car. It is equipped with two laser scanners and a long- and mid-range radar. A Velodyne HDL-64E and a DGPS-module are used for evaluation only. A robust bumper is mounted for crash scenarios

The radar sensor has two alternating modes. In the long-range mode maximum range is 174 m and the field of view is 20°, while in the mid-range mode maximum range is 60 m and the field of view is 90°. The cycle time for each mode is 50 ms. A camera is mounted for documentation and visualization purposes. A 64−layer Velodyne HDL-64E laser scanner is mounted for evaluation.

# 4 Framework Architecture

For the application, no global localization and mapping is necessary nor does a priori knowledge of the world exist. It is assumed that the vehicle starts at pose $X_0$ $[x\ y\ \Psi]^T = [0\ 0\ 0]^T$, i.e. at the origin of the map. The occupancy grid map (see Sect. 4.2.1) contains no information, which means every cell is occupied with probability 0.5, the so-called maximum entropy assumption.

SLAM is executed separately for each laser scanner. After that the scan data can be transferred into a common frame. Then, DATMO is performed simultaneously with measurements from both scanners.

Figure 2 shows an overview of the general framework architecture. The whole process is divided into four modules. For each module, several special types were implemented. Those that work best for the evaluated scenarios are presented.

SLAM is performed in the mapping and the localization module. DATMO is performed in the detection and the tracking module. The following section describes each module separately.

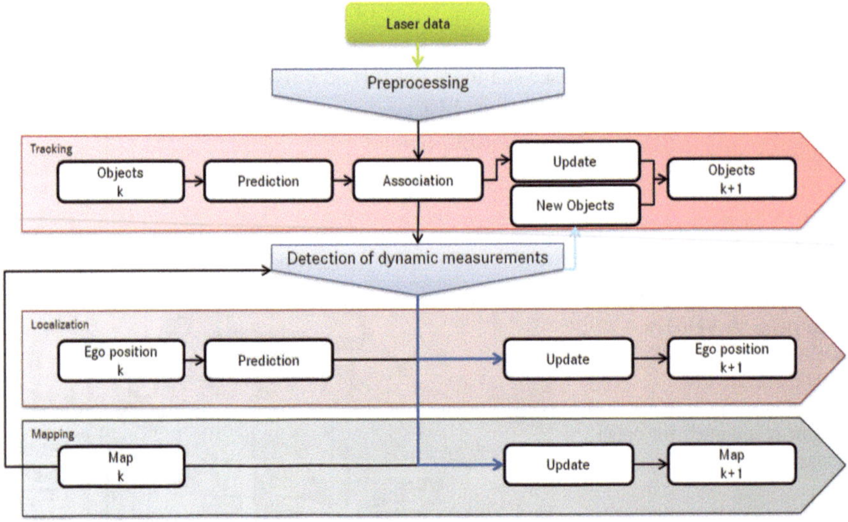


**Fig. 2** The figure shows how the laser sensor readings are processed and how the modules interact. Measurements classified as dynamic are shown in light blue, static measurements are dark blue

## *4.1 Localization*

Three localization modules with different complexities were implemented. Their appropriateness to the intended application in an outdoor environment turned out not to be equivalent. Since global localization doesn't have to be achieved in pre-crash scenarios, there must be a tradeoff between computational complexity and accuracy.

### 4.1.1 Odometry-Only Localization

A fast way to obtain the position and orientation of ego vehicle is dead–reckoning via vehicle's odometry sensor (velocity and yaw rate) from the starting position. This approach works well in standard scenarios with no extreme maneuvers. In those scenarios the drift can be neglected for the present application. The main advantage is the efficient computation of the algorithms using a standard Kalman filter.

### 4.1.2 Feature-Based Localization

Originally, feature maps are used in SLAM. A feature is a specific structure in the laser scan that can be found in consecutive scans such as points, lines, corners or other more abstract structures. Especially in indoor areas this approach provides robust maps that in turn allow achieving an accurate localization. In the present system, this method of localization allows creating an additional occupancy grid map with those "known" poses.

However, in outdoor areas obtaining robust features in every scan is challenging, especially when using range data that represents a slice of the world instead of real 3D information. A wide variety of environments with e.g. unstructured objects like bushes makes it difficult to find robust features. The approach proved to be inappropriate.

### 4.1.3 Occupancy Grid-Based Localization

In this method occupancy grid maps are used for localization instead of feature maps. The general advantage of those maps that contain the collected sensor information on a raw level also benefits the localization (see Sect. 4.2).

The present approach uses a particle filter. Each particle samples from an ego–vehicle model [4]. For each sampled pose, the laser scans are compared to the built occupancy map. The map is updated based on the probabilistic distribution represented by the particles. In contrast to approaches like the grid-based FastSLAM algorithm [4], it was found that not every particle needs to estimate its own map, since the accuracy of one common map is adequate. A slight drift error relative to the DGPS pose over time does not affect the performance of the system, which only depends on the position relative to the local surrounding map and its quality.

## *4.2 Mapping*

Since feature-based localization proved not to be robust enough, only a occupancy grid map is used. From this map, it is possible to extract static objects by searching clusters of occupied cells.

In occupancy grid maps the 2D space around the ego vehicle is divided into cells for which probabilities of occupation can be inferred via laser scanner measurements in combination with a probabilistic sensor model. Occupancy grid maps are first introduced by [5].

Occupancy grid maps store and combine the sensor readings on a low level. A beam model [4] is used as the sensor mode, a probabilistic model that incorporates free-space. Free-space is important to correct map errors resulting from slight localization drift. A decay factor helps to wipe out old and unnecessary occupation information over time.

## *4.3 Detection of Moving Objects*

The detection of moving objects step distinguishes between static and dynamic laser measurements. Radar data provides range rate information to extract moving objects out of laser data via a cluster-to-cluster association. Grid-based methods as proposed in [6] are currently evaluated and show a high noise level for 2D laser scanners.

As shown in Fig. 2 the mapping module includes static data, whereas dynamic data is processed in the tracking module.

## *4.4 Tracking*

### 4.4.1 Particle Tracking and Shape Estimation

Tracking is performed by a newly developed algorithm. In this algorithm the shape of the tracked object and its position are estimated simultaneously, comparable to the SLAM problem, since especially in pre-crash situations a box model of vehicles might not be appropriate. In Fig. 3 the curved front of a car can be seen which might be the reason that a crash will not take place. A pre-crash system however has to react conservatively and must not trigger mitigation algorithm

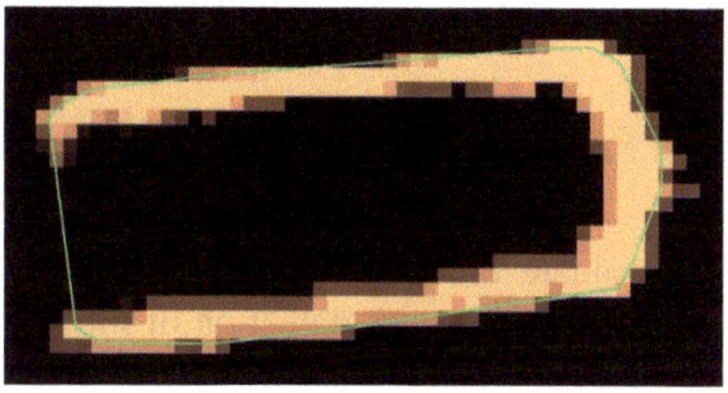

**Fig. 3** Local grid of a Mercedes C-class T-model. Orange cells denote the probability of occupation of each cell, black means unknown. The green line extracts the convex shape of occupied cells

when a crash is avoidable. Objects with a more complex shape can easily be imagined, where a box model would perform even worse.

A particle filter is used to track each object separately. Each particle has two parts (since it is *Rao-Blackwellized*). A dynamic state that is estimated by sampling and a shape that is represented by an object-local occupancy grid map containing occupation and an offset vector that describes the translation from a fixed anchor point in the map to the center of gravity of the car. The offset vector is modeled by a Gaussian random variable and filtered by a standard Kalman filter.

By applying this new algorithm, it is possible to avoid the known problem of the moving reference point on an object when changing the perspective, while simultaneously estimating its state and obtaining its shape. For details please see a separate paper by the authors.

Figure 3 shows the resulting local map of the estimation process of a Mercedes C−class. In most cases the obtained grid shape will not be as complete as shown in the picture. This is because of the special trajectory the target object drove for demonstration purposes on the test track.

The green line shows the extracted shape. The extraction is a post-processing step to achieve an abstract shape approximation for the following processing steps in the pre-crash system.

### 4.4.2 Decoupled Tracking

Wang [1] has shown that by estimating a joint probability over the ego-vehicles position, the mapping and the tracked targets, errors in localization directly influence the tracking accuracy. Since the localization error increases over time when no loop is closed (i.e. no area is visited again), tracking uncertainties will also increase over time.

Decoupling the tracking estimates from the ego-pose estimation and track in a relative frame avoids this problem. The ego-motion is compensated by the ego-velocity and yaw rate is improved by the localization module.

# 5 Evaluation

## *5.1 Static Objects*

To evaluate the performance for static objects, i.e. the evaluation of primarily SLAM modules with detection of moving objects that are filtered out, real world scenarios were recorded. Two of these are described here.

The first scenario is a test drive through a city. The distance traveled is about 2 km. Many moving objects such as pedestrians and cars are detected. Two crossroads are passed. The estimated trajectory is compared to a trajectory recorded by a DGPS module. The RMS error is 2 m compared to an odometry-only error of 36 m.

Since highly dynamic maneuvers would have been too dangerous to perform, a second scenario was evaluated. There the vehicle drove with high velocity around a 90° turn with a small turn radius of about 10 m. Shortly after the turn there was a narrow passage. The width of this passage was the car width plus 30 cm on each side and minus 20 cm on each side, respectively. Velocity was about 32 km/h during the turn. Measurements where mainly reflected by bushes at the side of the street. The algorithm presented in Sect. 4.1.3 could robustly distinguish between crash and no crash situations, whereas localization by odometry did not allow generating a map where the obstacle could be passed. RMS error with particle localization was 0.25 m, with odometry 0.7 m.

The absolute RMS error however doesn't have the effect that it would have in a map which is known a priori. Since the occupancy map is built on the basis of the estimated position, the map is better locally than the result would have led one to believe. This map correction is mainly achieved through the sensor model allowing the integration of free space and a decay factor for occupancy values. Figure 4.

**Fig. 4** Scenario 1: Typical urban scenario showing pedestrians with a baby buggy and cars

## 5.2 Tracking

The detection module successfully detects cars in the urban scenario. Pedestrians however are challenging since slow velocity results in a low signal-to-noise ratio for detecting movement from scan to scan. Also, the tracking algorithm works best for extended objects such as cars, while a standard point tracking algorithm may suffice to track pedestrians.

Figure 3 shows the result of the local map building process of the tracking module on a test track. The local occupancy grid is built with associated laser measurements. The sensor model does not represent free-space, so currently there is the differentiation between occupied and unknown space. This is to reduce the computational complexity, but might be integrated in the future. The target car is a Mercedes C-class T-model with a width of 1.77 m and a length of 4.59 m. The estimated size is 1.83 cm $\times$ 4.4 m. The occupied cells allow inferring a quality measure for these quantities. An initialization step for the dynamic state of the car is necessary to further decrease the number of particles from 40 particles.

# 6 Conclusion

A flexible probabilistic environment perception framework that can be applied to advanced driver assistance systems was presented. Further investigations concerning the reduction of computational complexity are necessary, but due to the particle filter approaches in the different module a highly parallelized implementation is possible.

# References

1. Wang CC, Thorpe C, Thrun S (2002) Simultaneous localization and mapping with detection and tracking of moving objects. IEEE Int Conf Robot Autom (ICRA) 3:2918–2924
2. Vu TD, Burlet J, Aycard O (2011) Grid-based localization and local mapping with moving object detection and tracking. Inf Fusion 12:58–69
3. Gambino F, Oriolo G, Ulivi G (1996) Comparison of three uncertainty calculus techniques for ultrasonic map building. In: SPIE international symposium on aerospace/defense sensing and control, vol 2761 pp 249–160,
4. Thrun S, Burgard W, Fox D (2005) Probabilistic robotics, MIT Press
5. Vu, TD (2009) Vehicle perception: Localization, mapping with detection, classification and tracking of moving objects, PhD Thesis, Grenoble Institute of Technology
6. Schütz M, Wiyogo Y, Schmid M, Dickmann J (2012) Laser-based hierarchical grid mapping for detection and tracking of moving objects. In: Meyer G (ed) Advanced microsystems for automotive applications 2012. Springer, Berlin

## 5.2 Tracking

The detection module successfully detects cars in the urban scenario. Pedestrians however [illegible]

[illegible]

## 6 Conclusions

[illegible]

## References

[illegible]

# Car2Pedestrian-Communication: Protection of Vulnerable Road Users Using Smartphones

**Sebastian Engel, Claudia Kratzsch and Klaus David**

**Abstract** The protection of vulnerable road users, like pedestrians and bicyclists, and the communication between vehicles and their environment, which is called Car2X-Communication, are two scopes in the development of the automotive industry that have gained much interest in the last years. This paper links together both topics by presenting a concept for the communication between vehicles and vulnerable road users, called Car2Pedestrian-Communication. A system architecture is presented that uses smartphones for the detection of vulnerable road users and for the transmission of data between the vehicle and the pedestrian. Based on the transmitted information, the collision risk between the vehicle and the pedestrian is estimated and a warning is given to the driver and the pedestrian.

**Keywords** Vehicle safety · Pedestrian protection · Car2X communication · Smartphones

## 1 Introduction

Actual developments in the automotive industry for enhancing traffic safety deal with the communication between vehicles among each other (Car2Car-Communication) and with the communication between vehicles and infrastructure

S. Engel (✉) · C. Kratzsch
Audi AG, Department of Vehicle Safety, 85045 Ingolstadt, Germany
e-mail: sebastian.engel@audi.de

C. Kratzsch
e-mail: claudia.kratzsch@audi.de

K. David
Chair of Communication Technology, University of Kassel, Wilhelmshöher Allee 73 34121 Kassel, Germany
e-mail: david@uni-kassel.de

J. Fischer-Wolfarth and G. Meyer (eds.), *Advanced Microsystems for Automotive Applications 2013*, Lecture Notes in Mobility, DOI: 10.1007/978-3-319-00476-1_4,

components like traffic lights (Car2Infrastructure-Communication). However, vulnerable road users (VRUs), like pedestrians and bicyclists, are not regarded in the development so far. In the German traffic situation of 2011 though, in every fourth traffic accident a pedestrian or bicyclist was involved [1]. This means that VRUs should be considered in the further development of Car2X-Systems.

This paper is organized as follows: In the next chapter the use of a radio-based pedestrian protection system is motivated based on an analysis of typical accident scenarios between pedestrians and vehicles in Germany. Afterwards the developed system architecture is presented. Based on different measurement setups, the suitability of the proposed concept for an automotive pedestrian protection is shown in chapter four. The last chapter concludes this paper and gives an outlook on further research and improvements.

## 2 Motivation for a Radio-Based Pedestrian Protection

To motivate the development of a radio-based pedestrian protection system the GIDAS (German In-Depth Accident Study) database [2] was analyzed regarding the most frequent accident scenarios between passenger cars and pedestrians/bicyclists. Since 1999 accidents in the areas of the German cities Dresden and Hannover have been analyzed in detail to obtain accident data. The database was filtered for accidents between passenger cars and pedestrians as well as bicyclists, where the pedestrian/bicyclist was injured heavily or killed. Afterwards the different scenarios, that look equivalent in the view of the sensor, were classified into categories. This led to three main categories for the pedestrian and the bicyclist case, which are shown in Table 1.

Regarding the pedestrian accidents, it can be seen that half of the accidents happen on straight roads without a visual obstruction of the pedestrian, which

**Table 1** Accident scenarios of passenger cars (→) with pedestrians/bicyclists (→)

| **Pedestrians** | | **Bicyclists** | |
|---|---|---|---|
| 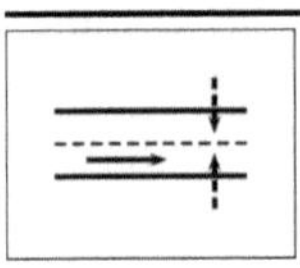 | Accident with a pedestrian crossing the street without visual obstruction **49.4 %** | 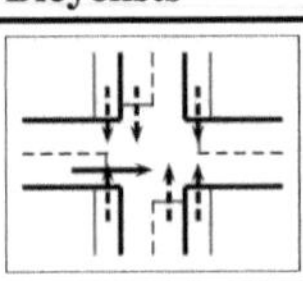 | Accident with a bicyclist at an intersection **65.1 %** |
| 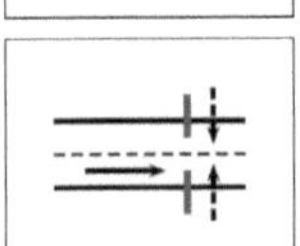 | Accident with a pedestrian crossing the street with visual obstruction **29.7 %** | 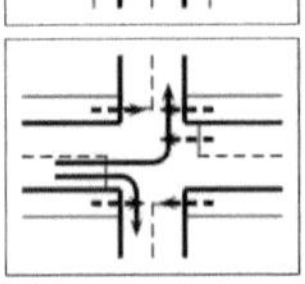 | Accident with a bicyclist crossing the street while turning into this street **16.5 %** |
| 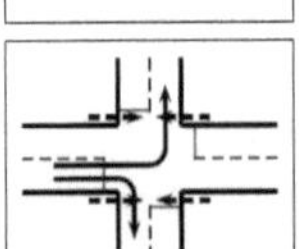 | Accident with a pedestrian crossing the street while turning into this street **7.6 %** | 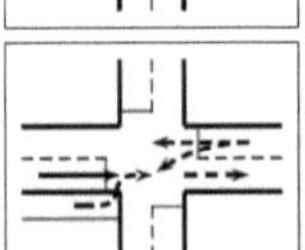 | Accident with a bicyclist in longitudinal traffic **9.1 %** |

means that these scenarios can be addressed well by e.g. cameras or laser scanners [3]. However, a great fraction of accidents happens if the pedestrian is not visible or if the vehicle turns into a street. In these cases, a sensor is needed that detects pedestrians that are visual obstructed and delivers a wide opening angle. In the case of bicyclist accidents the need of a sensor with such characteristics becomes even more obvious. This is because the main fraction of accidents between vehicles and bicyclists happens at intersections or in turning scenarios, where the bicyclist is likely to be concealed by cars or buildings and hits the vehicle from the side. The state of the art sensor systems for pedestrian and bicyclists protection like cameras or laser scanners can hardly fulfill these requirements. On the contrary, systems that work with radio waves can detect hidden VRUs because of the characteristics of radio waves and may deliver a virtual 360° opening angle, depending on the radiate characteristic of the antenna. This is why a radio-based approach is promising for the development of future pedestrian and bicyclist protection systems.

# 3 Car2Pedestrian System Architecture

In this section, the architecture of the developed system is presented. Smartphones are used to gather information about the position, speed and moving direction of the pedestrian and to transmit the data to vehicles in the environment. For the system only components-off-the-shelf are used, which means that no extra hardware had to be developed.

## *3.1 Communication Technology*

For establishing a communication between the vehicle and the pedestrian different technologies exist, especially if smartphones are used. Typical communication interfaces are WLAN, GSM/UMTS, Bluetooth or NFC (Near Field Communication). To identify the technology which fulfills the requirements of a Car2Pedestrian system the best, the GIDAS database was evaluated concerning the accident speeds of vehicles in crashes with pedestrians. The analysis showed that nearly 90 % off all accidents happened with vehicle speeds up to 70 km/h or approx. 20 m/s. Equation 1 describes the necessary braking distance of a vehicle with an initial speed $v_0$, a maximum brake acceleration $s_{brake}$, a brake inflation time $t_{fill}$ and a driver reaction time $t_{react}$:

$$s_{brake} = \frac{v_0^2}{2 \cdot a_{brake}} + v_0 \cdot t_{fill} + v_0 \cdot t_{react} \quad (1)$$

**Equation** 1. Braking distance of a vehicle, including the brake inflation time and the driver reaction time.

While the first summand describes the distance it takes the vehicle to decelerate from the initial speed to standstill, the second summand describes the distance it takes until the full break pressure is applied. In case of an automatic braking system, that does not need the input of the driver and brakes on its own, this is the total braking distance. If the driver is only warned and has to brake by himself, also the reaction time and its corresponding distance have to be considered, which is described by the third summand. The reaction time depends on several factors, e.g. the weather conditions or the necessity of the driver to turn its head, and is often assumed to be around one second [4]. To simplify further calculation, $a_{brake}$ is assumed to be 10 m/s$^2$, which equates to a maximum value that is only achievable with a strong break and optimal road grip. The brake inflation time is assumed to be 250 ms. In this case the total braking distance from an initial speed of 20 m/s adds up to 25 m for an automatic braking system and 45 m for a warning system.

Because of the required communication range that was deduced from the equation above, Bluetooth and NFC were disregarded in the development of the protection system, as the achievable communication range is too short.

Another criterion to evaluate the suitability of the different technologies is the communication latency. In [5] the latency of WLAN communication was compared to the latency of cell technology communication like GPRS and UMTS. For WLAN the latency was in the range of a few Milliseconds, for all cell technologies in the range of several hundred Milliseconds.

Therefore it was decided to use WLAN to establish the communication between the smartphone and the vehicle, as it delivers the best trade-off between availability, range and latency.

## 3.2 Hardware Description

To determine and communicate information about the pedestrian, Android smartphones are used. The GPS modules of the smartphones are used to determine the position of the pedestrian, which deliver this information with an update rate of 1 Hz. The connection between the smartphones and the vehicle is established by WLAN 802.11 b/g/n using external antennas and a D–Link DIR-825 router which is connected to a PC in the vehicle. The computer is connected to the vehicle CAN Bus (Controller Area Network) to gather information about the vehicle dynamics like the speed or the lateral and longitudinal acceleration with an update frequency of 50 Hz. To obtain the position of the vehicle in GPS coordinates, an uBlox EVK–6T evaluation kit is used which delivers the position information every 200 ms.

## 3.3 *Software Architecture*

A distributed system was developed to reduce the computational effort on the side of the smartphone and the vehicle. Different software modules are implemented in the smartphone and the vehicle PC, which complement one another.

### 3.3.1 Smartphone Software Architecture

On the side of the pedestrian, the GPS receiver of the smartphone determines the position, moving direction and speed and delivers these information over the NMEA 0183 (National Marine Electronics Association) protocol. The data are received by the smartphone software module and the position coordinates are transformed into Universal Transverse Mercator (UTM) coordinates. Therewith further calculation of other software modules (graphical representation, collision detection) can be executed easily by using Cartesian coordinates in a plane.

Moreover, the current movement state of the pedestrian is determined by an evaluation of the acceleration sensor and the gyroscope which are installed in the smartphone. The data are captured with a frequency of 40 Hz and stored in a window with a length of 128 samples and an overleap rate of 75 %. The following features are extracted out of the window:

- standard deviation and frequency entropy of the acceleration values (x-, y- and z-axis)
- standard deviation of the angular speed values (x-axis)
- frequency entropy of the angular speed values (x-, y- and z-axis).

Different activities (standing, walking, running, biking) were performed by 10 persons to generate a training set for a classification with a decision tree. The decision tree is coupled with a state machine to prevent impossible state changes (e.g. direct state transition from walking to biking). The activity information will be used as an input for different software modules in the future, e.g. for the correction of the pedestrian position (dead reckoning module) and the adaption of the warning algorithm, that is implemented in the vehicle.

In the last step, the UTM coordinates, speed, moving direction and activity of the pedestrian are broadcasted within the local network by UDP (User Datagram Protocol) packets. To distinguish between different smartphones, a randomized ID is attached to the packet.

### 3.3.2 Vehicle Software Architecture

A PC is installed in the vehicle that is connected to a router to receive the information of the smartphones. Furthermore the PC is connected to the uB-lox GPS Receiver and the vehicle CAN Bus, to obtain the position and driving direction information as well as the vehicle speed and the longitudinal and lateral acceleration data. To simplify further calculation steps, also the vehicle GPS data are transformed into UTM coordinates.

For the detection of a possible collision between the pedestrian and the vehicle, their positions and movement states have to predicted in each time step. For the vehicle, a constant acceleration model (CA) is assumed. The model predicts the position as well as the longitudinal and lateral velocity of the vehicle according to the current position, longitudinal and lateral velocity as well as the current longitudinal and lateral acceleration. The prediction is described in Eq. 2, where $y$ and $x$ describe the position of the vehicle in UTM coordinates (y: north value, x: east value) and $dt$ the step interval. Moreover $v_y$ and $v_x$ describe the velocities in the corresponding direction, $a_y$ and $a_x$ the acceleration in the corresponding direction. The longitudinal speed and acceleration as well as the lateral acceleration are delivered in reference to the vehicle coordinate system. Therefore these values are transformed into the corresponding north- and east values by a rotation according to the geographical driving direction of the vehicle, which is obtained by GPS.

$$\begin{pmatrix} \mathrm{y}(\mathrm{k}+1) \\ \mathrm{x}(\mathrm{k}+1) \\ \mathrm{v_y}(\mathrm{k}+1) \\ \mathrm{v_x}(\mathrm{k}+1) \end{pmatrix} = \begin{pmatrix} \mathrm{y}(\mathrm{k}) \\ \mathrm{x}(\mathrm{k}) \\ \mathrm{v_y}(\mathrm{k}) \\ \mathrm{v_x}(\mathrm{k}) \end{pmatrix} + \begin{pmatrix} \mathrm{v_y} \\ \mathrm{v_x} \\ \mathrm{a_y} \\ \mathrm{a_x} \end{pmatrix} \cdot \mathrm{dt} + \begin{pmatrix} \frac{\mathrm{a_y}}{2} \\ \frac{\mathrm{a_x}}{2} \\ 0 \\ 0 \end{pmatrix} \cdot \mathrm{dt}^2 \tag{2}$$

**Equation** 2. Prediction of the vehicle state from time step k to k + 1.

Similar to the prediction of the vehicle state the position of the pedestrian is predicted. This is done by a constant velocity model (CV), where it is assumed that the speed and the walking direction of the pedestrian do not change within the prediction period. The prediction is described in Equation 3, where $y$ and $x$ - describe the north- and east coordinates, $v$ describes the walking speed and $\varphi$ the geographical walking direction of the pedestrian.

$$\begin{pmatrix} y(k+1) \\ x(k+1) \\ v(k+1) \\ \varphi(k+1) \end{pmatrix} = \begin{pmatrix} y(k) \\ x(k) \\ v(k) \\ \varphi(k) \end{pmatrix} + \begin{pmatrix} v(k+1) \cdot \cos(\varphi(k)) \\ v(k+1) \cdot \sin(\varphi(k)) \\ 0 \\ 0 \end{pmatrix} \cdot dt \tag{3}$$

**Equation** 3. Prediction of the pedestrian state from time step k to k + 1.

It has to be noted that the speed and walking direction, which are obtained by GPS, can be erroneous, especially if the pedestrian changes his movement behavior suddenly. This can lead to a falsified prediction which might cause a false alarm afterwards. Therefore movement recognition algorithms will be integrated in the position estimation and prediction in the future development, which use the smartphone sensor data like a magnetic compass, the accelerometer sensor or the gyroscope.

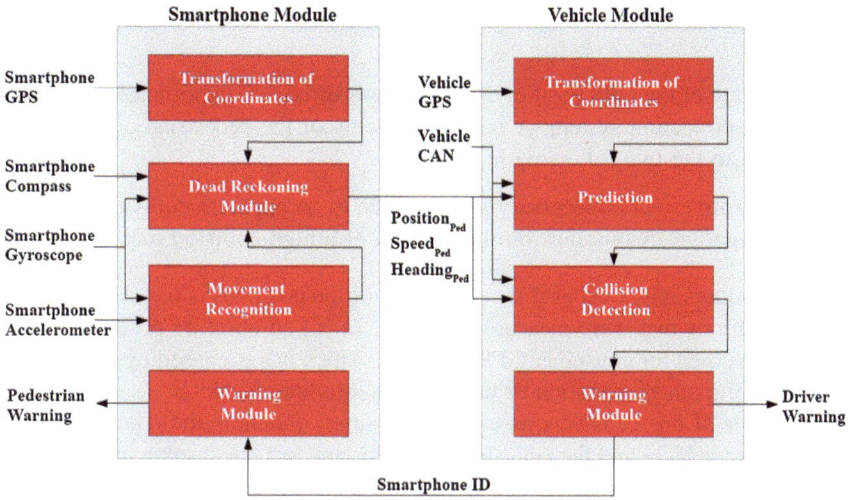


**Fig. 1** Software architecture of the developed protection system

To evaluate the criticality of a possible collision, the positions of the vehicle and the pedestrian are predicted according to the formulas given above. If the predicted positions of the vehicle and the pedestrian intersect, the criticality is assessed as a function of the necessary time prediction interval and the geometric constellation.

If the criticality exceeds a certain threshold a warning is given to the driver. Furthermore an UDP packet is sent to all smartphones within the local network that contains the ID of the smartphone that belongs to the endangered pedestrian. The software module of each smartphone in the network evaluates the received message and triggers a warning sound if the received ID matches the own ID. With this method, a two-way protection function is implemented. The whole system concept is shown in Fig. 1.

## 4 Performance Measurements

To evaluate the suitability of the architecture for a pedestrian protection system, different measurement series were carried out. These included the analysis of the smartphone GPS accuracy, the communication range as well as the energy consumption. The results will be described in the following sections.

## *4.1 GPS Accuracy*

The accuracy of the GPS position information is of crucial importance for a robust pedestrian protection system. A false positioning of the pedestrian or the vehicle could lead to the two following malfunctions:

- False positive: An unnecessary warning due to an apparent collision
- False negative: A missing warning in case of a high collision risk.

Both failures might occur if the prediction of the positions is based on incorrect initial positions that were measured by GPS. In [6] an absolute position error of 10 m is assumed for consumer GPS receivers, which is e.g. caused by atmospheric effects, multipath propagation or drifts in the satellite clocks.

However, if one considers an automotive safety function, the relative distance between the vehicle and the pedestrian is important, not the absolute position with reference to the GPS coordinate system. If the pedestrian and the vehicle are close to each other, one can assume that the GPS receivers evaluate the signals of similar GPS satellites, suffer from the same atmospheric disturbances and underlie similar multipath effects due to the same buildings in the environment. The short distance between the vehicle and the pedestrian is not a heavy assumption in this case, because this is exactly the field of application for a Car2Pedestrian protection system.

Different static measurements were performed in urban and rural scenarios to assess the absolute and relative positioning accuracy of smartphone GPS receivers. Some of the results are shown in Fig. 2:

While the measurement points 1 and 5 were located in a rural area, the measurement points 3–5 were located in an urban area. Due to multipath effects, the

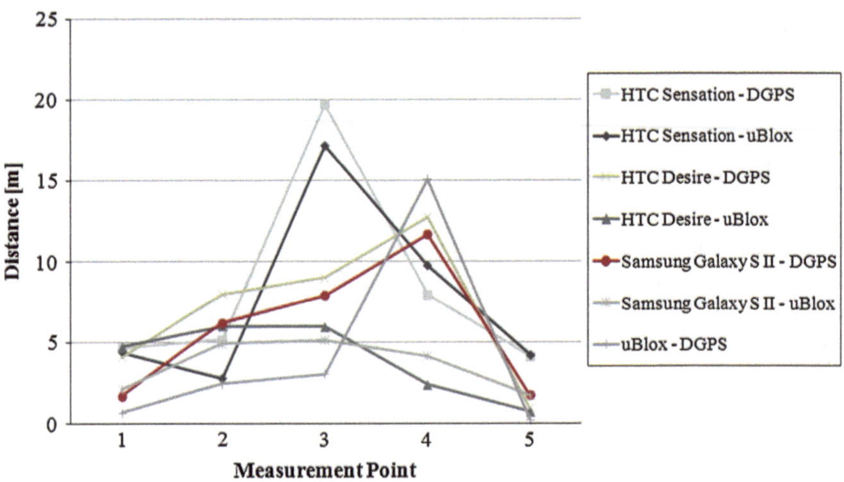


**Fig. 2** Accuracies of smartphone GPS measurements against DGPS and uBlox measurements

error is greater there. Figure 2 shows the distance between the smartphone GPS position and the reference position as well as the distance between the smartphone GPS position and the uBlox GPS position. The reference position was measured by DGPS (Differential GPS) and defines the point where the receivers were placed. As it is to be seen, the error between the smartphones and the vehicle uBlox GPS receiver is usually 2 m smaller than the error between the smartphone and the reference position. This is because of similar error sources which affect the measurements of the smartphone and the vehicle GPS receivers. However, for a warning system the average error between the smartphones and the uBlox GPS receiver of approx. 5 m is still too high and could cause false warnings and missing warnings. Therefore, further improvements regarding the positioning of smartphones is necessary.

## *4.2 Communication Range*

To evaluate the performance of the WLAN communication regarding the possible range, two different series of measurements were carried out. Both series were performed statically, where the test subjects carried different smartphones in the pockets of their trousers and moved away from the standing vehicle. The smartphones broadcasted UDP packets with a frequency of 40 Hz. The tests were performed in a free field scenario without any obstacles and on a crowded parking area with shadowing effects by other cars. Afterwards the received packet rate as a function of the distance of the pedestrian to the vehicle was evaluated. Some of the measurements are shown in Fig. 3.

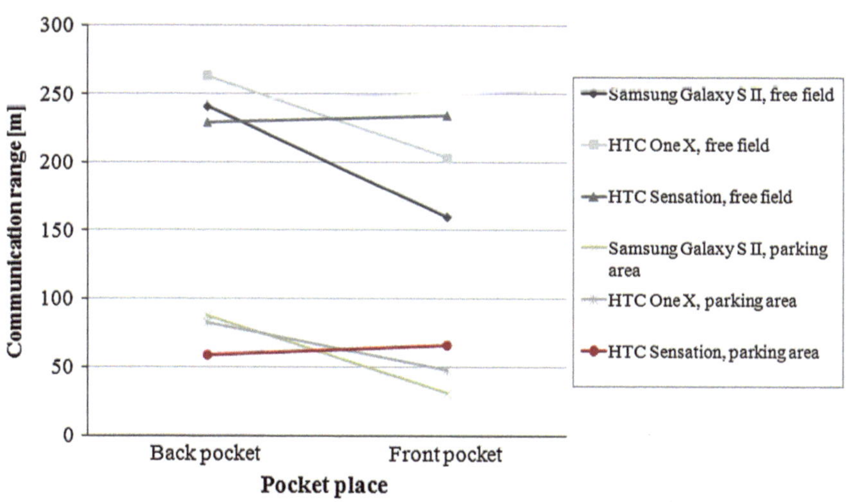


**Fig. 3** Measurements of the WLAN range with different smartphones

The rate of successful received packets did not depend on the distance between the pedestrian and the vehicle. However, in nearly all measurements a shorter range could be noticed if the smartphone was worn in the front pocket. This is because the test subjects moved away from the vehicle forwards and the human body shadowed the antenna of the vehicle in this case.

As it is to be seen, in the free field scenario a range up to 200 m was achieved. In the parking scenario however, the range reduced to approx. 60 m. Nonetheless the necessary communication range of 25 m and 45 m, that were deduced from the accident speeds in Sect. 3.1 were achieved. It has to be said that the communication range depends on several physical factors like the transmitting power, the antenna gain and the geometry of the vehicle, but even without special hardware the necessary range with respect to the accident analysis was attainable.

### *4.3 Energy Consumption*

In contrast to passive Radio Frequency Identification (RFID) tags, that are stimulated and powered by electromagnetic waves of the reading device, the Car2Pedestrian system uses an active scheme. That means that the smartphones have to transmit the information on their own and are not detected passively.

The whole developed software architecture that is implemented in the smartphones was evaluated regarding the energy consumption. Therefore the smartphones were carried by a test subject and all software modules were activated. The tests were performed outdoors to obtain a GPS fix and the data were transmitted via WLAN. The average overall operating time during the tests was about 5 h. This value mainly depends on the battery capacity, which is different for each smartphone and the lifetime of the battery. Therefore no generalized statements regarding the energy consumption can be made. However, one can say that the energy consumption is not higher compared to other regular utilizations like surfing the internet or using an application.

## 5 Conclusion and further research

An architecture was presented that uses smartphones as sensor devices for a radio-based protection system. Because of the characteristics of radio waves this system has the capability to address the major accident scenarios of pedestrians and bicyclists with vehicles. No special hardware was used to demonstrate the feasibility of the proposed architecture. The system detects critical traffic situations between passenger cars and pedestrians and delivers a warning to the driver as well as the pedestrian to prevent a possible collision.

The main challenges for a successful application of the system remain in the GPS positioning accuracy and the suitable communication technology. For a warning system, the relative distance between the vehicle and the pedestrian has to be determined exactly. For a braking system, this requirement is even stronger, to prevent false warnings and, as a consequence thereof, wrong brake applications. Therefore the motion sensors of the smartphones will be evaluated by a dead reckoning module in the future development to deliver higher accuracy and a positioning solution if no GPS is available. Regarding the communication it could be shown that the achievable range is sufficient in terms of accident speeds. However, a fast connection establishment is necessary to transmit the information to an early point of time. Therefore ad hoc communication solutions for consumer WLAN are required. Another possibility would be to use future fast cell technologies like Long Term Evolution (LTE) in combination with a server. In this way a connection would be given instantly and the connection time could be disregarded. In this way the Car2Pedestrian system would offer a benefit for pedestrian safety.

## References

1. Statistisches Bundesamt, "Verkehrsunfälle - Zeitreihen 2011", 2012
2. German In-Depth Accident Study, "Unfalldatenbank", last access: 22.11.2011. [Online]. Available: http://www.gidas.org
3. Gandhi T, Trivedi MM (2007) Pedestrian protection systems: issues, survey, and challenges. IEEE Trans Intell Transp Syst 8(3):413–430
4. Dettinger J (2008) Reaktionsdauer bei Notbremsungen - Entwicklung und Status quo des Erkenntnisstandes. Verkehrsunfall und Fahrzeugtechnik 6:180–187
5. Flach A, David K (2009) A physical analysis of an accident scenario between cars and pedestrians. In: IEEE 70th vehicular technology conference fall, pp 1–5
6. Mansfeld W (2010) Satellitenortung und Navigation. Vieweg + Teubner Verlag, Wiesbaden

# Retrieving Human Control After Situations of Automated Driving: How to Measure Situation Awareness

**Arie P. van den Beukel and Mascha C. van der Voort**

**Abstract** When applying automated driving as a means for congestion assistance, developers need to account for regular and sudden transitions between automation and manual control. To assess then the ability to retrieve human control, we tested within a driver simulator experiment how Situation Awareness (SA) gained during time-critical take-overs, and based on a simulation freeze technique (SAGAT) and SA queries (SART), could be measured. The results show that drivers are able to build up SA in these time-critical situations. With regard to measurement method, only SART showed positive correlation with successfulness in retrieving control. Nonetheless, the applied measurement method showed to allow future selection of support types drivers are in need for when retrieving control after automation.

**Keywords** Automated driving · Congestion assistance · Situation awareness · Take-over control

## 1 Introduction

Demos with prototype vehicles show far reached technical capabilities for automated driving [1]. The application of automated driving enhances comfort under monotonous and tiresome driving conditions. Moreover, automated driving is expected to offer a major contribution to more efficient and environmental-friendly road transport [2]. The automotive industry's current focus for automated driving

A. P. van den Beukel (✉) · M. C. van der Voort
University of Twente, Laboratory of Design, Production and Management,
Postbus 217 7500 AE ENSCHEDE, The Netherlands
e-mail: a.p.vandenbeukel@utwente.nl

M. C. van der Voort
e-mail: m.c.vandervoort@utwente.nl

J. Fischer-Wolfarth and G. Meyer (eds.), *Advanced Microsystems for Automotive Applications 2013*, Lecture Notes in Mobility, DOI: 10.1007/978-3-319-00476-1_5,

is on developing congestion assistance providing automated lateral and longitudinal control during low speeds on motorways [1]. Due to this application, transitions between automation and manual driving occur regularly: For example when traffic speed exceeds the low speed threshold, a take-over will be requested by the system. However, during automation a driver is placed remote from the control loop. Likely, this causes slower reaction times and reduces driver's awareness of the situation. This in turn causes a misunderstanding of what corrective actions need to be taken [3]. Successful application of congestion assistance is therefore related to the effectiveness of the developed interface to support drivers in taking over control. To assess the effectiveness of such interface the driver's ability to gain Situation Awareness (SA) is an informative measure. The ability to take-over control is also influenced by timing aspects, i.e., the frequency and urgency of requested take-overs. Therefore, this paper presents the outcome of an experiment which aim is to investigate to what respect drivers build up SA in time-critical situations of taking over control after automation. Based on the results of the experiment, this paper provides conclusions with respect to the applicability of the methods tested to assess SA. Therewith, the results also support the development of (future) interfaces that provide appropriate assistance to successfully gain human control after automation.

## 2 Definition and Possible Methods to Measure Driver's SA

The first aim of this research is to assess how Situation Awareness (SA) gained during take-over of control can be measured. A reliably insight in SA gained during take-over will help to determine the support drivers are in need for as well as to assess the quality of possible interfaces providing this support. In this paper we will first determine which elements comprise operator SA within our take-over task to achieve a better understanding what we want to assess. Then we will review relevant existing measuring methods and choose applicable methods for our aimed SA-assessment.

### *2.1 Understanding How Drivers Achieve Situation Awareness*

The driving task basically involves five processing steps: (1) perception; (2) comprehension; (3) projection, (4) decision making and (5) implementing. Based on information detected on the state of the environment, drivers select courses of action that may or may not change their driving task. Driver actions include accelerating, decelerating, taking over, etc. With automated driving the driver is placed "out-of-the-loop", i.e., the driver is not, or less, involved in before mentioned processing steps. To determine which elements comprise driver's Situation Awareness, we first recall how SA is psychologically be defined. According to the most widely adopted definition [3] Situation Awareness is a psychological construct, containing three levels: (1) The observed presence or absence of elements in the situation; (2) The participants' comprehension of the meaning of these elements; and (3) The anticipated future state of the elements. Because Situation Awareness contains the first three of before mentioned processing steps (perception, comprehension and projection), SA is theoretically a prerequisite for the remaining steps and a necessity to enable performance of the complete driving task [5].

Understanding how task performance proceeds in sudden take-over situations is however encompassed with assumptions. It could be that with time-critical and unexpected take-overs control is more based on intuition, assuming a more or less impulsive step from 'perception' directly to 'implementation'. Nonetheless, Matthews et al. [5] related the three levels of SA to the three tasks (i.e., operational, tactical, strategic task) from Michon's generally accepted taxonomy of the driving task. Accident avoidance involves tactical tasks, requiring short span projection of the driving environment. According to Matthews' model, tactical tasks in turn involve especially level 1 and level 2 SA. As we are interested in measuring SA during take-over (accident avoidance) it seems plausible to especially focus on the first two levels of SA.

## 2.2 *Available Methods and Choice to Measure Situation Awareness*

Based on Salmon et al. [6] we consider 4 categories of SA measurement methods applicable within driving simulator experiments: Freeze probe techniques; Real-time probe techniques; Self-rating techniques; Performance measures.

Freeze probe techniques involve the administration of queries during 'freezes' in a simulation. The queries relate to probes and their answers are assumed to be representative for SA, e.g.: "Is there currently (i.e.: during 'freeze') a vehicle on the neighbouring lane?" During these 'freezes' the screens of a simulation are typically blanked and the queries are randomly chosen from a set of relevant probes. The primary advantage of freeze-probe techniques is their direct nature. A disadvantage, however, is their intrusiveness upon primary task performance. Within this category, the Situation Awareness Global Assessment Technique (SAGAT) [7] is most commonly used. Although especially developed for the aviation domain, the method is also applicable for SA assessment in driving [5].

Real-time probe techniques also use probes. The technique is intended to be less intrusive, while an expert administrates the real-time probes during task performance. Typical application is for time-taking supervisory tasks, like video-surveillance. As our application (emergency take-over) requires quick responses this method is not considered appropriate.

Self-rating techniques involve self-assessment by participants based on standardized queries. The main advantage is their non-intrusiveness, since they are typically administrated post-trial. However, self-rating techniques are criticized, due to interference of SA with task performance, their dependence on memory capacities and because of participant's inability to rate levels of SA—especially with regard to accurately measuring poor SA as participants may not realise that they have inadequate SA in the first place [7]. Due to the ease of application (fast and low cost) self-rating techniques are nonetheless wide-spread used and the Situation Awareness Rating Technique (SART) [8] is most popular.

Performance measures allow an indirect assessment of SA, for example lane position or crash avoidance. Their advantages are that they are objective and non-intrusive. The main problem when applying performance measures for SA-assessment is however the underlying assumption that efficient performance is achieved as a result of efficient SA. However, it may be that efficient performance is achieved despite an inadequate level of SA, or that poor performance is achieved regardless of a high level of SA (e.g. a novice driver with good SA, but inferior driving skills). However, since performance measures are normally taken during task performance anyway, they are valuable as a back-up and extra reference with regard to the success of performing a task.

Existing studies which focus on measuring SA during driving, mostly relate to SA while being involved in some secondary task (e.g. telephoning) versus 'normal' driving [4]. No studies have been found specifically focussing on drivers' SA during transition from higher to lower levels of automation. Based on afore review of techniques, we consider both SAGAT and SART most suitable for our intended SA assessment. Hence, both SAGAT and SART are most widely used and validated. SAGAT has demonstrated reliability and validity within driving task studies [6] and SART's main advantage is its ease of application. SAGAT's disadvantage of primary task intrusiveness will be accounted for in the specific set up of our experiment; Because fast response are linked to the nature of our experiment (emergency take-over) and therefore SA-measurement will take place directly after primary task performance. Moreover, by comparing correlation of the two methods with successfulness of taking over control, we will assess to what respect the disadvantages of SART (accuracy of self-assessment) come into play for our specific situation (take-over scenarios).

## 3 Method

### *3.1 Task*

The task used in this study was to avoid an accident, within a short period of time, after automation of the driving task was suddenly being terminated, due to a critical road situation. Participants were seated in a mocked-up vehicle placed in a

**Fig. 1** Driving simulator used for the experiment. Photo on the *right* shows secondary task on the touch screen at the centre console

simulated motorway environment (see Fig. 1), while driving automatically. For taking over control user inputs occurred through common automobile control interfaces, including a physical steering wheel and physical gas and brake pedals. Other vehicles drove in front and behind the simulated vehicle, as well as on the neighbouring lane. To simulated congested driving all vehicles were driving with time headways between 1 and 1.5 s at approximately 60 km/h. The vehicles differed in type and colour and the environment included exits, emergency lanes, grass, road markings and buildings, offering a realistic driving scenery. During automation no input from the driver was required and to simulate a driver being out-of-the-loop the driver was asked to perform a secondary task. This task was a motoric-visual task executed with a touch-screen located in the centre console below the dashboard (see Fig. 1). The task involved manipulation of a blue dot by moving it over the screen and avoiding collision with moving red coloured dots. This secondary task required right hand operation and to take the eyes off the road.

The critical road situation consisted of the car in front braking by $-8$ m/s$^2$, which corresponds with emergency braking and is far beyond the boundaries within which typical systems for automated driving can brake by themselves, thus requiring take over. The system's detection of the critical situation was announced by an alarming sound (A 3-tone sound of 82 Hz, exceeding the simulated engine and road roar with about 12 Hz). The time to react (i.e.: the time between the alarming sound and the moment an accident happens if the driver would not intervene) was varied to simulate three different levels of criticality. The participants received the instruction to take over control as soon as the situation requires, based upon their own judgement. Moreover, participants were advised to avoid an accident by swerving out if possible, but any kind of accident avoidance was allowed. The critical situations (i.e. the lead vehicle makes an emergency brake, marked with an alarming sound) were distinct from the automated driving situation. Hence, none of the participants took over control at another moment.

### 3.2 Experimental Design

The independent variables for the experiment comprised of 'criticality' (i.e. the time between the moment at which the front vehicle starts braking and the moment an accident would happen, if the driver does not react) and 'method' (i.e. SA assessment by using SART and SAGAT). 'Criticality' was manipulated within subject: Each participant was confronted with three different time conditions in which take-over was required. To avoid habituation, one condition was added in which no take-over was required because no critical road situation occurred. Each time condition was duplicated to make two sets: one set with 4 conditions while driving on the left lane, and one set while driving on the right lane. To increase data generation, all conditions were also repeated once, resulting in a $4 \times 2 \times 2$ design (16 trials totally per participant). The order of trials was randomized for each participant to avoid carry-over effects. The take-over required only a few seconds and the time beforehand (in which the vehicle was driving automatically and the participant performed the secondary task) varied between 15 and 20 s. After successful take-over the simulation was shut down. If an accident happened, the simulation also ended. Hereafter, SA was measured using both the Situation Awareness Global Assessment Technique (SAGAT) and Situation Awareness Rating Technique (SART). While the simulation stopped, the screens were blacked out and the experimenter subjected the participant to a SAGAT questionnaire. Each questionnaire presented a sample of three SAGAT questions from a pool of 11 questions. Within these queries, participants were required to e.g. recall car type, location and colour and give the answers promptly to the experimenter. After participants completed this questionnaire, they were required to fill out a SART questionnaire themselves, using paper and pencil. The applied SART method involves a 3-dimensions subjective, self-assessment of Situational Awareness. Based on Taylor [8], the questionnaire referred to (1) the involved cognitive *demand*, (2) the involved mental resources (*"supply"*), and (3) the gained *understanding* of the situation. After completing both the SAGAT and SART questionnaires, a new trial was started. Then, the driver (participant) was again 'launched' into the road situation, already driving automatically, with a new set of surrounding vehicles on the road. The successfulness of take-over was assessed through observation (i.e. the experimenter rated the successfulness with: "successful" and "unsuccessful").

### 3.3 Simulator Environment

The experiment was conducted with a high fidelity driving simulation, using a stationary (non-moving) driving simulator with 180° viewing angle. See also Fig. 1. The simulation was programmed using a modified version of "Drive v2.5" (developed by Re-lion) and Lua script. The simulator-vehicle's interior consisted

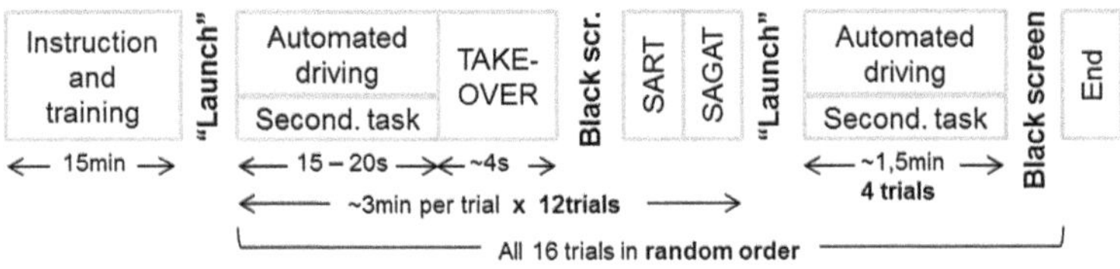


**Fig. 2** Schematically presentation of the experiment's procedure

of a complete dashboard. The interior was mounted in a neutrally designed car mock-up. The 180° viewing angle was realized by placing the mock-up in front of three connected screens each 2.5 m high and 3 m wide, illuminated by three opposing video projectors. The simulated vehicle has an automatic clutch and the mirrors were projected onto the screens. The elapsed time, vehicle speed and following headway were recorded at 5 Hz. Between experiments the position of the neighbouring vehicles was similar per time-condition. This is important to ensure that participants got the same chance of resolving the situation.

### 3.4 Participants and Procedure

Thirty-four university students were recruited for the simulator experiment, who had no prior knowledge of the study. All participants had normal or corrected to normal vision and were required to have at least 1 year of driving experience. Participation was voluntary. The average age was 21, 3 years.

Per participant the experiment lasted 1 h. The time schedule is shown in Fig. 2 and consisted of 16 times an approximately 3 min trial per participant. Per trial the automated driving lasted between 15 and 20 s until a take-over was requested. The take-over required only a few seconds. After take-over the questionnaires took approximately 2½ min. 4 Out of 16 trials required no take-over, each lasting ca. 1½ min. The participant started each trial manually via an ignition handle, then they were directly driving automatically.

### 3.5 Hypotheses

The aim of this experiment is to investigate to what respect drivers build up SA in time-critical situations of taking over control after automation and to assess what method for measuring SA is most suitable for these situations. Within various existing studies the SAGAT technique has proven to be superior in terms of reliability and sensitivity when compared to SART [6]. Also in view of the expected inaccuracy of self-assessment involved in SART, we therefore hypothesized in particular a positive correlation between SA and successfulness of retrieving control based on SAGAT measurement. Furthermore, we expected that the successful take-overs would correspond with the least critical time conditions, as these conditions allow more time to become situation aware.

# 4 Results and Discussion

Figure 3 presents the percentage of successful and non-successful take-overs per time condition (criticality). The graph shows that unsuccessful take-over ("Accident") occurs most often during the most critical time condition (0.5 s) and that the time-conditions relate positively to success-rate, i.e. the least critical time-condition has the highest success-rates. This is in line with expectations. A Pearson correlation coefficient revealed a significant correlation between the percentage of successful takeovers and criticality ($r = 0.541$, $p < 0.001$). Analysis of variance (ANOVA) indicated that the mean success-rates for each of the three conditions differ significantly [$F(2, 89) = 20.417$, $p < 0.001$], confirming the effect of criticality on success-rate.

Figure 4 presents the SA scores per time condition (criticality) provided by the SAGAT and SART questionnaires. Because the SART scores range from 1 to 14 [8] and the SAGAT questionnaire we used ranged from 0 to 15 (With respectively 14 and 15 denoting high SA), the scores of SART have been corrected to match the range of scores for SAGAT. (Further analysis was performed without any correction to avoid inaccuracies.) The plot reveals that based on SAGAT participants have on average best SA at the intermediate time condition (1.0 s) and worst SA for the least critical ('easiest') time-condition, which is contrary to expectations.

Analysis of variance (ANOVA) showed, based on either SART or SAGAT, that there was an interaction effect of criticality (time-condition) on SA [SAGAT: $F$(2,

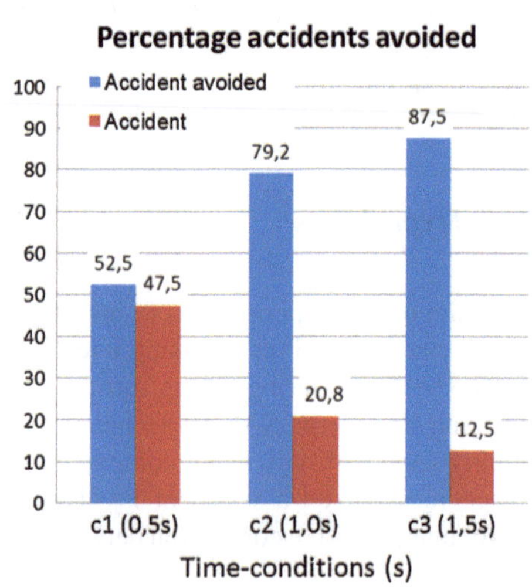


**Fig. 3** Percentage successful (accident avoided) and non-successful take-overs per time-condition

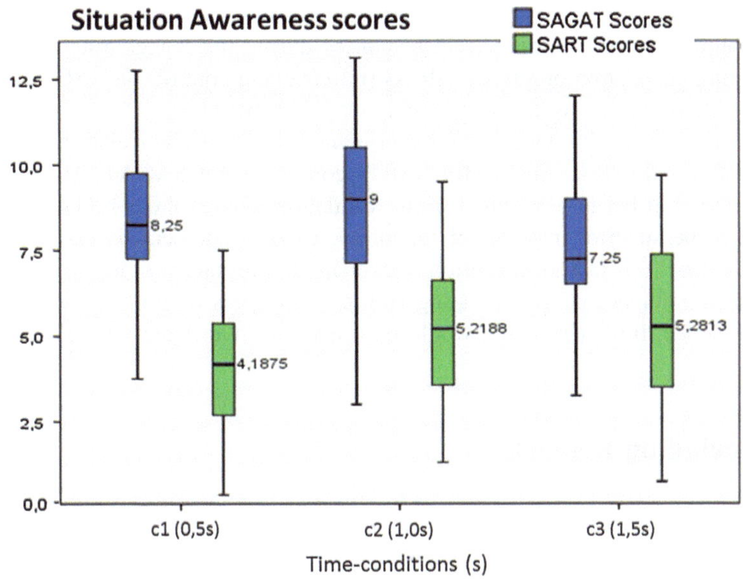


**Fig. 4** SA-scores based on SAGAT and SART per time-condition

101) = 3.391, $p = 0.038$; SART: $F(2, 101) = 4.721$, $p = 0.011$]. Pearson correlation coefficients showed an insignificant correlation between criticality and mean SAGAT scores ($r = -0.169$, $p = 0.089$). However, a significant correlation was found between criticality and the mean SART-scores ($r = 0.284$, $p = 0.004$). The correlation between SART-scores and criticality is according to expectations. Al-though insignificant, the effect of SAGAT-score on criticality (time-condition) is remarkable because of its direction: the more critical the time-condition is, the better SAGAT scores. This result with respect to SAGAT is contrary to the expectation that situation awareness would increase as the criticality decreases. An explanation for this effect could bc that within the least critical time condition, the traffic situation is most changeable. The SAGAT quires were taken directly after the simulation stopped and there were approximately 4 s between take-over request and termination. The lower SAGAT scores are therefore not likely to result from remembrance effects, but rather from confusion associated with the more interchangeable traffic during the longer time conditions. In other words, when there was more time for take-over, participants encountered during take-over more changeable traffic, making it more difficult to precisely refer to type or colour of the neighbouring vehicle at the very moment the take-over was requested. This might have resulted in more wrong answers. Hence, reduced SAGAT-scores.

We also expected a positive correlation between SA and successfulness of take-over, as SA could be assumed a prerequisite for good driving performance. However, Pearson correlation coefficients show that only SART-scores are positively correlated with successfulness ($r = 0.323$, $p = 0.002$). The SAGAT-scores show no correlation ($r = 0.020$, $p = 0.852$). As successfulness has shown to be correlated with criticality and the SAGAT-scores have presumably be influenced by ambiguity in giving the correct responses during the longer time-conditions, this failing correlation of successfulness (in retrieving control) on SAGAT-scores is most likely caused by the same ambiguity in probe-taking.

Overall, the results within this experiment show better correlation with successfulness and time-criticality for SART-scores then for SAGAT. These results are remarkable because within various validation studies the *SAGAT* technique proved to be superior in terms of reliability, validity and sensitivity when compared to SART [6]. The exception shown in this research seems to indicate a significant influence of probe-taking (timing and types of probes) on SAGAT-scores. Hence, further research would be needed to investigate this relation.

## 5 Concluding Remarks

Following the purposes of the current study (i.e.: Investigating to what respect drivers build up SA in time-critical situations of taking over control after automation, and assessing what method for measuring SA is most suitable for these situations), we draw the following conclusions and recommendations.

The SART questionnaire has shown some positive correlation with the successfulness of take-over. This seems promising: when SA increases, so does the chance for a successful take-over. However, in this experiment, SA measurement based on SAGAT technique does not correlate with successfulness of taking over control. When comparing SA-scores for their relation with criticality (i.e. time available for taking over control), the contradiction between SAGAT and SART is even stronger: SART-scores showed positive correlation again, whereas SAGAT-scores showed to be negatively correlated with criticality. The latter is not only contrary to expectations, but also opposing previous research since previous research showed SAGAT technique to be superior with respect to validity and sensitivity in comparison with SART. As explained in the previous chapter, the reason for this contradictory influence between SART and SAGAT method on successfulness of retrieving control is likely to be the result from ambiguity in probe-taking involved in the SAGAT technique we applied. In general, the probes involved in SAGAT are tailored for each experiment. Hence, we need to point at the precarious driving situation applied in our research: accident avoidance within time-critical situations after automation. For such situations SA measurement has hardly been examined. We therefore recommend to further improve probe-taking within SAGAT for our future experiments, e.g. by looking at the precise contents probes refer to and their relevance for accident avoidance.

In view of our results and previous research we cannot draw a final conclusion yet which method is to be preferred for measuring SA during time-critical situations of retrieving control after automation. Nonetheless, we do cautiously conclude the SART method to be at least a secure consideration for our application, while undermining the precariousness of probe-taking involved in SAGAT.

Finally, the results of this experiment show that the chosen levels of criticality influenced driving performance. Meaning that although criticality was high in all conditions, the less critical conditions were more manageable to the participants than the more critical ones. Based on this result, we conclude that applying these time conditions in future interface-tests will allow to distinguish in support for taking over control provided by different interface-types. This is important for the development of future interfaces which provide appropriate support in retrieving human control after automation.

## References

1. Viehmann S (2011) Assistenzsysteme: Entlastung im Stau. www.zeit.de
2. European Commission (2006) Raising awareness of ICT for smarter, safer and cleaner vehicles. Intelligent Car Initiative, Brussels, pp 59-final
3. Endsley MR, Kaber DB (1999) Level of automation effects on performance, situation awareness and workload in a dynamic control task. Ergonomics 42(3):462–492
4. Ma R, Kaber DB (2005) Situation awareness and workload in driving while using adaptive cruise control and a cell phone. Int J Ind Ergon 35:939–953
5. Matthews ML et al (1779) Model for situation awareness, driving. Transp Res Rec 26–32:2001
6. Salmon P et al (2006) Situation awareness measurement: A review of applicability for C4i environments. Appl Ergon 37:225–238
7. Endsley MR (1995) Measurement of situation awareness in dynamic systems. Hum Factors 37:65–84
8. Charlton SG (2002) Measurement of cognitive states in test and evaluation. Handbook of Human Factors Testing and Evaluation, London, pp 115–122

# Multi-Modal Cooperative Intelligent Transport Systems to Improve Safety

**Jugdutt Singh, Anirudha Desai, Felix Acker, Stanley Ding, Aashath Abdoul Rachide and Peter Nelson-Furnell**

**Abstract** Population increases globally have put considerable strain on transport infrastructure and transport management resulting in large social, environmental and economic costs. Cooperative Intelligent Transport Systems (ITS), based on Information and Communication Technologies (ICT), can help focus on problems associated with traffic management, infrastructure management and security, as well as enhanced driver safety and logistics support for transport operations. It will improve drivers' commutes, provide better information to city planners, increase the productivity of businesses and raise citizens' quality of life, reduce congestion, shrink fuel use and cut $CO_2$ emissions. This paper presents a multi-modal Cooperative Intelligent Transport Systems based solution that has been designed as a more efficient and effective approach to prevent and potentially eliminate collisions at rail/road crossings. Large scale field trials of the system at level crossing interfaces have produced promising outcomes is a major step towards improving safety at level crossings.

J. Singh (✉) · A. Desai · F. Acker · S. Ding · A. A. Rachide
Centre for Technology Infusion, La Trobe University, Melbourne, VIC 3086, Australia
e-mail: Jack.Singh@latrobe.edu.au

A. Desai
e-mail: a.desai@latrobe.edu.au

F. Acker
e-mail: f.acker@latrobe.edu.au

S. Ding
e-mail: s.ding@latrobe.edu.au

A. A. Rachide
e-mail: a.abdoulrachide@latrobe.edu.au

P. Nelson-Furnell
Public Transport Victoria, State Government of Victoria, Melbourne 3086, VIC, Australia
e-mail: peter.nelson-furnell@ptv.vic.gov.au

J. Fischer-Wolfarth and G. Meyer (eds.), *Advanced Microsystems for Automotive Applications 2013*, Lecture Notes in Mobility, DOI: 10.1007/978-3-319-00476-1_6,

**Keywords** Intelligent transport systems · Cooperative ITS · DSRC

# 1 Introduction

Australia faces major transport challenges through increasing traffic congestion, greenhouse emissions, strong public demands for enhanced safety, and the urgent need for an integrated transport system, namely:

- the national costs of traffic congestion in Australian cities will exceed $20 billion by 2020;
- stop-start congested traffic increases fuel consumption and greenhouse gas emissions (GHG) by around 30 %, contributing to transport being the nation's 3rd largest and 2nd fastest growing GHG sector;
- the National Road Safety Strategy 2011–2020 aims to reduce deaths and serious injuries by approximately 30 % by 2020;
- road accidents cost the nation around $27B annually and,
- a latent demand exists for greater transport efficiency.

According to the Australian Transport Safety Bureau (ATSB) between 2001 and 2009 over 630 road vehicle collisions occurred at level crossings in Australia with the state of Victoria alone accounting for 236 of these incidences [1]. In addition to the financial loss of level crossing accidents, they have also resulted in high level of tragic fatalities (over 70 deaths between 1997 and 2002 [2]). Investigation by the ATSB into 12 level crossing accidents between April 2006 and December 2007 observed that nine of these incidences involved heavy road vehicles [3]. These incidences resulted in 19 deaths and an estimated financial loss of over $100 million, including a single incident near Karang, Victoria, where 11 people died.

The major factors which have accounted for collisions have included driver behaviour (and errors), poor communications, signalling difficulties and environmental conditions such as visibility [4–6]. Existing approaches to improve safety include both passive and train activated warning signs and signalling systems. Although passive signs are economical to deploy, they can be less effective in addressing common causes of crossing incidences e.g. poor visibility, human errors due to fatigue or inappropriate driver behaviour [7, 8]. On the other hand, active warning systems such as boom gates can be expensive to deploy.

Furthermore the absence of active warning systems or the presence of adverse weather or road conditions are not always the primary factors behind level crossing incidences. According to a study of 87 level crossing accidents conducted by ATSB [9], over 80 % of the fatal accidents at level crossings occurred in excellent driving conditions that included daylight (excluding dawn and dusk), fine weather and straight and dry road. In addition to this, at least half of the level crossings involved in these collisions employed some form of active warning system such as boom gates, flashing lights or barriers. Driver error was the most common cause of level crossing collisions accounting for 46 % of all incidents. Errors, however, are not monocausal and range from attentional lapses to false expectations, inappropriate decision making, driver mental overload and perceptual misjudgement [4, 5].

Over the past few years, the Federal and State Governments of Australia have made several recommendations [10, 11] to improve safety at rail-road crossings. Seven of these recommendations relate to the development and use of new Cooperative Intelligent Transport System (ITS) to improve safety at level crossings. The State Government of Victoria through the Public Transport Victoria and the Federal Government of Australia through the Cooperative Research Centre for Advanced Automotive Technology (AutoCRC) along with a number of key Universities and industry partners have cooperatively funded a new multi-million dollar project which aims at developing a co-operative Intelligent Transport System to improve safety at rail-road crossings. The specific aims within the scope of the project are to develop an Intelligent Transport System simulation platform for rail-road crossings based on 5.9 GHz DSRC technology; to develop an ITS demonstrator system and to implement field trials with up to 100 vehicles at several level crossing interfaces. This paper presents the research, system architecture and the outcomes of this project.

## 2 Intelligent Transport Systems

Existing technologies, such as Global Positioning System (GPS), can be deployed to establish communication between vehicles. Such systems offer communication platform to facilitate sharing of information within the transportation infrastructure to allow realisation of a range of safety and mobility applications. The development of the Dedicated Short Range Communication (DSRC) standard by the IEEE has paved the way to efficiently implement a wireless ad-hoc network, which can be used for vehicle-to-vehicle (V2V) and vehicles-to-infrastructure (V2I) communication to facilitate the development of ITS. ITS enables applications across three main categories, with the potential to offer significant social, economic and environmental benefits, namely:

- Safety: ITS safety applications use the communication mechanism within DSRC to create complete situation awareness for vehicles.
- Mobility: ITS mobility applications include travel and route planning, traffic and congestion management including public transport, transport network productivity and reliability enhancement etc.
- Commercial: ITS can allow a very wide range of commercial applications ranging from next generation electronic toll collection (ETC) and fleet management to variety of data and infotainment applications.

## *2.1 Dedicated Short Range Communication*

In late 1999 United States Federal Communications Commission (FCC) allocated a 75 MHz spectrum for DSRC at 5.9 GHz for use in Intelligent Transport System applications. The Australian Communication and Media Authority (ACMA) is also planning allocating a similar frequency spectrum in Australia [12]. The Dedicated Short Range Communication (DSRC) technology allows rapid and high speed communication between multiple vehicles and between vehicles and infrastructure to enable both safety and non-safety applications. This technology at 5.9 GHz offers 7 channels of 10 MHz each, allowing communication data rates of up to 27 Mbps, with communication capabilities at vehicle speeds up to 200 km/h. This next generation 5.9 GHz DSRC technology is based around a set of standards being developed by the IEEE. Following is a summary of the key standards that are being developed to support this technology:

- IEEE 802.11p (Wireless LAN medium access control [MAC] and physical layer [PHY] specification for WAVE): The WAVE architecture uses an adaption of IEEE 802.11a Wireless LAN standard for supporting wireless communication in vehicular environments (initially developed in ASTM standard E2213-03 [13]). The IEEE 802.11p amendment [14] to extend the IEEE 802.11 standard is intended specifically towards addressing harsh vehicular ad-hoc network communication requirements—to enable rapid and reliable communication mechanism for dissemination of safety messages.
- 1609.1 (WAVE Resource Manager): This standard [15] defines the WAVE "Resource Manager" application and specifies its services and interfaces. It allows multiplexing communications of multiple remote applications, each of which can interact with radio units in many vehicles.
- 1609.2 (WAVE Security Services): This standard [16] defines the DSRC security services for applications and management messages (including secure message formats and processing) which are intended to provide anonymity, authenticity and confidentiality for the messages exchanged.
- 1609.3 (WAVE Networking Services): This standard [17] defines network and transport layer services (including addressing and routing) in support of secure WAVE data exchange.
- 1609.4 (WAVE Multi-Channel Operation): This standard [18] defines the multichannel wireless radio operation of WAVE. It is intended to tightly integrate with the IEEE 802.11p to manage the lower layer usage of DSRC channels.

# 3 The Project: Multi-Modal Cooperative Intelligent Transport Systems to Improve Safety at Level Crossing

The overall aim of the project was to research, develop, implement and trial a DSRC based ITS solution to improve safety at level crossings. The solution aims to reduce and potentially eliminate level crossing accidents by enabling dynamic V2V and V2I communications using DSRC technology. The specific aims within the scope of the project are:

- Phase 1: To develop an Intelligent Transport System (ITS) simulation platform for rail-road crossings based on 5.9 GHz DSRC technology.
- Phase 2: To develop an ITS demonstrator and Proof-of-Concept system.
- Phase 3: To carry-out field trials at level crossing interfaces.

Figure 1 illustrates the system architecture of the proposed DSRC based Cooperative Intelligent Transport System deployed at a level crossing. The system is composed of DSRC enabled roadside units (RSUs) and on-board units (OBUs). The RSUs are placed at locations such as rail-road crossing active signs while the OBUs are installed in road vehicles and trains.

A DSRC based ITS at level crossing will enable communication between infrastructure nodes and vehicles in the vicinity of the crossing including trains and road vehicles. This communication enables sharing abundant data between vehicles including basic information such as vehicle size and type, position and motion and other control information such as brake status, throttle, steering angle etc. Depending on various parameters and conditions, the information can be shared directly using V2V communication or indirectly using V2I communication. The information will be used by a specialised level crossing safety application to generating warning messages such as advice of an approaching train, advance indication of a potentially faulty crossing, expected delay at the level crossing, suggestion of alternate routes etc. The Human–Machine–Interface (HMI) used to communicate safety messages to drivers was developed to not only address the immediate safety requirements but also to promote overall long term behavioural change of drivers towards safety consciousness.

## 3.1 Simulation Platform

Existing simulation platforms/tools such as CarSim/TruckSim, Automated Highway Simulations (AHS), are generally domain specific. Most high-end simulation platforms are commercial, closed-source projects and co-simulation can be extremely computation-intensive in nature. The overall result is a limited capacity for simulation systems to perform simulations on a larger scale in a respectable time that will allow real empirical evaluation of a system.

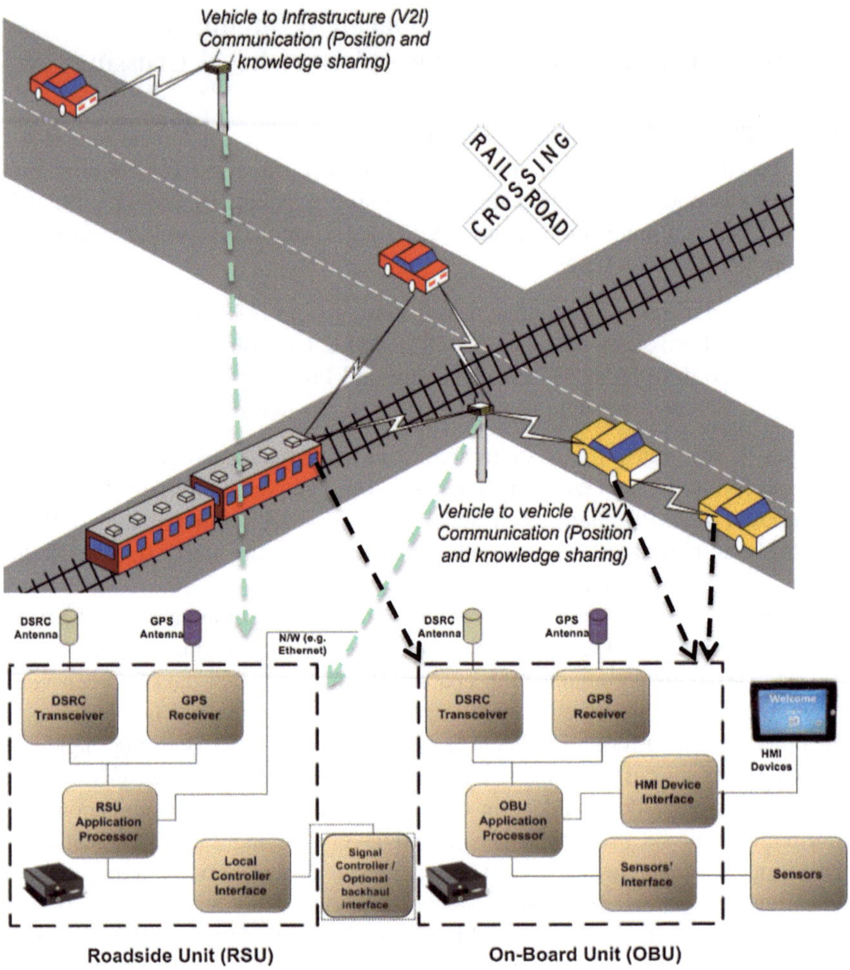


**Fig. 1** Architecture of DSRC-based ITS for level crossings

To address the limitations of existing solutions, this project involved the development of a DSRC based ITS simulation platform which can accurately simulate the mobility, DSRC communication and complete interaction of trains and road vehicles at a level crossing. Figure 2 illustrates the architecture and the elements of the simulation platform. The platform has been developed to allow simulation of all domains in a unified simulation environment that is specialised for level crossings.

The features of the simulation platform include:

- Distributed High Performance Computing (HPC) cluster based simulation environment.

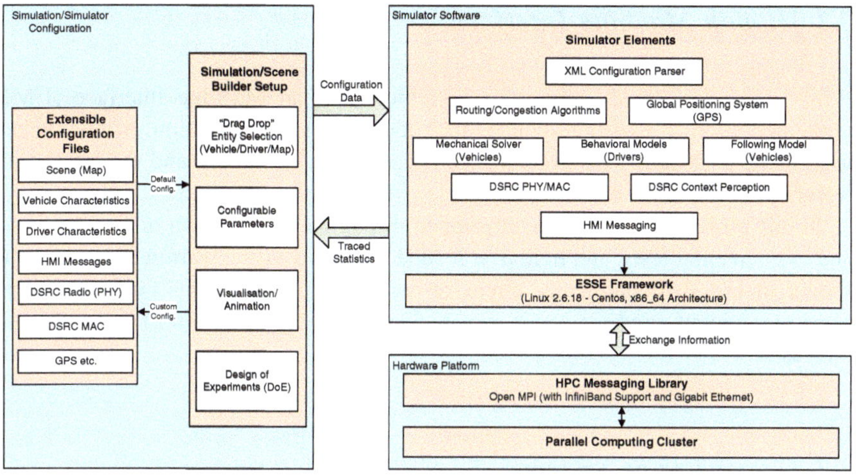


**Fig. 2** Simulation platform architecture

- Scalable models of DSRC communication, GPS, Intelligent Collision Avoidance Applications, Vehicle models (articulated trains, buses, trucks, cars), Mapping system models, and Human behaviour models (driving behaviour and HMI).
- Flexible parameter configurability and efficient simulation trace/debugging/event filtering mechanisms for all models to allow Design of Experiment (DoE) and heuristic search.
- A complete simulation set-up/visualisation engine with graphical user interfaces that includes real-world map loading, drag and drop of element placements, optimal traffic routing and parameterisation of various elements.

The simulation platform allows detailed modelling and forecasting of the typical interactions between rail and road traffic. A simulator normed with complex empirically collected field trial data can be used to further the design of intelligent collision avoidance algorithms, and visually and intuitively evaluate system boundary conditions for safety factors such as the predicted motion paths of vehicles, different configurations of collision warning thresholds, and human behaviour variations on attributes such as reaction time, driving style or attentiveness. With its high level of customisability, this platform has the potential to direct the design and implementation of practical DSRC road-safety infrastructures and mobile devices.

### 3.2 Human Machine Interface

While there are already numerous in-vehicle Human Machine Interface (HMI) implementation in existing vehicles that address forward collision or lane departure, there is currently no system dedicated to provide timely and effective train warnings at rail level crossings.

In this project, the HMI was developed through a combination of focus groups, and experimental tests, and field trial feedback. Audio-visual warning messages, as shown in Fig. 3, dynamically intensify on multiple dimensions (volume, pitch, semantic content, contrast, static versus dynamic image animation) as driver risk increases.

### 3.3 Demonstrator System

A demonstrator system was developed using Cohda Wireless's 5.9 GHz DSRC radio platform. The main objective of this work was to establish a proof of concept through the development of a technology demonstrator based on 5.9 GHz DSRC technology. The most significant aspects of this work include:

- development of optimised embedded software for DSRC-based safety applications based on intelligent new algorithms for proactive collision avoidance, human factor research outcomes, and with work leading towards congestion and emission reduction.
- the identification, development, and implementation of core technologies for HMI, including both functionality and display characteristics of HMIs and drivers' behaviour issues. The resultant HMI interface was designed to be scalable.

(a) (b)

**Fig. 3** **a** Warning sequence on a direct, **b** parallel approach to level crossing

### *3.4 Field Trials*

Phase 3 of the project involved conducting real-world field trials of the system with a large number of vehicles at level crossings in urban and rural locations to verify:

- radio communication performance under adverse conditions including vehicular congestion, 'high' speed vehicular interactions, and built environment operation;
- key system performance factors including data rate, application response, multipath fading, Doppler effect, on-time reception of safety critical messages, and GPS positioning accuracy;
- driver reaction (risk reduction) factors such as message effectiveness, usability of the HMI, and effect of a passenger on the driver.

In addition to the orchestrated trials, a 6 week longitudinal trial was conducted at the two urban trial sites. Ten local drivers were recruited to drive across the level crossing as they normally would and 16 radios were fitted to eight live trains that passed through the DSRC controlled level crossing. The aim of this project phase was to gather data on alert adaptation over time, system acceptance and, if applicable, changes to driver behaviour as a result of exposure to warnings.

## 4 Outcomes

The final stage of the trials was being completed at the time of writing this paper. The key results available are related to the driver's interaction with the system during first part of the orchestrated field trials. The evaluation of the driver experience focused on drivers' perceived ease of use of the system, perception and assessment of the individual alerts with respect to both nuisance and safety, and, most importantly, a global assessment of system effectiveness.

Across all field trial conditions participants reported a high level of effectiveness for the alerts. As presented in Fig. 4, ratings were highest for the most

**Fig. 4** Average effectiveness rating (with standard deviations) for each of the three alert levels (n = 85)

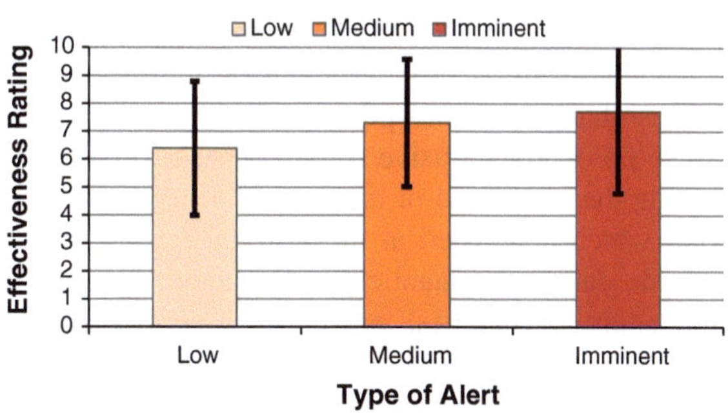

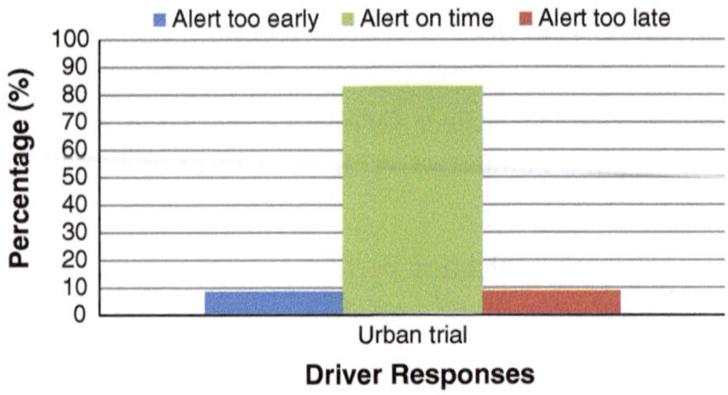


**Fig. 5** Driver assessment of alert timings (n = 70)

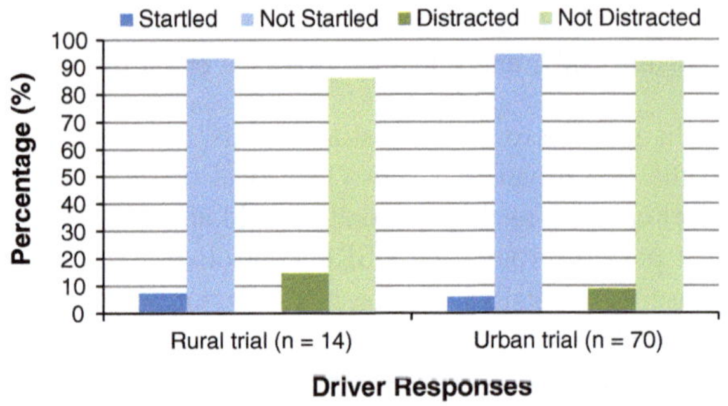


**Fig. 6** Driver assessment of distraction and startle

noticeable and intrusive alert, and lowest for the low level advisory notification. This result was in keeping with the objective of design, that alerts should be non-intrusive in non-threating situations.

A second important factor for the overall assessment of effectiveness is the appropriateness of alert timing. Early alerts are likely to lead to driver frustration, while late alerts present a significant safety hazard. Figure 5 illustrates the driver's response in the urban field trial.

A third important aspect was drivers' perceived level of distraction and startle. Both of these are safety–critical in emergency situation as they lengthen reaction time and impair appropriate decision making. Aggregated across all alerts, participant drivers reported low levels of both distraction and startle as presented in Fig. 6.

Overall, the data collected in this project provide some understanding of the behavioural effects brought about by the in-vehicle alerts. Importantly, the data suggest that participants perceive the system to be effective, while at the same time not to be overly intrusive or otherwise adverse. This is a crucial aspect in determining the likelihood of uptake and acceptance by drivers. The longitudinal field trial, which concluded at the time of writing this paper, on the other hand, did provide data from unanticipated encounters between road- and rail vehicles, yet the nature of the trial prevented a similar rigor of data collection (e.g., no in-vehicle video recording) that was implemented during the orchestrated trial.

Future projects will address this aspect specifically by adding to the existing data and through tests of additional scenarios both under laboratory and field conditions.

## 5 Conclusion

Level crossing collisions can be catastrophic in nature and lead to tragic deaths and significant financial losses. Study and analyses carried out by the Australian Transport and Safety Bureau have indicated that driver error is one of the most significant factors that lead to collisions at level crossings. DSRC technology has the potential to offer a cost effective means of deploying an Intelligent Transport Systems offering social, economic and environmental benefits.

This paper has presented the overview of the research, system architecture and the preliminary outcomes of a $5.5 million Intelligent Transport System to Improve Safety at Level Crossing project. This project has resulted in development of a complete proof-of-concept safety solution for improving safety at level crossings. The outcomes indicate that the DSRC technology has performed under adverse conditions including vehicular congestion and preliminary results from large scale field trials of the safety solution indicate that overall, drivers perceive the system to be effective and largely resilient to adverse effects. Broadly, the solution will not only result in immediate safety outcomes in terms of collision avoidance, it may also help in long term behaviour change of drivers and result in an overall improvement in safety consciousness on Australian roads.

## References

1. Australian Transport and Safety Bureau (ATSB) (2009) Australian rail safety occurrence data 1, January 2001–30 June 2009. ISBN: 978-1-921602-99-3
2. Australian Transport and Safety Bureau (ATSB) (2003) Level crossing accident fatalities. Rail Saf Stat
3. Australian Transport Safety Bureau (ATSB) (2008) Railway level crossing safety bulletin. ISBN 978-1-921490-11-8
4. Caird JK, Creaser JI, Edwards CJ, Dewar RE (2002) A human factors analysis of highway-railway grade crossing accidents in Canada. Cognitive Ergonomics Research Laboratory, Calgary, Alberta
5. Edquist J, Stephan K, Wigglesworth E, Lenné M (2009) A literature review of human factors safety issues at Australian level crossings. MUARC, Melbourne
6. Kosky L (2009) Towards zero: a strategy for improved level crossing safety in Victoria. State of Victoria, Melbourne
7. Cairney P (2003) Prospects for improving the conspicuity of trains at passive railway crossings. ARRB Transport Research Ltd, Melbourne
8. Rudin-Brown M, Lenne M, Edquist J, Trotter M, Navarro J, Tomasevic N (2010) Driver compliance with, and understanding of, level crossing controls. J Australas Coll Road Saf 21(2):50–55

9. Australian Transport and Safety Bureau (ATSB). Monograph 10: Level Crossing Accidents - Fatal accidents at level crossings. ISSN 1444-3503. Jan 2002
10. Parliamentary Road Safety Committee (2008) State Government of Victoria, Inquiry into improving safety at level crossings, ISBN: 978-0-9751534-6-8
11. House of representatives standing committee on infrastructure, transport, regional development and local government, commonwealth of Australia. Level crossing safety, ISBN: 978-0-642-79188-7, 2009
12. Australian Communications and Media Authority (ACMA) (2009) Planning for intelligent transport systems: proposals for the introduction of intelligent transport systems into the 5.9 GHz band in Australia
13. ASTM International (2003) Standard specification for telecommunications and information exchange between roadside and vehicle systems-5 GHz band dedicated short range communications (DSRC) medium access control (MAC) and physical layer (PHY) specifications. ASTM E2213-03
14. IEEE Draft Standard for Information Technology (2010) Telecommunications and information exchange between systems—Local and metropolitan area networks—Specific requirements, part 11: wireless LAN medium access control (MAC) and physical layer (PHY) specifications, Amendment 6: wireless access in vehicular environments. IEEE Std 802.11p-2010, July 15 2010, pp 1–51
15. IEEE Trial-Use Standard for Wireless Access in Vehicular Environments (WAVE) (2006) Resource manager. IEEE Std. 1609.1-2006
16. IEEE Trial-Use Standard for Wireless Access in Vehicular Environments (2006) Security services for applications and management messages. IEEE Std. 1609.2-2006
17. IEEE Trial-Use Standard for Wireless Access in Vehicular Environments (WAVE) (2007) Networking services. IEEE Std. 1609.3-2007
18. IEEE Trial-Use Standard for Wireless Access in Vehicular Environments (WAVE) (2006) Multi-channel operation. IEEE Std. 1609.4-2006

# Vision Zero: Technologies and Limitation

**Klaus Krumbiegel, Hadj Hamma Tadjine, Benedikt Schonlau and Robert Schwaiger**

**Abstract** In order to meet the growing requirements on vehicle safety, additional safety systems are usually integrated that then fulfill a certain function. The function only becomes active in the defined use cases, since it could have a negative effect on the vehicle safety in other situations, or the necessity of intervention is not at all recognized. For the implementation of a traffic system without traffic fatalities, it is very difficult to implement this method, since an infinitely large number of situations have to be taken into consideration. In an integral safety concept, the individual safety systems are closely networked and act interdependently with each other. This paper examines in simple scenarios whether it is possible to ensure accident-free driving such that the so-called Vision Zero can be realized in any case.

**Keywords** Vision zero · Integral safety · Accident-free driving · Appropriate driving

## 1 Introduction

In 2011, 395,934 people were involved in accidents on public roads in Germany. 4,002 of them died [1]. In 2012, for the first time in 30 years, both of these numbers have risen compared to the previous year [2].

K. Krumbiegel (✉) · H. H. Tadjine · B. Schonlau · R. Schwaiger
IAV GmbH Ingenieurgesellschaft Auto und Verkehr, Carnotstraße 1, 10587 Berlin, Germany
e-mail: klaus.krumbiegel@iav.de

H. H. Tadjine
e-mail: hadj.hamma.tadjine@iav.de

B. Schonlau
e-mail: benedikt.schonlau@iav.de

R. Schwaiger
e-mail: robert.schwaiger@iav.de

J. Fischer-Wolfarth and G. Meyer (eds.), *Advanced Microsystems for Automotive Applications 2013*, Lecture Notes in Mobility, DOI: 10.1007/978-3-319-00476-1_7, 

The White Paper on Transport [3], published by the European Union, associates the year 2050 with the Vision Zero—the vision of no fatalities or serious injuries in road traffic [4]. In regards to accident statistics, it becomes obvious that this vision will not come true with today's efforts. Therefore, the Vision Zero documents include a set of requirements divided into three fields of action: the human being, the car, and the infrastructure. In this context, 'human being' refers to the driver of a car. In 95 % of all accidents in road traffic, he or she is at least partially responsible. Nevertheless, Vision Zero accepts that humans are fallible. And if they do not adapt to the road traffic system, it has to adapt to them. The automobile manufacturers and their suppliers have to make a contribution to fulfill the four fundamental rules of Vision Zero: to correct human mistakes, to make the occupants' survival non-negotiable, to regard man's biological "toughness" as the limit, and to give people the right to be part of a safe road traffic system. This paper addresses the question of whether it is possible to develop a suitable safety concept. It has to prevent accidents caused by human errors that the Vision Zero rules are not violated at any time. Initially, the examination is limited to the longitudinal traffic with multiple lanes—the entire analysis is based on this scenario. Further investigations extended to include the entire road traffic system are conceivable after initial positive results. Furthermore, accidents should be incorporated in the analysis provided that these accidents do not result in serious injuries violating the rules of the Vision Zero.

To solve the problem, a comprehensive analysis of the state of the technology in vehicle safety is performed. Only comprehensive knowledge about all safety systems, their mode of functioning, constraints and prospects make possible the drawing of conclusions about why we are so far away from accident-free driving despite our advanced technology.

Out of the variety of active, passive and assistance safety systems, only those which are significant for the implementation of Vision Zero should be carried over to the integral safety concept. Furthermore, the concept contains strict specifications on detecting hazards and on who has which rights and duties to intervene. Also, current laws will be ignored if this is necessary to achieve positive results.

## 2 Preliminary Results

This chapter concerns about classification and specification of vehicle systems affecting safety. Furthermore, simple driving scenarios are defined in which the investigation takes place. Safety can be considered as a condition in which no unacceptable risks occur. A risk in turn is "the expected frequency of occurrence of an event that leads to damage and the expected extent of damages" [5]. According to this, vehicle safety systems have the task of lowering the risk to a reasonable degree while the vehicle is in use.

In technical literature, vehicle safety systems are usually divided into active and passive systems. This does not refer to the mode of operation but to their placement in the chronological sequence of an accident or a dangerous situation. According to this clear classification, the driver assistance systems (DAS) and driver information systems (DIS) are to be counted as active systems since they already take effect before $T_0$, the occurrence of an accident. However, a further division is logical for this paper. As well as the intervention point, the authority to act is to be used to make a comparison. Besides the passive and active safety systems, this results in the driver information and assistance systems as individual categories. Thus, four categories were created in the results of the status analysis.

Passive safety systems are those that reduce the consequences of an accident independent of $T_0$. For example, airbags or seatbelts have to be listed in this category. Active safety systems were defined as those systems that function to prevent accidents and whose function cannot be overridden by the driver, such as the Electronic Stability Program or Antilock Braking System. All systems in the investigation that also fulfill a safety function but which can be overridden by the driver at any time were categorized as driver assistance systems. The Lane Departure Warning System or Adaptive Cruise Control are classic examples. The last category consists of the driver and surroundings information systems. This includes all systems that serve the exchange of driver, vehicle and surroundings information. This class includes simple systems, like the car-lighting as well as the technical advanced Car-to-X Communication.

The vehicle that was created using the information compiled was equipped with all currently available safety systems. This vehicle then had to run through the scenario described in the following. In this way it was to be determined if accident-free driving could currently be implemented such that the Vision Zero is fulfilled.

The driving scenario was initially designed as simple as possible but at the same time representative of real traffic. The driving task is to be managed on an asphalt road with two lanes and a shoulder. The model for the following specifications was a highway in Germany in accordance with the German Highway Traffic Code, see [6].

The roadway of the oncoming traffic is structurally separate and is not further considered for the driving scenario. There is also a structural separation on the right side of the road shoulder that is only interrupted in special cases. A highway does not have intersections on the same level, and on and off ramps are not part of the examined scenario.

All vehicles have to fulfill certain requirements in order to be allowed on the road. In the scenario, all vehicles correspond to the category of vehicles with a permitted total weight of up to 3.5 metric tons and an attainable maximum speed of at least 60 km/h. A listing of the comprehensive rules for driving in cases of limited functionality, e.g. in road works, was not considered since these are not used in the scenario.

The partial scenarios developed in the following represented the basis for the analysis of the feasibility of Vision Zero and the subsequent concept development, see Table 1. Note that in this paper no specific simulation tool was used to represent the scenarios or the links between them.

**Table 1** Different scenarios for the analysis

| | |
|---|---|
| Scenario 1 | A slightly slower car ahead on right lane |
| Scenario 2 | Adjusting the vehicle speed to follow a car |
| Scenario 3 | Lane-change to the left due to an obstacle |
| Scenario 4 | Driving on the left lane, obstacle ahead |
| Scenario 5 | Following a car on the left lane |
| Scenario 6 | Lane-change to the right after overtaking |
| Scenario 7 | Nearly stationary object on right lane ahead |
| Scenario 8 | Emergency brake to avoid a collision |
| Scenario 9 | Collision with an obstacle |

## 3 Analysis Regarding Feasibility of Vision Zero

This paper concentrates on accident-free driving since it is generally not possible to estimate the consequences of an imminent accident precisely. Of course, the majority of collisions are non-fatal such that the Vision Zero is not violated. Nevertheless, avoiding accidents lead to a strict fulfillment of the Vision Zero. It will be part of ongoing work, how it is possible to allow also for non-fatal collisions in the present investigation since for certain crash scenarios the consequences can be estimated. By linking the presented scenarios, this section will analyze whether or not accident-free driving is realizable.

A program flowchart was generated from the different determined partial scenarios and by linking them with each other. This schedule serves to illustrate the driving task in the scenario and should demonstrate at which points violations regarding accident-free driving take place. Due to lack of space Fig. 1 shows only a section of this flowchart, see [7].

All connectors that represent a violation of the assumption that can no longer be averted are marked red in the diagram. The analysis starts backwards from the end marked red. This end represents the violation of the stated requirement and is

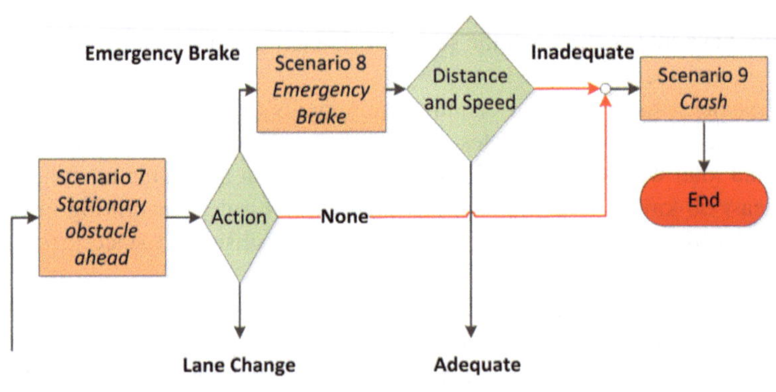


**Fig. 1** Detail of the program flowchart, describing a stationary obstacle ahead and resulting options

reached via the partial scenario 9, the accident. Since at that time there is no possibility to avoid the collision, the arrows are followed backwards further. It is clear that, in these scenarios, two causes are responsible for the collision of the examined vehicle with another road user. The first cause is insufficient scanning of the surrounding area.

For instance during a lane change, the driver steers into the desired lane and collides directly with another vehicle or interferes with it in such a way that it causes an accident. The relevant vehicle safety systems such as Blind spot warning and Lane keeping assistant that are to counteract this situation can simply be ignored or overridden by the driver. The following critical scenario can also be attributed to insufficient scanning of the surroundings. The driver does not react to an emerging oncoming obstacle. Since intentional wrongdoing can be ruled out, it is assumed that the driver does not detect the obstacle due to distraction or other outside influences.

The other cause for the unavoidable accident is inappropriate driving behavior, which in most cases expresses itself as driving too fast or not keeping a safe distance to the vehicle ahead. Regarded to the emergency braking scenario, a full stop can no longer be achieved in time. The safe speed, distance and visibility are directly affected by environmental influences such as rain, snowfall or ice on the road thus having influence on the appropriateness of driving.

As a consequence there are two reasons that Vision Zero cannot be fulfilled: insufficient scanning of surroundings and inappropriate driving behavior. Both of these are driver weaknesses. In turn, this means that Vision Zero would also be possible without safety systems if there was a perfect driver in every vehicle. Since this cannot be assumed and Vision Zero even explicitly states that human error is natural and allowed, the listed causes have to be addressed in another way.

As the link between the driver and the road, the vehicle lends itself to intervention in risk situations through assistance and active safety functions.

# 4 Concept Proposal

## *4.1 Legislative Adjustments*

Since Vision Zero states that man is to be accepted as a source of errors, the Vienna Commission on Road Traffic [8] can no longer apply in its current wording. As long as the driver has to always be the last resort in a vehicle and a system cannot autonomously intervene before the unavoidableness of an accident, human errors will continue to result in fatal accidents. In order to fulfill Vision Zero, the authority to act in the last resort has to therefore be extended to the vehicle safety systems. If one of the safety systems detects an action of the driver that will lead to an accident, this driving command is not to be executed. The adjustments have to contain extensive term definitions and sequences. When and why the driver is relieved of the control over the vehicle has to be comprehensible. If technical systems are allowed as the last resort and can therefore no longer be overridden by the driver, this will also have effects on function validation and tests.

## 4.2 Function Validation

Several necessary adjustments will be described under the generic term of function validation. It is fundamental that the technical vehicle safety systems that are able to intervene as a last resort must always work. They cannot be deactivated or malfunction without the driver being informed of this. The driver of the vehicle must not at any time have the possibility to deactivate safety systems for emergency intervention. Since vehicles with inactive safety systems are controlled solely by a human being who is considered fallible, a Vision Zero without this requirement is not enforceable. People deactivate systems frequently for the simple reason that they believe the systems trigger erroneously, intervene uncomfortably, or are disturbing. If drivers no longer have the option of switching the systems off, their acceptance has to be increased, because only when safety systems are accepted and drivers do not feel restricted by them will they stop searching for alternative ways to bypass the systems.

## 4.3 Scanning the Surroundings

A further point that caused today's vehicle to fail the requirements in this work was the insufficient surroundings detection. The variety of sensors used in the vehicle already record the majority of the surroundings. This information is sometimes only available to certain systems, or sensor types are not suitable for all objects that are to be detected. Accordingly, it is necessary to adjust the sensors used and the processing of the data.

In order to enable accident-free driving, the driving speed has to be adjusted to the visual range of the vehicle in which all objects can be detected. The goals of surroundings detection in the discussed scenarios are therefore to implement complete object detection with as high a range as possible.

The state of the technology in the field of vehicle sensors compiled in the run-up shows the use of radar, video, infrared and ultrasonic sensors. The examination shows that only the information of the radar sensors is available for long ranges, but they only reflect metal surfaces. Non-metal obstacles such as animals, humans or lost cargo cannot be detected. The video sensors for the front end are currently primarily used for the detection of lanes and traffic signs. In terms of Vision Zero, video sensors possess several advantages over the other sensors used. The information content of the recorded signal is larger and can be used, given adequate video processing, for all safety functions with environmental scanning.

Furthermore, the "perception" of the video sensors is the most similar to human vision. Since the existing surroundings sensors have already been adjusted to the requirements of the corresponding functions and have advantages over imaging systems in some areas, it is imperative that they are maintained. An overlapping of the monitored areas of some sensors is advantageous, since disadvantages can in this way be compensated for and the results can be checked against each other for plausibility.

Concealed targets cannot be detected by the sensors used actually. The possibilities of the C2X communication could supplement the data of the surroundings scanning in this case. If one's own vehicle receives the C2X signals from three vehicles in the right lane, but only detects two of these with its own sensors, the obvious concealment can be reacted to.

The vehicle speed permissible for the realization of Vision Zero depends directly on how far in front of the vehicle objects can be detected. The rear end and side areas have to cover the lanes in order enable lane changes with adequate surroundings scanning.

## *4.4 Appropriateness*

This section addresses the definition of the appropriate driving behavior on the basis of the considered scenario. Even under the assumption that the legislation allows autonomous driving interventions, that these systems always work, and that environmental sensors detect all obstacles reliably, an accident resulting from inappropriate driving behavior cannot be reliably prevented. The reason for this can be a speed at which the braking distance is too long for an emergency braking action to prevent a collision. The parameters reflecting appropriate driving behavior have to be continuously calculated by the vehicle data and data provided by surrounding scanning sensors. It has to be ensured that the limiting values are not exceeded at any time or under any circumstances.

Related to the considered scenario it is only necessary to compare two variables in order to ensure that one's own vehicle comes to a complete stop on time in front of an immobile obstacle when travelling unobstructedly. One's own braking distance ($l_B$) must never be larger than the visual range ($l_S$) of the sensors used for the front area. As long as $l_B < l_S$ is valid, an emergency stop can prevent a collision. Figure 2 graphically illustrates the correlation. The delay between the obstacle detection and start of braking, i.e. the reaction time, is still disregarded here.

Even though the figure in this case shows a vehicle travelling at its own speed as the obstacle, in the worst case a standing obstacle has to also be taken into consideration for the investigation. The figure depicts precisely the point when the obstacle was detected. Since the braking distance here is shorter than the visual range, the green car can stop on time. Obstacles that have a negative relative speed, i.e. oncoming vehicles, have not been dealt with in accordance with the

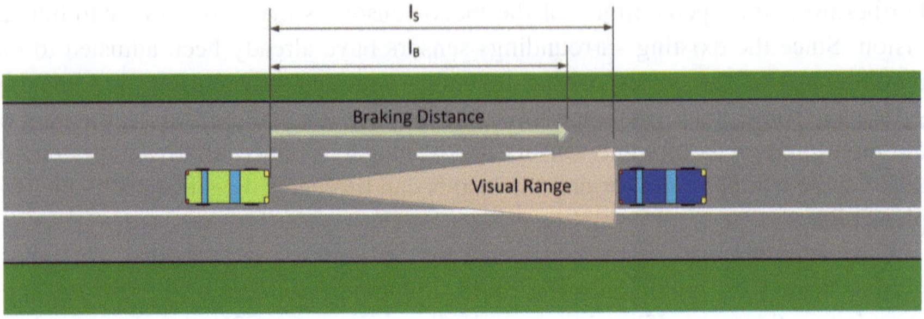


**Fig. 2** Appropriateness shown via relation of braking distance and visual range

scenario basis. The input values of the braking distance and visual range are not directly recordable by the sensors.

Methods for detecting the own braking distance in real-time, often compare estimated longitudinal forces with values from estimated road conditions. The quality of the estimated values and the simplifications strongly influence the result. Thus, the available methods for predicting one's own braking distance using the traction potential are not yet accurate enough to fulfill the requirements of ensuring accident free driving [8]. For detecting the own visual range more than one already working method is available. For example Winner [9] explains that LiDAR sensors can be used to determine an atmospheric visual range. Another way of detecting the own visual range is described by Charbonniér [10]. This method uses a stereo camera which generates a three-dimensional image. The algorithms used make an exact assertion about the distance at which object detection is still possible. The additional costs are lower than for the LiDAR detection, since the stereo camera is already used for other tasks in the vehicle.

When the visual range of the sensors ($l_S$) is determined, it can be compared with the calculated current braking distance. Test calculations showed that permissible speeds over the valid recommended speed are possible. For the acceptance of the concept it is important that all limitations, especially with greater visual ranges, are minimal. Therefore, provided that the assumptions for the calculation are close to reality, this requirement can be considered as fulfilled.

This approach to appropriate driving behavior is based on and is also valid only for the driving scenario of this paper. Approximations of real traffic result in additional situations. The assumptions made here have to be supplemented in order to be able to react to this.

## 5 Conclusions

The objective of the investigations was to determine if and under what circumstances the vision of zero traffic fatalities can be achieved. For this purpose a uniform driving scenario was defined. A vehicle corresponding to the previously researched state of the technology had to perform different tasks in the scenario. The ability of the driver-vehicle alliance to complete the individual partial scenarios without an accident was investigated. The result of the analysis was that the requirement of accident free driving cannot be achieved if the driver has the final authority to act, since he or she always has to be considered fallible. In turn, it was therefore certain that the vehicle has to prevent accidents itself using suitable sensors. From these and further results, the requirements for the integral safety concept were derived with the help of which Vision Zero is to be made possible. As far as the necessary sensors for the detection of the vehicle's surroundings are concerned, it could be proven using the previously determined state of the technology that the technical prerequisites are in large part already available. Modern driver assistance and safety systems already scan large areas around the vehicle. On the other hand, it was determined that the adjustments of the authority to act and function validation that are theoretically easy to carry out are not compatible with current legislation. The Vienna Convention on Road Traffic that is valid in many European countries currently prohibits necessary intervention in the driving task. Although the legal requirements were ignored for the continuation of the work, this means that accidents can only be avoided to a limited degree. Accidents in which the physical limits of driving dynamics prevent early enough obstacle avoidance and braking maneuver must be averted by constantly appropriate driving behavior. This results in the driver only being able to freely decide about the driving parameters within a framework of determined limiting values. The requirements on this appropriate driving behavior in the considered scenarios were easily defined. However, it became clear that driving dynamic parameters that are currently not available with sufficient precision in the vehicle are to be determined for the implementation of the requirements. Since it involves important data—but data that is very complex to collect—only approaches were made whose implementation could be the topic of further papers. For future studies it is still conceivable to expand the scenario in order to thus come closer to the complex actual infrastructure. Some possibilities would be, for example, to expand to three lanes or road construction situations. Furthermore, accidents and passive safety systems should be incorporated in the analysis provided that these accidents do not result in serious injuries violating the rules of the Vision Zero. Therefore, it is necessary to deal with the estimation of the severity of accidents and evaluation of these results concerning accuracy.

In conclusion, for the investigations it can be stated that the question of implementing Vision Zero has to be considered under different aspects. On the one hand, the investigations showed that the necessary adjustments to our current system are less complex than they first appear. On the other hand, the presumably extensive effects on economic aspects and the important acceptance by the automobile manufacturers' customers could not be determined. Statements about the implementation before 2050 could also not be made with the current state of knowledge. However, it turned out that the limitations resulting from the integral safety concept will strongly affect public road traffic. Humans as the drivers of vehicles would possess absolute freedom to act only in a certain framework at whose limits they would lose all authority. The necessary intervention of safety systems could possibly be, in certain situations, considered an unjustified limitation of one's freedom to act, which would reduce the acceptance of the concept. Therefore the question remains as to how much freedom a person is willing to forgo for his or her own safety, because Vision Zero cannot be enforced without the stated limitations.

## References


1. Statistisches Bundesamt (Hrsg.) (2012) Verkehr aktuell 04/12, erschienen. Wiesbaden, 20 Apr 2012
2. Statistisches Bundesamt (2011) Unfallstatistik—Verkehrsmittel im Risikovergleich. Wiesbaden
3. European Commission (2011) White paper on transport. Luxemburg
4. Deutscher Verkehrssicherheitsrat (2012) Schriftenreihe Verkehrssicherheit—vision zero, Bonn
5. Kramer F (2009) Passive Sicherheit von Kraftfahrzeugen, 3. Auflage, Vieweg + Teubner, Wiesbaden
6. Bundesministerium der Justiz (2012) STVO—nichtamtliches Inhaltsverzeichnis, abgerufen am 18 Dec 2012. http://www.gesetze-im-internet.de/stvo
7. Schwaiger R (2012) Integrales Fahrzeugsicherheitskonzept zur Umsetzung der Vision Null im Längsverkehr. Masterthesis, HTW Berlin, Oktober
8. Lex C, Eichberger A, Hirschberg H (2011) Methoden zur Ermittlung des Reifen-Fahrbahn-Kraftschlusspotenzials für Fahrerassistenzsysteme. in: ATZ, Dezember
9. Winner H, Hakuli S, Wolf G (2012) Handbuch Fahrerassistenzsysteme, 2. Auflage, Vieweg + Teubner Verlag, Wiesbaden
10. Charbonnier P, Muzet V, Nicolle P, Hautiére P, Tarel J-P, Aubert D (2008) Stereovision applied to road scene analysis. In: Bulletin des Laboratoires des Ponts et Chaussées

# Advanced Headlight System: 3D High Beam

**Robert Büthorn, Hadj Hamma Tadjine, Bert Auerbach and Karsten Schulze**

**Abstract** In this paper, a new concept for an advanced headlight system is presented. The weather or the road geometry cause a reduced illumination range or glare which is critical for driver safety. An optimal headlamp system should be able to adapt on a multitude of parameters from the ego vehicle and the environment to ensure a sufficient illumination of the street without glaring other road users. A perfect road illumination without providing glare for other traffic participants and giving the maximum additional information about the environment to the driver is still a hard task for these systems. The improvement of such systems will be achieved only with the combination of lane data and a list of detected objects provided from the camera system with the PSD format data of the GPS (Global PS) and the navigation system. The most part of the technology which is needed for the realization of an optimal adaptive system is still available.

**Keywords** Vehicle lighting · Bending light · Adaptive main beam · Prediction · Camera based · Global positioning system · PSD-data

R. Büthorn (✉) · H. H. Tadjine · B. Auerbach · K. Schulze
IAV GmbH, Carnotstraße 1, 10587 Berlin, Germany
e-mail: robert.buethorn@iav.de

H. H. Tadjine
e-mail: dr.hadj.hamma.tadjine@iav.de

B. Auerbach
e-mail: bert.auerbach@iav.de

K. Schulze
e-mail: Karsten.schulze@iav.de

J. Fischer-Wolfarth and G. Meyer (eds.), *Advanced Microsystems for Automotive Applications 2013*, Lecture Notes in Mobility, DOI: 10.1007/978-3-319-00476-1_8,

## 1 Introduction

The first vehicles were left parked in darkness because the drivers could barely see where they were driving. Light is an important factor of a person's comfort and well-being, especially for the elderly. Having appropriate lighting conditions in a household is both crucial and difficult to achieve. The only choice a car driver had for a long time was to switch the lights on or off. This is different today: The headlights think with the driver, sometimes even faster than him. They can illuminate corners, automatically switch between high beam and low beam and even leave out or highlight certain areas.

The first car with adaptive lights was already built in 1918 [1]. Electric lights were not invented until the 1920s of the last century. The adaptive light entered series production in the 1960s [2]. The headlights were connected to the steering axle and followed the steering motions through a simple pulley system. Today this is all managed electronically: Either using additional headlights which switch on simultaneously with the indicator lights or swiveling headlights [3].

In the last years the development of driver assistant systems which can improve the visibility range have hardly increased. Reasons for this development are the usage of the new technologies such as the Xenon head-lights, new adequate optical sensors, High dynamic range cameras and very faster ECUs.

Studies on the well-being of drivers at night show that 48 % of drivers feel stressed when driving in low-light conditions [4] and 34 % of drivers are using light assistance systems [5–7].

Accident statistics show that the current task of the automotive industry is to guarantee for the driver more safety, more comfort and more security. One of the most important topics is the improvement of street visibility at night. In 2011 died 4,002 peoples due to traffic accidents in Germany. Approximately 38 % of these accidents are happened at night. In contrast the driving performance amounts only 25 % at night. Furthermore 29 % of all seriously injured and 36 % of all killed people are listed at night [8] (see Fig. 1).

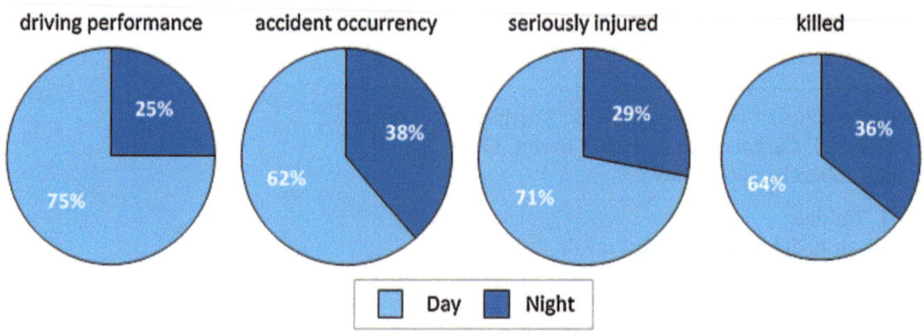


**Fig. 1** Accident statistic—comparison between day and night

# 2 State of the Art Light Functions

## 2.1 Evolution of Light Functions

It has been a long way from a headlamp that normally illuminates the area in front of the car to modern light based driver assistance systems. The first system was integrated at the beginning of the 20th century, when two differently angled headlamps were integrated and installed that enabled low beam and high beam settings. Almost 40 years later, the first type approval for headlamps with asymmetrical light distribution was granted. Although different approaches for cornering light and bend lighting have been carried out over the time, it took until the beginning of this century to establish these dynamic light functions. Moreover, in 2006, legislation allowed the use of adaptive front lighting systems that provide specialized light distributions for different driving situations.

Currently, more than 25 % of all new German cars were equipped with adaptive lights, compared to 2009 where only 15 % were equipped [9]. In the current automotive market, there are a variety of technical solutions to facilitate these light distributions like: dynamic bending light, high beam assist, adaptive main beam and the adaptive high beam. All of these systems could improve the visibility for the driver. However, all these systems are technically limited and suffer from the problem of the package and the high costs [10].

## 2.2 Dynamic Bend Lighting

Dynamic bend lighting is a well visible lighting function with great benefit for the end-customer, because the entire headlamp light distribution is swiveled up to $\pm 15°$. This dynamic movement of the light distribution represents an active gain in safety due to the increased range of illumination along the road. On winding roads, visibility can therefore be increased by up to 50 %, depending on the curve radius. The calculation of the swivel angle depends on the vehicle speed, steering angle and yaw angle [10, 11].

The proposed system suffers from some limitation. Due to the CAN Vehicle data, the swivel angle could be calculated only in one time and the system could not allowed any prediction. In several scenarios the light distribution due to the main beam will not be distributed on the street, where the driver needs it. As seen in the Figs. 2 and 3 at the beginning of a curve the main beam shines straight forward, not into the curve. Also, in the end of a curve, the main beam is bended to the left or right side but the lane in front of the car is already straight.

Fig. 2 Dynamic bending light at the beginning of a curve

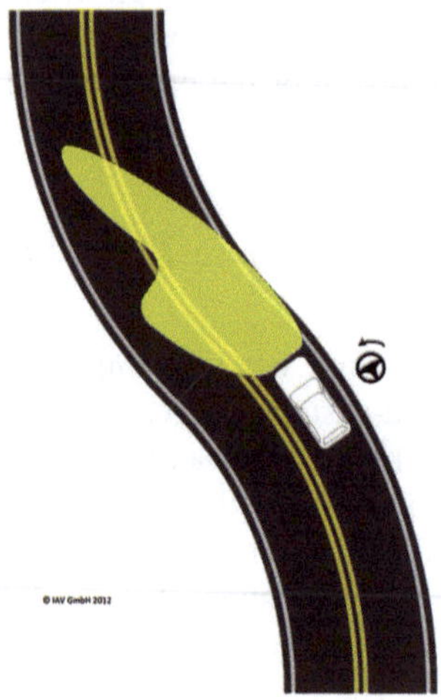

Fig. 3 Dynamic bending light at an alternating curve

## 2.3 *High Beam Assist*

High-Beam Assistant System manages the operation of dipped and main beam lights whilst driving at night. The system comprises primarily of a camera located in the interior rear-view mirror housing. The camera monitors the light from any source at the front of the vehicle (e.g. rear lights of a vehicle in front, street lights, and headlights of an approaching vehicle) which then triggers the system to automatically operate the dipped or main beam function of the Xenon headlights.

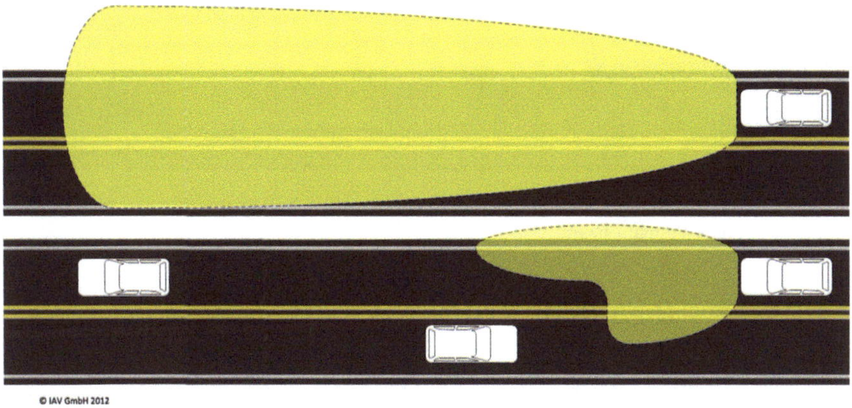


**Fig. 4** The high beam assist

This system can be overridden by either manually selecting the 'On' position for the headlights or by using the dipped beam selector on the indicator stalk.

As benefits, an optimum light control and visibility at all times with improved focus on the road and a reduction of actions required from the driver will be enabled. Also, the possibility of dazzling approaching drivers by automatically selecting dipped or main beam headlights will be reduced. This system has a main limitation. In case of automatic deactivation due to oncoming vehicles the area up to the car in front is not illuminated and obstacles especially in this area can be missed (Fig. 4).

## *2.4 Adaptive Main Beam*

The adaptive main beam assistant is a camera based system which adapts the range of the main beam to the distance to the other cars. The headlamps work in conjunction with a camera fitted on the inside of the windscreen that detects oncoming vehicles and vehicles in front. The system automatically adjusts the headlamps to prevent other drivers from being dazzled by gradually transitioning to high-beam mode if the road ahead is clear (see Fig. 5).

A front camera recognizes continuously the oncoming traffic. Blobs of the head lights or the rear lights will be automatically detected and classified using an image processing algorithms. The cone of the main beam always ends at the nearest detected light to the ego-car. With a presence of other cars near to the ego-car, the main beam will be in the normal position or in another height without any assistance.

Obstacles at the roadside will not be marked und could be overlooked or missed.

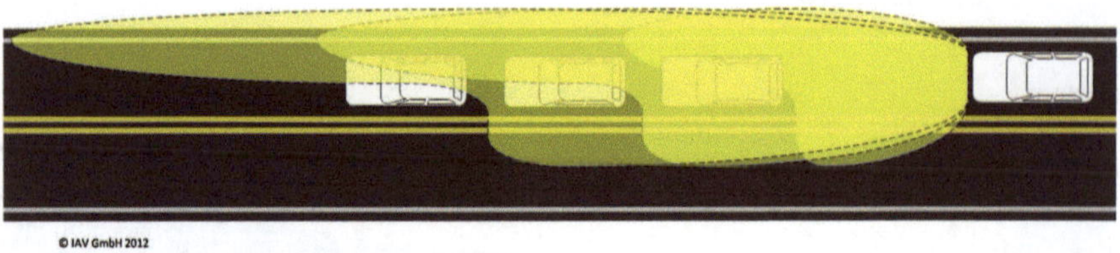


**Fig. 5** The adaptive main beam

## 2.5 Adaptive High Beam Assist

A camera system mounted on the front windscreen is used to detect an oncoming vehicle in the opposite lane or another vehicle in front and could automatically adjusts the headlamp range to the appropriate distance—in contrast to traditional systems which only switch between dipped and main beam. This allows the range of the dipped beam to be increased from around 65 to approaching 300 m without dazzling other drivers (Adaption of the illumination range dependent on the position of other road users). Consequently, the driver always benefits from the optimum illumination range, providing earlier and better recognition of the road ahead, pedestrians and danger spots [3].

With the current adaptive System, the high beam is almost active. An optimum light distribution needed for the driver will be calculated and generated from the System in order to reduce the dazzling effect. This is pretty up-to-date, so the current camera system and the headlamps adapted for the system are real expensive. Furthermore the system is set inactive, if the traffic density is high and a lot of turn ups and dips of the headlamps is needed (Fig. 6).

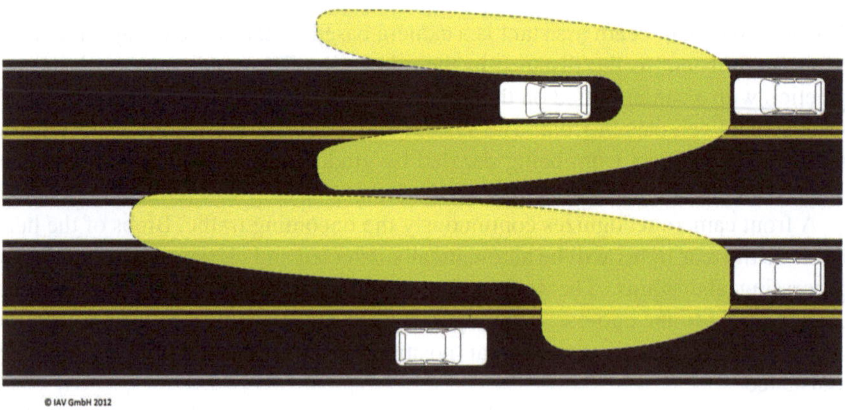


**Fig. 6** The adaptive high beam assist

## 3 Adaptive Bending Light

During night time driving it is more demanding to keep the vehicle within the lane boundaries. The automotive lighting equipment illuminates the driver's viewing field and has to deliver the appropriate information for lane keeping under many different driving situations.

Regarding the driver's needs for visual information about the road ahead, present headlight technology insufficiently lights up bends in the road ahead. The advantage of using an adaptive lighting system, where one observes a country road lit correctly. Instead of illuminating some fields or trees, the available light is directed dynamically in the direction where it is actually needed, enhancing the visual information for the driver. The adaptive bending light is not realized due to vehicle speed, steering-, and yaw angle, like the dynamic bending light.

The new systems are controlled by means of path prediction based on vehicle dynamics, lane information from the camera system and route vectors of the navigation system.

With the output of the path prediction model it is possible to control the movable headlamps appropriately. To assure a best illumination of the street, the mid-position of the light distribution is placed on the left lane of the road (see Figs. 7 and 8).

The prediction consists of an estimation of the lateral accelerations acting on the vehicle body and a prediction of the future path of the vehicle based on the current vehicle dynamic states in combination with a simple model of the vehicle. Extra filtering has been added to improve the quality of the prediction. If no PSD- and GPS-data are available, the adaptive system will fall back automatically to the normal lighting function such as the dynamic bending light.

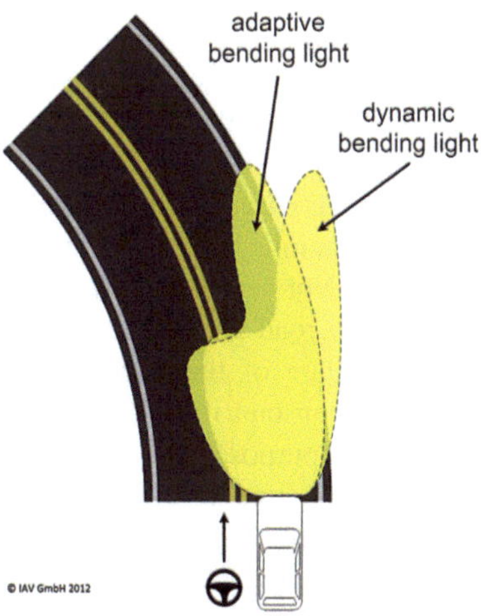


**Fig. 7** Comparison between dynamic and adaptive bending light at beginning of a *curve*

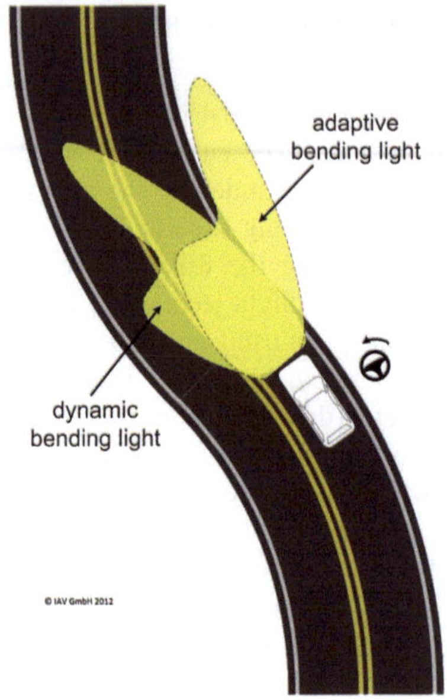


**Fig. 8** Comparison between dynamic and adaptive bending light at an alternating *curve*

## 4 Adaptive Main Beam at Hilltops and Dips

Classic headlamps are static systems which emit the light always in the same direction independent from the road geometry. The optimal range can be achieved for a straight, horizontal street geometry. In case of cambers, dips and curves the illuminated range on the street can decrease while the glare of the oncoming traffic can increase. According to this an optimal headlamp should be able to adapt on the horizontal (curves) and vertical (cambers and dips) street geometry to assure a better illumination, a constant visibility range and a minimized glare.

One possibility to realize this concept is a swiveling light module. Some modern headlamps, especially from cars of the upper class are already equipped with this system which is called dynamic bending light. They offer an adaption on curves. The potential of these systems to increase the road safety is classified as very high. The risk to cause an accident in curves increases significant at night especially in curves with a small radius [4].

The dynamic bending light offers only an adaption on the horizontal road geometry. Not yet realized in production cars but also important is an adaption on the vertical road geometry. Cambers and dips can cause a reduced visibility distance or glare of the oncoming traffic, too. A vertical swiveling light module ensures a constant illumination range.

For this purpose, the headlight of oncoming traffic and the traffic backlights ahead will be used as reference for the distance estimation between the ego vehicle and other oncoming vehicles. The system will be activated if no other vehicles in the relevant area where other drivers could be dazzled are detected. Else, the system will be switched to the adaptive mode. That means, the cut-off of the main beam is lowered infinitely variable up until the headlights or backlights of the other car. A maximal high is not intended (see Figs. 9 and 10).

## 5 Proposed Approach

As extension for the existing system, a combination of the adaptive main beam with the adaptive high beam assist will assure the best possible illumination of the road is proposed. This improves the headlamp illumination by means of continuous adaptation of the headlamps according to the current driving situation and current environment. For this reason the vehicle trajectory must be analyzed three-dimensional. To assure a good functionality of the proposed system and to become a close view of the real lane, an adequate prediction is required.

First road estimation is achieved based on the camera system. The road vertical curvature estimation is performed, due to the fact that at far distances the road may be not plane. Once, this road curvature is estimated, a vertical offset correction is

**Fig. 9** Adaptive main beam at a hilltop and a dip with traffic ahead

**Fig. 10** Adaptive main beam at a hilltop and a dip with oncoming traffic

only done for those objects located close to the horizon line, at the infinity. A representative features are calculated for each tracked object, and then objects are classified (main nuisance light source due to the reflections of the own vehicle's light over the road-traffic signs) or vehicles, using a special classification. To estimate the distance between the ego vehicle and the detected vehicles using monocular camera, the Dickmanns Model is used [12].

By using the road vertical curvature estimation camera based, a correction of the vertical image coordinate of all the objects located at the infinity where the flat road assumption is not valid is performed [13, 14]. Vertical and horizontal mapping models can be carried out. However, this distance is only an approximation, since some sources of error can affect this estimated distance, such as: elevation of the vehicle's light, road unevenness, object's centroid, etc. In order to minimize these errors, a vertical coordinate correction is proposed [15].

Generally, the road is assumed as plane. However, this is a dangerous assumption, since at far distances (more than 500 m) the road vertical curvature may influence the vertical position at which the objects appear in the image. As a simplifying assumption, the horizontal road curvature is assumed to be so small

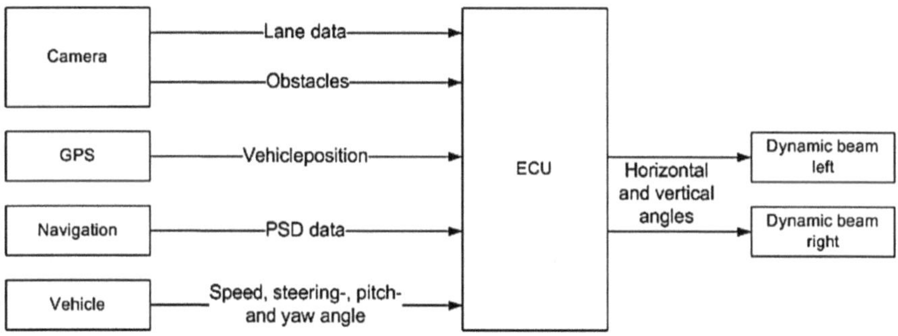


**Fig. 11** Block diagram

that is not considered in the perspective model [16, 17]. If the road vertical curvature is significant, some mistakes can arise in the classification process due to the flat road assumption, so road vertical curvature estimation must be included into our camera perspective model. For these reasons additional information from other systems are needed.

In addition to the vehicle dynamic parameters, route vectors from the navigation system are used. The navigation system permanently keeps track of the precise location of the car, and with its built in digital road map it knows the road characteristics ahead. Finally, the path prediction, the navigation data, vehicle CAN data and an objects list provided from the camera system will be coupled to ensure consistent control of the headlamps (see Fig. 11).

## 6 Conclusion

In this paper, a new concept for an advanced headlight system is presented. The weather or the road geometry cause a reduced illumination range or glare which is critical for the driver safety. An optimal headlamp system should be able to adapt on a multitude of parameters from the ego vehicle and the environment to ensure a sufficient illumination of the street without glaring other road users. A perfect road illumination without providing glare for other traffic participants and giving the maximum additional information about the environment to the driver still a hard task for these systems.

The improvement of such systems will be achieved only with the combination of the lane data and the list of detected objects provided from the camera system with the PSD format data of the GPS (Global Positioning System) and the navigation system. The most part of the technology which is needed for the realization of an optimal adaptive system is still available. But these systems are not mandatory by the Economic Commission for Europe (ECE).

# References

1. Roslak J, Kauschke R, Wallaschek J (2005) Active headlamps to increase traffic safety. ISAL proceedings 11, p 1
2. Schönbeck S, Ellmers U, Gail J, Krautscheid R, Tewis R (2005) Abschätzung möglicher Auswirkungen von Fahren mit Licht am Tag (Tagfahrleuchten/Abblendlicht) in Deutschland, closing report 1, German Federal Ministry of Transport, Bergisch Gladbach, p 1
3. Von Hofmann A (2003) Lichttechnische Anforderungen an adaptive Kraftfahrzeugscheinwerfer für trockene und nasse Fahrbahnoberflächen. PhD, Technical University Ilmenau, Germany, p 1
4. Frost and Sullivan: Night Light. http://www.frost.com/
5. Puls Market Research (2011) Fahrerassistenzsysteme
6. Konrad R (2010) Fahrstabilisierungssysteme und Fahrerassistenzsysteme. ISBN: 978-3-8348-1314-5
7. Konrad R (2010) Bosch Autoelektrik und Autoelektronik: Bordnetze, Sensoren und elektronische Systeme. ISBN-10: 3834812749
8. German statistics. https://www.destatis.de
9. DAT-Report (2011) http://www.dat.de/products/
10. Akashi Y, Rea M (2001) The effect of oncoming headlight glare on peripheral detection under a mesopic light level. In: Progress in automotive lighting
11. GENTEX: Vehicle lamp control (2005) http://www.patentstorm.us/patents/6947577-fulltext.html
12. Dickmanns ED, Mysliwetz BD (1992) Recursive 3-D road and relative ego-state recognition. IEEE Trans Pattern Anal Mach Intell 14(2):199–213
13. Chen YL (2009) Nighttime vehicle light detection on a moving vehicle using image segmentation and analysis techniques. WSEAS Trans Comput 8(3):506–515
14. O'Malley R, Glavin M, Jones E (2008) Vehicle detection at night based on tail-light detection. In: 1st international symposium on vehicular computing systems, Trinity College Dublin
15. López A, Hilgenstock J, Busse A, Baldrich R, Lumbreras F, Serrat J (2008) Nighttime vehicle detection for intelligent headlight control. In: ACIVS 08: proceedings of the 10th international conference on advanced concepts for intelligent vision systems, Springer, Berlin, pp 113–124
16. Alcantarilla PF, Bergasa LM, Jiménez P, Sotelo MA, Parra I, Fernández D, Mayoral SS (2008) Night time vehicle detection for driving assistance. In: IEEE intelligent vehicles symposium (IV)
17. Alcantarilla PF, Sotelo MA, Bergasa LM (2008) Automatic daytime road traffic control and monitoring system. IEEE intelligent transportations systems conference (ITSC), pp 944–949

# A Prototyping ITS Station for Advanced Driver Assistance Systems and Pedestrian Safety

**Robin Streiter, Christian Adam, Sven Bauer, Marcus Obst, Timo Pech, Pierre Reisdorf, Robin Schubert, Jan Thomanek, André Welzel and Gerd Wanielik**

**Abstract** The steadily increasing complexity of cooperative Advanced Driver Assistance Systems (ADASs) requires efficient development and prototyping strategies for new ADAS applications. In this paper the prototyping road side unit (RSU) *PROTECT1* from Chemnitz University of Technology is presented as a generic intelligent transportation system (ITS) with a variety of sensors, computing units, and different human–machine interfaces (HMIs) in order to allow the fast implementation and evaluation of new ADASs and safety applications for

---

R. Streiter (✉) · S. Bauer · M. Obst · T. Pech · P. Reisdorf · A. Welzel · G. Wanielik
Chemnitz University of Technology, Chemnitz, Germany
e-mail: robin.streiter@etit.tu-chemnitz.de

S. Bauer
e-mail: sven.bauer@etit.tu-chemnitz.de

M. Obst
e-mail: marcus.obst@etit.tu-chemnitz.de

T. Pech
e-mail: timo.pech@etit.tu-chemnitz.de

P. Reisdorf
e-mail: pierre.reisdorf@etit.tu-chemnitz.de

A. Welzel
e-mail: andre.welzel@etit.tu-chemnitz.de

G. Wanielik
e-mail: gerd.wanielik@etit.tu-chemnitz.de

C. Adam · R. Schubert
BASELABS GmbH, Chemnitz, Germany
e-mail: christian.adam@baselabs.de

R. Schubert
e-mail: robin.schubert@baselabs.de

J. Thomanek
IAV GmbH, Chemnitz, Germany
e-mail: jan.thomanek@iav.de

J. Fischer-Wolfarth and G. Meyer (eds.), *Advanced Microsystems for Automotive Applications 2013*, Lecture Notes in Mobility, DOI: 10.1007/978-3-319-00476-1_9, 

vulnerable road users (VRUs). It is an integral part of the rapid prototyping framework for ADAS applications of the Professorship for Communications Engineering—which furthermore consists of two multisensory equipped vehicles *CARAI 1/2* and the modular software prototyping framework BASELABS Suite.

**Keywords** Road side unit · ADAS development · VRU protection · Precise positioning · Classification · Tracking · Localization · Intelligent transportation systems

## 1 Introduction

The idea was to set up a generic ITS research station in order to complement the existing ADAS prototyping vehicles to expand the scope from Car2Car (V2V) to Car2Infrastructure (V2X) applications. In addition to the description of the technical components, the utilized software prototyping framework and the usage of *PROTECT1* is demonstrated on the example of different ADAS applications.

With the help of the complete test bed, new and innovative ADAS applications will be implemented as proposed. Based on the integrated positioning sensors, communication and visual sensors, different applications can be derived. On the one hand there will be all ADAS applications based on low cost precise localization (e.g. relative positioning). In order to do that, *PROTECT1* will determine and distribute DGPS corrections via the integrated V2X communication equipment or mobile data networks (LTE) to V2V enabled vehicles and pedestrians. New Global Navigation Satellite System (GNSS) augmentation strategies, like the distribution of urban satellite shadow maps will be evaluated. On the other hand there are VRU safety applications. The main objectives are the evaluation of tracking algorithms for crowded pedestrians basing on data fusion algorithms for the integrated stereo vision- and a far infrared camera (FIR). Here new approaches to classify crowds of pedestrians will be evaluated. The pedestrian safety applications will directly benefit from the increased positioning performance for mobile devices—augmented by the ITS station.

The processing of different data types originating from several sources is managed by the software framework BASELABS Connect and Create which handle data acquisition from sensors, data recording and replay, rapid prototyping of algorithmic components with focus on probabilistic data fusion and data distribution. The paper is structured as follows. In the first part the test bed and the equipment of the components will be illustrated. In the second part of the paper some applications are introduced that have already been implemented and some future scenarios are described that are based on existing components, including how these applications will be set up within the test bed.

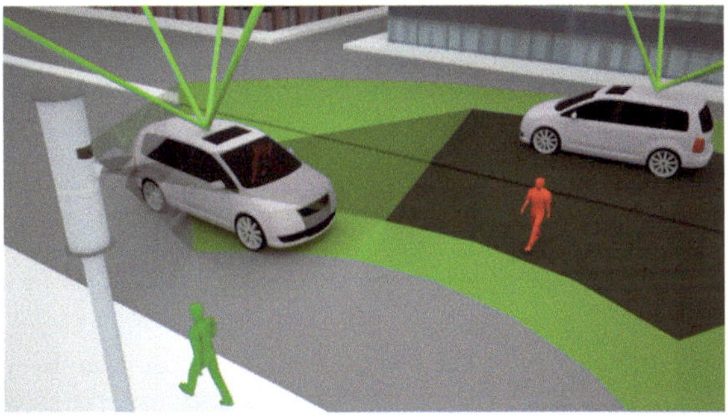

**Fig. 1** ADAS research and rapid prototyping Framework with ITS Station *PROTECT1* (*left*) and test vehicles "CARAI 1&2" (*right*)

## 2 Prototyping Test Bed and Sensor Equipment

Figure 1 shows the components of the prototyping test bed. Since the measurement vehicles introduced in detail in [1] the focus in that paper will be on *PROTECT1*—also named RSU. It is equipped with an amount of sensor types, able to observe the vicinity and to communicate via different channels. There are vision sensors, localization sensors and communication devices which are utilized depending on the current application. Figure 2 shows the sensor configuration of the whole test bed—while the following list shows the detailed configuration of *PROTECT1.*

Vision Systems:

- Far Infrared Camera (FLIR Photon 320)
- 3-sensor multi-baseline IEEE-1394b compliant stereo vision system (Bumblebee XB3).

Communication Devices:

- V2X and V2V compliant communication device (IEEE 802.11p) (Denso Wireless Safety Unit)
- Wi-Fi/3G/4G enabled network.

Other Components:

- Embedded PC
- u-blox GNSS localization module LEA-6T with precision timing
- Further sensor modules and components within the vehicles are shown in Fig. 2.

The ITS station is able to receive and to redistribute information from participants on various information channels, simultaneously. The communication devices are providing the infrastructure to set up various broadcasting and data distribution applications. Basing on DENSO Wireless Safety Units a V2X infrastructure can easily be established between the RSU and the vehicles. In order to address further road users like pedestrians or cyclists communication via Wi-Fi is enabled. In order to receive information from external sources (traffic center

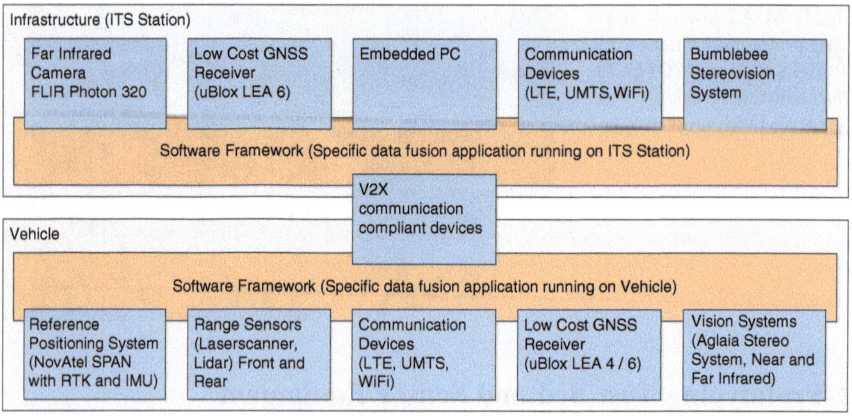


**Fig. 2** The test bed hardware architecture is divided into infrastructure represented by the ITS station (*top*) and the vehicles (*bottom*)

information, weather, GPS augmentation) 3G and 4G can also be used. The position of road users within the sensor coverage who are not equipped with appropriate devices will also be distributed. The detailed description about the implemented algorithms is given in the following sections.

# 3 ADAS Prototyping Software Framework

The software framework can be broken down in two parts. On the one hand there is the BASELABS software framework where the whole application is set up at component level and on the other hand there are the concrete sensor and algorithmic components. Both are described in this section.

## *3.1 Multi-Sensor Application Design Using BASELABS Connect*

In order to facilitate a rapid implementation of system prototypes, the development software tool BASELABS Connect is applied [2]. It provides a library of ready-to-use sensor components, including various cameras, Radar, LIDAR, GPS, and others. In addition, different communication components including TCP, UDP, or CAN are provided. For evaluating the advantages of such a software-supported design methodology, some core features of Connect will be described in the following.

The software facilitates a rapid top down approach of complex multi-sensor applications. The first step is the definition of encapsulated software modules with defined data interfaces (called components). Such components can be designed quickly by using the provided software assistants. After that, the components can be used via drag and drop within the development environment. In particular, components can be connected with each other given that the output data type of the data source complies with the input data type of the data sink.

Figure 3 illustrates an example of this methodology. The image shows that BASELABS Connect can be used seamlessly in existing development environments which facilitates simultaneous application design and debugging. In the lower right part, the so-called toolbox contains all components which are linked to the project (including components created by the user). The upper left part illustrates the graphical usage of these components and, in particular, their connections. These connections specify the data flow within the designed application. The control flow is based on a data-driven paradigm and requires no further attention in standard scenarios. For complex applications, however, the synchronization behavior of the system can be adjusted in detail.

An additional feature of BASELABS Connect is the automatic recording and replaying of sensor data. The term automatic refers to the fact that no manual interaction such as serializing/deserializing or timestamping is required for this functionality as long as the data interfaces between the modules are appropriately defined. The software supports a proprietary, efficient binary format as well as the possibility to include user-specific formats. In addition, the usage of databases for sensor data management is supported.

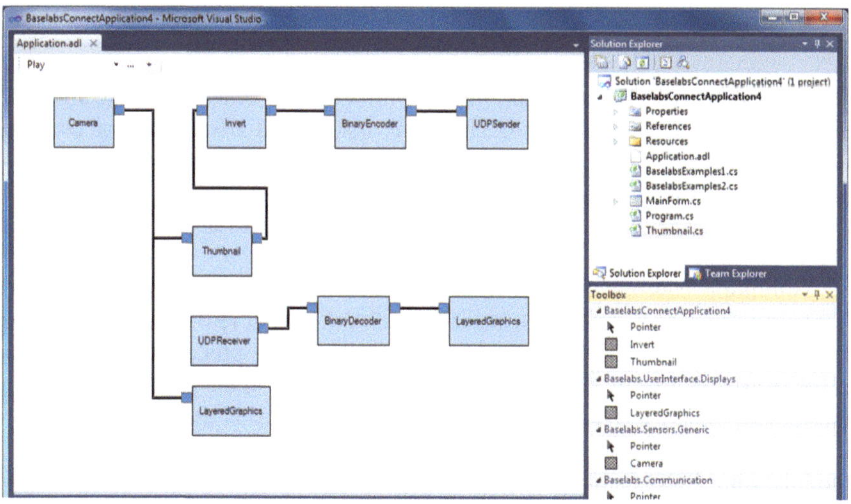


**Fig. 3** Screenshot of BASELABS connect illustrating the rapid application design

## *3.2 Algorithmic Components*

This section introduces several algorithmic components that are used in the described software framework. Based on these components the concrete application can be set up.

### 3.2.1 Global Satellite Navigation

GNSSs, e.g. GPS, GLONASS or the emerging GALILEO, are an appropriate solution for many absolute positioning tasks like vehicle localization. Typically, accuracy between 2 and 10 m is achievable under good conditions with standard low-cost hardware like a u-blox LEA6T GPS receiver.

The prototyping vehicle *CARAI* and the ITS station are equipped with different algorithms appropriate for single and multi-constellation satellite navigation. This includes standard least square solvers as well as loosely and tightly coupled Bayes filter variants which are especially useful for dynamic applications like moving vehicles [3].

Moreover, the ITS station broadcasts corrections for some of the GNSS errors by relaying information provided by EGNOS/EDAS to the connected vehicles. Another approach, similar to differential GPS, called relative positioning which was developed inside of the European research project CoVeL [4] is implemented in the ITS station as well. Here, the ITS stations constantly observes the measurements of the built-in u-blox GPS receiver and compares them to the well-known position of the ITS station itself. The error between the observation and the a priori knowledge is then broadcasted to the vehicle with V2X.

### 3.2.2 Generic Classifier

To detect objects like pedestrians, vehicles or other features at an image a generic classifier has been implemented. Therefore, normalized *Histograms of Oriented Gradients* (HOGs) [5] are used as feature descriptors. The extracted feature vector is classified using a trained *Classifier Cascade* [6]. It consists of a series of *AdaBoost classifiers* which contains several decision trees as weak learner. The classifier was trained for vehicles and pedestrians [7].

As multiple cameras are used for environment recognition, a high-performance implementation was required to manage the workload in real-time. To meet this requirement, the parallel nature of the necessary image processing and classification was exploited by offloading the complete process to the graphics processing unit (GPU). The CUDA [8] based implementation uploads the grayscale camera image to the graphic cards video memory, calculates integral images to speed up the following HOG feature computation and computes the classification result for numerous regions of interests. Positive detections are then returned to the host for further processing, e.g. by the tracker described in the following.

### 3.2.3 Tracking

The tracking algorithm estimates the motion of all objects detected by the image classification over time. That includes the direction and the velocity of VRUs and vehicles. With this information, the likely position of all objects can be predicted for a few seconds in future, which enables the system to detect possible collisions and to broadcast suitable warnings to the concerned traffic participants.

The tracking also compensates the unreliable detection performance of the image classifier. In some cases the classifier can miss an object, that is, it provides no detection for this object. In other cases, the classifier may provide multiple detections for the same object. The reason for that is that the object is small enough to fit into several regions of interest of the classifier, but is great enough to cause a positive classification. In that case, each investigated region becomes a valid detection. One tracking algorithm that especially manages multiple detections from one object is the General Probabilistic Data Association (GPDA) [7].

The tracking is implemented with BASELABS Create. Other Bayesian filtering algorithms are also provided—like the Unscented Kalman filter (UKF), various tracking algorithms like SPRT, IPDA, and GPDA, and ready-to-use sensor and motion models. Furthermore, it provides graphical assistance for the user-friendly implementation of multi-sensorial systems and reduces the effort for the implementation significantly. The modular architecture for Create enables quick interchangeability of algorithm parts. For example, the same tracking component can be used with different association algorithms and motion models without the need to change the code significantly. Thereby, different algorithms can be implemented in short time and evaluated against each other to find the best one for the intended application.

# 4 Applications

The ITS prototyping station is used in various research projects in order to provide the required V2X infrastructure. In this chapter several ADAS and VRU protection applications are introduced.

## *4.1 Vulnerable Road User Protection*

Our team develops ADAS applications enhanced by visual human machine interfaces (HMIs). This includes a system to display detected pedestrians and traffic related information within a head up display in the driver's line of sight on the windshield. The ITS station contributes to increase the safety of VRUs. Vehicles in the range of the RSU receive the position and movement information of pedestrians approaching the road. The early warning system detects whether a collision of the vehicle and the pedestrian is imminent and visually alerts the driver. Using *PROTECT1*, drivers whose vehicle is equipped with communication devices only, can be made aware of VRUs as well. In order to support the in-vehicle intersection assistant, the ITS station broadcasts information regarding the right of way and traffic signs in the surrounding intersections. The visual ADAS application will generate the projection for the head up display.

#### 4.1.1 Reliable Pedestrian Detection

The development of pedestrian protection systems is in the main focus of research in the field of ADASs. The automated identification of pedestrians has the objective to reduce the number of collisions and to improve the safety of regular vehicles. However, pedestrian detection is a challenging task due to different illumination and weather conditions, and the large variations of the poses of the pedestrians. Various types of sensors, for instance cameras, Radars and LIDARs have been examined in order to find the most suitable sensor for a reliable and robust detection of pedestrians [9]. But one input given by a single sensor may often not be capable of reproducing a complete view of the environment. Therefore, computer-processing tasks use the process of combining substantial information from several sensors to provide a more accurate description from observed scenes. Particularly for imaging sensors, e.g. visible light cameras or infrared cameras, such multi-sensor fusion can be done on pixel-level. The output of the data fusion is a single composite image that will be more comprehensive and thus, more useful for further object detection tasks [10].

Generally, FIR cameras seem to be more suited for the detection of pedestrians than the visible light ones. However, the FIR camera would fail to correctly detect pedestrians in the case of hot and sunny weather. Therefore, the proposed application is based on a combination of far-infrared and visible light sensors to improve the capability of the detection system with respect to single and crowded people in different conditions of illumination and weather.

The pixel-based fusion is the precise spatial alignment of the infrared and visual image. Figure 1 depicts the aligned images (dark green is FIR image light green is stereo image). Finally, the extracted features are passed to the classifier described in Sect. 3.2.2. The algorithm flow is depicted in Fig. 4.

### *4.2 Low Cost Vehicle Localization in Urban Areas*

Reliable and precise positioning is a key technology for various ADAS applications. Erroneous pseudorange measurements are a major problem for GNSS based positioning. Large errors are caused by so called multipath and non-line-of-sight (NLOS) effects due to signal reflection and blockage. This is especially a problem in dense urban areas, where foliage and large buildings are usually nearby. In order

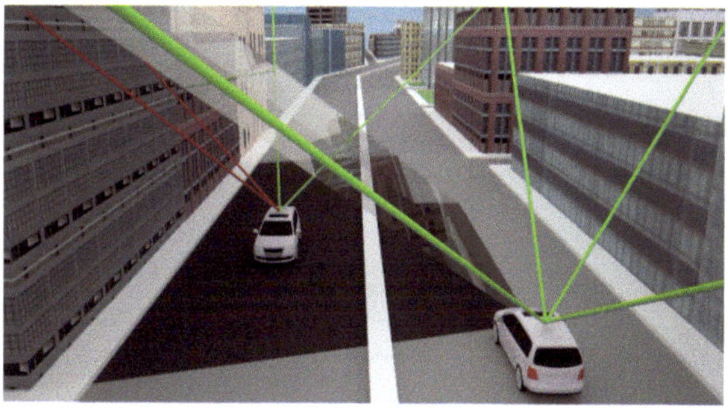

**Fig. 4** Pedestrian detection system based on sensor image fusion of far-infrared and video data

to overcome the limitations of existing positioning solutions *PROTECT1* will help to transfer state-of-the-art positioning algorithms to road applications. Figure 5 shows the approach to mitigate multipath errors is the creation and distribution of so called shadow maps [11]. Using a 3D environment model consisting of an elevation map and building models the areas where a satellites ranging signal can be directly received and where only NLOS reception is possible can be computed. This can be done by calculating the satellites 3D position at a certain point in time and casting the buildings shapes to the ground as seen from this position. As shadow casting is a classical problem in 3D computer graphics. GPUs can be utilized to optimize this algorithm. The resulting maps can then be broadcasted to the vehicles using the ITS station's V2X interface or by using cellular networks to allow the in-vehicle receiver to weight down pseudorange measurements. By that approach, observations potentially affected by NLOS conditions can be excluded.

Other error sources like tropospheric and ionospheric influences cannot fully be compensated by correction services like EGNOS. Therefore, *PROTECT1* implements another approach called relative positioning (Fig. 6). As the location of *PROTECT1* is precisely known, it can estimate errors in measured satellite observations and broadcast them using the 3G network or its V2X interface to the connected vehicles. One step further, it can also broadcast its own pseudorange measurements to nearby vehicles. Assuming that the ITS station and the vehicle measurements are affected by the same global errors—which is likely in close proximity—a more precise relative location can be computed by single

**Fig. 5** Precise positioning through satellite shadow map exchange

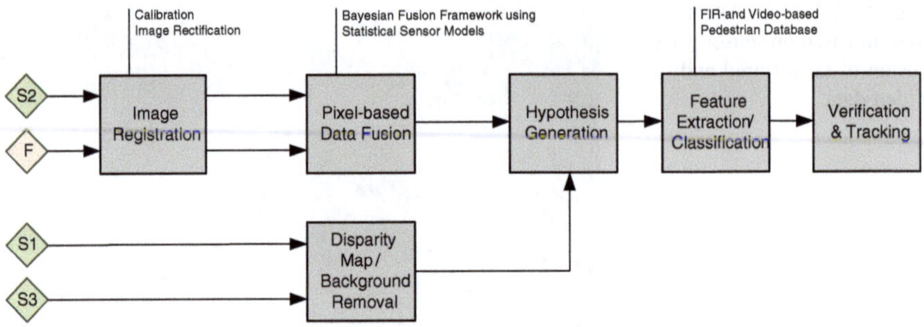


**Fig. 6** Precise positioning through low cost differential GPS via V2X

differencing as shown in [12]. Furthermore, the described ITS station extends the concept of relative positioning between vehicles to mobile devices of pedestrians. The smart phones of the connected pedestrians may use a cellular network or even Wi-Fi to receive precise correction information from *PROTECT1*. In a next step, it cross-checks the received pedestrian position with its internal sensors and retransmits this to the nearby vehicles to leverage an enhanced perception.

## References


1. Schubert R, Richter E, Mattern N, Lindner P, Wanielik G (2010) Advanced microsystems for automotive applications: smart systems for green cars and safe mobility. In: Meyer G, Valldorf J (eds). Springer, Berlin
2. BASELABS GmbH (2012) ADAS rapid prototyping and data handling. Website http://www.baselabs.eu. Last visit 30 Jan 2013
3. Obst M, Bauer S, Wanielik G (2001) Urban multipath detection and mitigation with dynamic 3D maps for reliable land vehicle localization. In: Proceedings of the IEEE/ION PLANS., Myrtle Beach, South Carolina
4. Obst M, Mattern N, Schubert R, Wanielik G (2012) Car-to-Car communication for accurate vehicle localization—the CoVeL approach. In: Proceedings of the 9th international multi-conference on systems, signals and devices, 2012, Chemnitz, Germany
5. Dalal N, Triggs B (2005) Histograms of oriented gradients for human detection. In: Proceedings of the IEEE conference on computer vision and pattern recognition, vol II
6. Viola P, Jones M (2001) Rapid object detection using a boosted cascade of simple features. In: Proceedings of the IEEE computer society conference on computer vision and pattern recognition, vol 1, p 511
7. Schubert R et al (2012) Generalized probabilistic data association for vehicle tracking under clutter. In: Proceedings of the IEEE intelligent vehicles symposium
8. Nvidia Corporation (2013) Parallel programming and computing platform | CUDA. Website http://www.nvidia.com/object/cuda_home_new.html. Last visit 1 Feb 2013
9. Gerónimo D, López AM, Sappa AD, Graf T (2010) Survey of pedestrian detection for advanced driver assistance systems. IEEE Trans Pattern Anal Mach Intell, vol 32
10. Blum RS et al (2006) Multi-sensor image fusion and its applications. Taylor & Francis Group, London

11. Bauer S, Obst M, Streiter R, Wanielik G (2013) Evaluation of shadow maps for non-line-of-sight detection in Urban GNSS vehicle localization with VANETs—The GAIN approach. Accepted for publication at IEEE Conference on Vehicular Technology
12. Obst M, Richter E, Wanielik G (2011) Accurate relative localization for land vehicles with SBAS corrected GPS/INS integration and V2V communication. ION GNSS, 2011 Portland, Oregon

# A Comprehensive Approach for Modeling, Simulation and Virtual Validation of Integrated Safety Systems

**Michael Karner, Martin Krammer, Markus Schratter, Peter Wimmer, Daniel Watzenig and Christian Michael Gruber**

**Abstract** One of the main contributors to the massively rising complexity within automotive systems are the huge improvements made in the area of integrated safety systems. This enormous complexity stimulates the demand for methodologies supporting the efficient development of such systems. Within this work, a comprehensive approach for modeling, simulation and virtual validation of integrated safety systems is proposed. Based on the idea of co-simulation, a process is defined which integrates the development and parameterization of simulation models, the creation of a suitable co-simulation configuration and the according virtual validation of integrated safety systems. This helps the developer to get a comprehensive view about the whole integrated safety system.

**Keywords** Co-simulation · Integrated safety · Active safety · Safety system · Pedestrian detection · Co-simulation configuration

M. Karner (✉) · M. Krammer · M. Schratter · P. Wimmer · D. Watzenig
Virtual Vehicle Research and Test Center, Inffeldgasse 21a/1,,
8010 Graz, Austria
e-mail: michael.karner@v2c2.at

M. Krammer
e-mail: martin.krammer@v2c2.at

M. Schratter
e-mail: markus.schratter@v2c2.at

P. Wimmer
e-mail: peter.wimmer@v2c2.at

D. Watzenig
e-mail: daniel.watzenig@v2c2.at

C. M. Gruber
BMW Group, Knorrstrasse 147, 80788 Munich, Germany
e-mail: christian-michael.gruber@bmw.de

J. Fischer-Wolfarth and G. Meyer (eds.), *Advanced Microsystems for Automotive Applications 2013*, Lecture Notes in Mobility, DOI: 10.1007/978-3-319-00476-1_10,

# 1 Introduction

In recent years, enhancements of sensors led to substantial progress of integrated vehicle safety. The goal of integrated safety is the avoidance of accidents and the reduction of severity of accidents as well as a further reduction of the risk of injuries. For this purpose, active and passive safety systems are combined, leading to extensive cross-linking of safety functions in the vehicle. This, however, means an increasing complexity in the development and integration of such distributed safety systems. The enormous complexity represents a major challenge: currently, safety functions often can only be developed and examined individually and not as a distributed system. This makes it very difficult to get a comprehensive view about the whole system.

An essential utility to understand the whole system is co-simulation. Co-simulation means to simulate a system using different modeling languages, abstraction levels, solvers, tools or computing systems in one common simulation [1, 2]. Different simulators and models of various disciplines are combined to represent the properties of the complete system. As a result, cross-disciplinary issues can be answered and dependencies as well as interactions between individual subsystems can be easily analyzed. Based on the idea of co-simulation, a process is defined within this work which integrates the development and parameterization of simulation models, the creation of a suitable co-simulation configuration and the according virtual validation of integrated safety systems.

# 2 Modeling, Simulation and Virtual Validation of Integrated Safety Systems

Within this chapter, the proposed approach for modeling, simulation and virtual validation of integrated safety systems is described. First, an overview about the approach is given, followed by details about selected individual steps.

## *2.1 Approach Description*

Based on the well-known V-model, we propose an approach for modeling, simulation and virtual validation of integrated safety systems. It is shown in Fig. 1. Consisting of six main steps, it describes the overall way from the requirements level up to the concluding HW/SW test and verification levels.

In the first step, the requirements for the integrated safety systems have to be captured. This can be done by using specialized tools (e.g. Doors) and by using techniques for requirement specification (e.g. boiler plates, semi-formal notation…) or by using common tools like Microsoft Word or Excel. The accurate

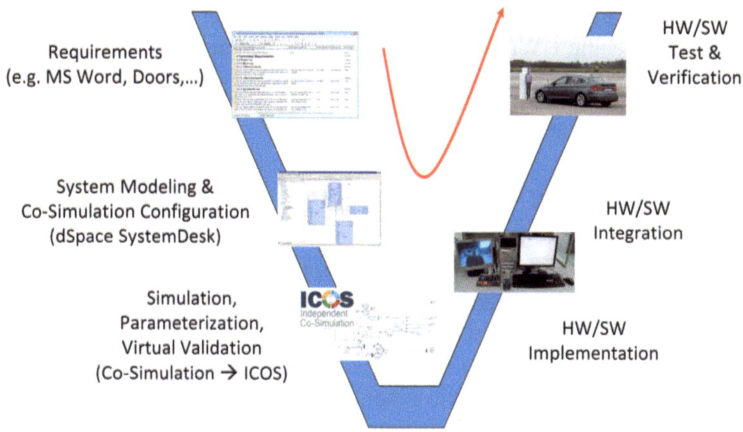


**Fig. 1** Proposed approach for modeling, simulation and virtual validation of integrated safety systems

specification of requirements is very important because they will be used again later on for the validation, verification and test of the system.

The next step of the proposed approach is the system modeling and co-simulation configuration phase. During that phase, based on requirements a hierarchical system model of the integrated safety system and its environment is created. All required interfaces and external influences have to be represented in this system model. Out of this, a corresponding co-simulation configuration (mapping of components to simulation models and tools, distributed over different platforms) can be created automatically. This step is described in detail in Sect. 2.2.

Following, detailed simulation models for the different system components are created and parameterized. Together with the resulting co-simulation configuration from the previous step, an executable co-simulation environment is available which can be used to perform a virtual validation of the integrated safety system. With co-simulation different simulators and models of various disciplines can be combined to represent the whole system. Dependencies and interactions between different models can easily be analyzed. The whole system includes all relevant effects and dependencies between the different models. With this approach it is possible to virtually validate the integrated safety system by performing a number of co-simulation based dynamic test runs with different parameters. The developer is able to change parameters in one model and immediately gets the result. This helps to get a comprehensive view about the complete system, massively increasing the understanding of the overall mechatronic system. More in-depth details about this step are described in [3, 4].

Within this paper, we are focusing on the left side of the V-model shown in Fig. 1, hence not demonstrating the actual implementation. Based on the simulation models and the results of the virtual validation process the actual HW/SW implementation can be done. Following, the according HW/SW integration is performed and specific HW/SW tests and verification can be accomplished (e.g. to cover important corner cases).

## 2.2 System Modeling and Co-simulation Configuration

In the automotive domain, the environment for developers is usually inhomogeneous. Engineering domains like mechanics or electrics follow diverse approaches and use different types of models. For example, finite element or software models are of different nature, but might still interact. The discipline of system engineering tries to deal with such environments. In this context, a number of tools are available which comply with certain standards. In this work, we want to exploit these facts in order to (a) generate a high level system view and (b) use this view to set up a system level co-simulation environment for virtual system verification and validation.

### 2.2.1 System Modeling

Model-based system design is a key issue in the development of automotive E/E systems. There are several reasons for that. Models are often more suitable to deliver a rough overview, due to their limited size and complexity. Usually they can be structured according to several properties. Furthermore, their increased re-usability and testability account for decreased development time and cost. A meta-model unifying automotive development effort is AUTOSAR. The AUTOSAR standard provides a common meta-model for the development of automotive software and systems. It is founded on model-based approaches, supporting the automotive industry's need for a data exchange format between OEMs and suppliers. Software components are used to group and encapsulate single functionalities. Compositions are structuring elements including software components. The interconnection between software components and compositions is defined by using ports. In the AUTOSAR context, this represents a logical grouping. The final distribution of software components to ECUs is not affected and may differ from that. Each software component implements one or more certain behaviors, which are called Runnables in AUTOSAR.

Additionally to the architectural information, all elements are enhanced with information relevant for simulation. For example, a software component may include information linking to a simulation tool and corresponding model, which are capable of reproducing the component's behavior.

These architectural elements are used to build a system's description. We decided to use dSPACE SystemDesk for that matter. Its project library enables the management of architectural elements to enforce re-usability and structuring of information. The result of this process step is an AUTOSAR ARXML file.

### 2.2.2 Model Conversion

In the next step we want to take the previously generated system description file, and convert it to a co-simulation configuration file. For this task, a model converter was written. It performs the following steps.

First of all, all software components, compositions, ports and interconnections from the AUTOSAR model are read. The interconnections are analyzed and evaluated. All software components are then converted to so-called wrappers, which are the basic building blocks within the co-simulation framework used, where one wrapper connects to one solver. Each of the software component's ports is converted to one parameter. Parameters determine the data interface for co-simulation. They include information about physical units, data types, interpolation and more. Using this mapping, it can easily be seen that the interconnection of wrappers is represented by the system's architectural design.

After that, the resulting data structure is written in XML file format as a co-simulation configuration file.

### 2.2.3 Independent Co-simulation

The co-simulation software "Independent Co-Simulation" (ICOS) [5] is an in-house development by Virtual Vehicle. Its core feature is the ability of coupling solvers of different simulation tools. The connecting element between a simulation tool and the coordinating simulation engine is called wrapper. ICOS currently supports more than 30 industrial simulation tools by providing wrappers for each of them. Typically, the workflow of setting up an ICOS based co-simulation involves the manual assignment of solver quantities to wrappers and also between wrappers. This process can now be automated by taking the generated configuration file. This file can be imported to ICOS. Assumed that all wrappers are running on their dedicated machines, the co-simulation may be started instantly.

### 2.2.4 Co-simulation Configuration Process Overview

Figure 2 shows the complete process from graphical system description to the execution of the proposed co-simulation experiment. A graphical block representation is linking to its simulation model. This information is included within the block itself, using customized parameters. This information is also preserved during serialization. A standard compliant ARXML is exported and fed to the XML-converter. It transforms the ARXML file to a so called ICOSXML file. This file defines the necessary data types and parameters of simulation models, which are coupled in the next step. ICOS reads this ICOSXML file and prepares all required solvers. Note that ICOS is capable of linking solvers from different models as well as solvers running on very different computers.

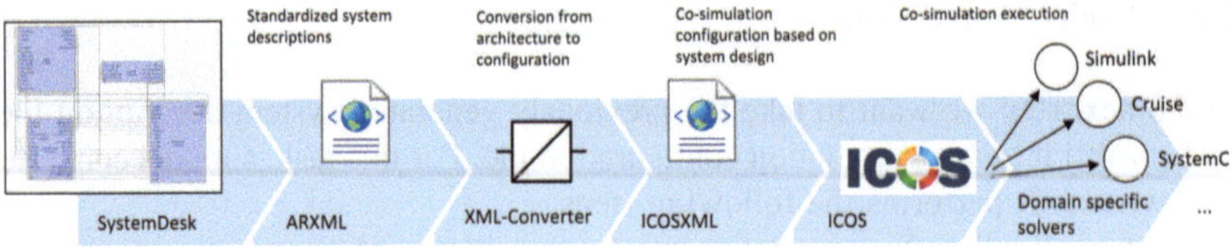


**Fig. 2** Proposed process for co-simulation configuration generation

This process is fully automated and is flexible related to changes of the system architecture, since all necessary information is stored directly in the used components and compositions.

## 3 Case Study: Camera-Based Pedestrian Detection

Crossing the road is the most frequent accident scenario for a pedestrian. In [6], it is shown that about 65 % of all pedestrian accidents happen in the front of the car. For this reason we analyzed a scenario where a pedestrian is crossing the road.

Within this case study, we show how our proposed co-simulation based approach is applied to a pedestrian detection system. The focus is on modeling of a camera-based pedestrian detection and the associated safety function algorithm (warning and braking logic) within the control units of the vehicle. The area in front of the vehicle is analyzed by the safety function algorithm. In case of a critical situation with a pedestrian in front of the car, the algorithm calculates the situation. Depending of the level of the situation, it initiates an acoustic warning to the driver or starts an autonomous braking intervention to reduce the accident severity. For the parameterization of simulation models, tests on a pedestrian test bed with different scenarios were carried out to get a realistic baseline for the models.

The case study which we want to simulate is shown in Fig. 3. The pedestrian starts behind the vehicle on the roadside and is hidden until he walks onto the road.

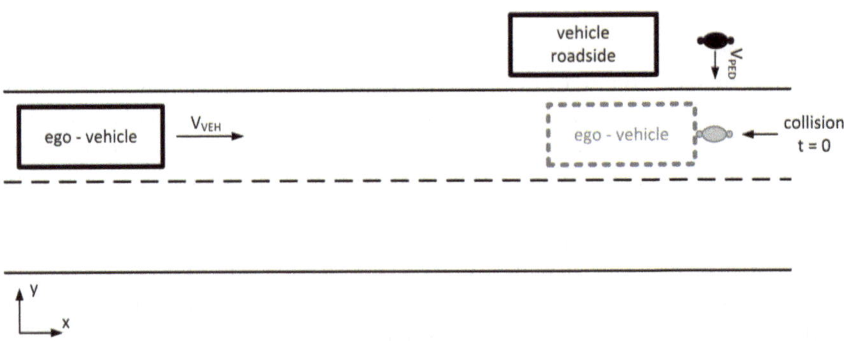


**Fig. 3** Case-study for the camera-based pedestrian detection

The ego-vehicle drives with a constant speed in the direction of the pedestrian. To simulate the scenario shown in Fig. 3, the following models are needed:

- Ego-vehicle plus integrated safety system (vehicle, safety function algorithm, sensor)
- Environment (pedestrian model, vehicle on the roadside)
- Simulation control (visibility model, collision detection).

We will now demonstrate the individual steps as shown on the left side of Fig. 1. After the requirements have been captured (not shown here, trivial task), the hierarchical system model is created by using SystemDesk. An excerpt is shown in Fig. 4, also already showing the resulting co-simulation configuration in ICOS.

It mainly consists of the different components mentioned before (ego-vehicle plus integrated safety system, environment…) including all interfaces and information relevant for co-simulation configuration (e.g. modeling languages and simulators/solvers used for different parts of the system model, configuration data for distributed co-simulation…).

Out of this, the co-simulation configuration is automatically created according to the process shown Fig. 2. After the different simulation models have been implemented and parameterized (Fig. 5 is showing an example for the camera model), the created co-simulation configuration can be executed and used for virtual validation of the integrated safety system.

Within the co-simulation based environment, it is easily possible to change the parameters e.g. for the safety function algorithm used in the model. For example a reduced time-to-collision for the braking zone. With this possibility it is easy to get

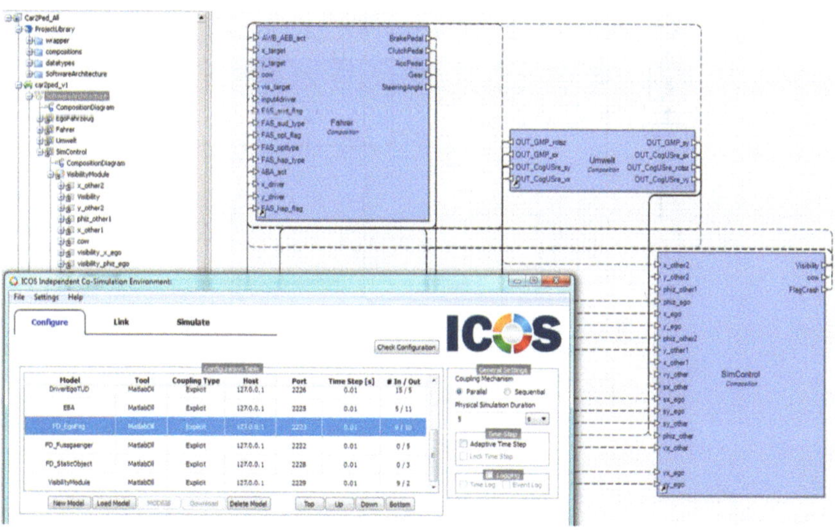


**Fig. 4** System model with resulting co-simulation configuration

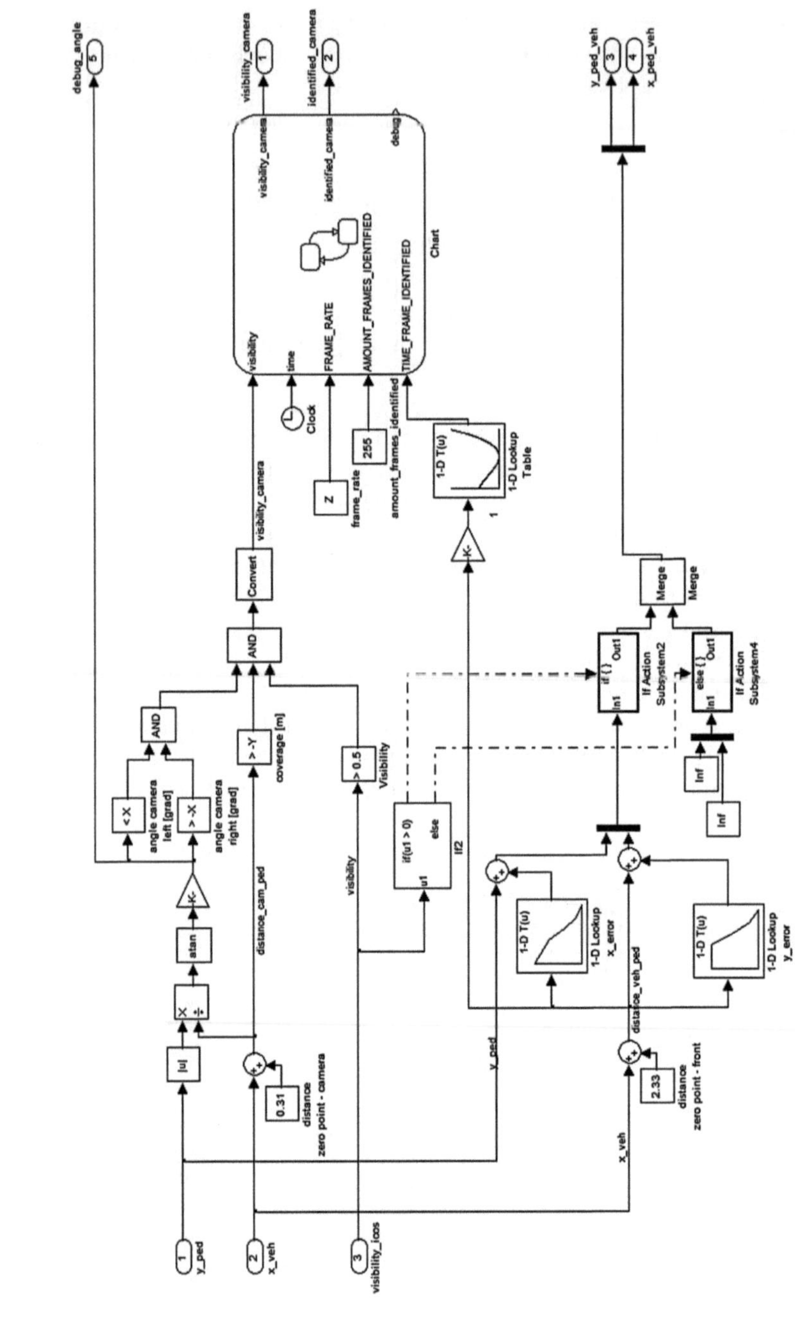


**Fig. 5** Camera model implemented in Matlab Simulink

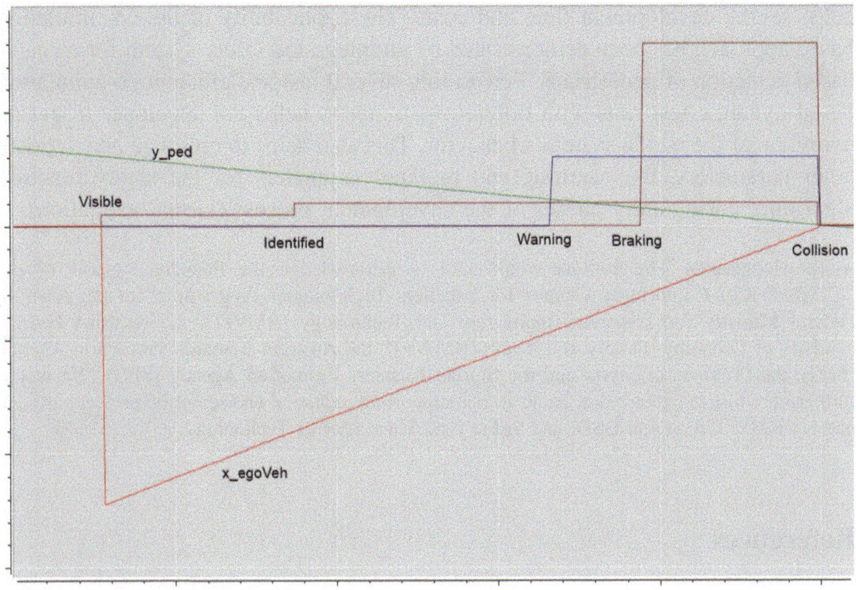


**Fig. 6** Co-simulation result using one specific set of parameters

results for a new set of parameters in the simulation without doing tests on the test bed. If changes have the desired effect, parameters can be transferred to the vehicle and tested on the test bed.

A co-simulation result of the preventive pedestrian protection using one specific set of parameters is shown in Fig. 6. Here, the different states of the integrated safety function can be seen. First the pedestrian is not visible because of the vehicle on the roadside. If the pedestrian is visible, the camera needs some time to identify the pedestrian and to estimate the distance to the pedestrian. After identification, the driver is notified by a warning signal. If there is no reaction by the driver, the system starts an autonomous braking action.

More in-depth details about the implemented camera-based pedestrian detection system can be found in [3].

## 4 Summary

In early phases during development of an integrated safety system, it is difficult to get a comprehensive view on the whole system. This stimulates the demand for methodologies supporting the efficient development of such systems. For the first time, a comprehensive virtual assessment of integrated safety systems is now possible. By using co-simulation, we demonstrated a workflow for the efficient modeling, simulation and virtual validation of integrated safety systems. This helps saving development time and costs. The applicability of the co-simulation based approach has been demonstrated by an integrated safety system for camera-based detection of pedestrians. Performing several fast and efficient co-simulation based dynamic test runs with different parameters helps the developer to get an overview of the whole system's behavior. This also helps to optimize and validate other parameters, like warning and braking parameters for the safety function algorithm, during early phases of the development process (virtual validation).

**Acknowledgments** The authors would like to acknowledge the financial support of the "COMET K2—Competence Centres for Excellent Technologies Programme" of the Austrian Federal Ministry for Transport, Innovation and Technology (BMVIT), the Austrian Federal Ministry of Economy, Family and Youth (BMWFJ), the Austrian Research Promotion Agency (FFG), the Province of Styria and the Styrian Business Promotion Agency (SFG). We would furthermore like to express our thanks to our supporting industrial and scientific project partners, namely BMW Group and DSD, and to the Graz University of Technology.

## References

1. Karner M, Steger Ch, Weiss R (2010) A cross domain co-simulation platform for the efficient analysis of mechatronic systems. In: SAE Paper No. 2010-01-0239 (SAE World Congress 2010, Detroit, 2010-04-15), SAE, 2010
2. Geimer M, Krueger T, Linsel P (2006) Co-Simulation: gekoppelte Simulation oder Simulatorkopplung. O+P Zeitschrift für Fluidtechnik 50(11–12):572–576
3. Schratter M, Karner M, Wimmer P, Watzenig D, Gruber C (2013) A co-simulation based approach for the validation of integrated safety systems. In: SAE Technical Paper No. 2013-01-0201 (SAE World Congress 2013, Detroit, 2013-04-16), SAE, 2013
4. Wimmer P, Rieser A, Gruber C, Sinz W (2012) Effectiveness assessment of vulnerable road user protection systems via numeric simulation. In: Proceedings of the ICRASH 2012
5. Virtual Vehicle Research and Test Center (2013) ICOS—Independent Co-Simulation. http://www.v2c2.at/en/products/icos. Accessed 27 Jan 2013
6. Unselt T, Breuer J, Eckstein L (2004) Fußgängerschutz durch Bremsassistenz. http://www.ftm.mw.tum.de. Accessed 27 Jan 2013

# Enhanced Low-Cost Sensing Technologies for Vehicle On-Board Safety Applications (ADOSE Project)

**Erwin Schoitsch, Christoph Sulzbachner and Jürgen Kogler**

**Abstract** A key objective of the EU Transport Policy is to reduce considerably (50 % and more in the future) casualties due to road accidents. Various programs have been established to investigate new sensor technologies and the resulting applications as basis for advanced driver assistance systems. The project ADOSE (Reliable **A**pplication Specific **D**etection of Road Users with Vehicle **O**n-board **Se**nsors, FP7 ICT 216049) addressed the enhancement of preventive and active safety functions through the development of high performance and low cost sensing technologies suitable for detection and classification of vulnerable road users under harsh environmental, weather and light conditions by combining and evaluating complementary technologies. Partners were Fiat CRF (co-ordinator), BOSCH, Magneti Marelli, ST Microelectronics, TRIAD, UMICORE, Paragon, IMEC, VTT, Fraunhofer IZM, Uppsala University and AIT Austrian Institute of Technology. The paper will shortly describe the ADOSE approach, the scenarios that were taken into account (and why), major results and provide some comments on the most unusual sensor validated, the SRS sensor from AIT.

**Keywords** ADOSE · ADAS · On-board safety · Sensor technology · Active and preventive safety functions · Collision avoidance · Pre-crash warning · Pre-crash preparation

E. Schoitsch (✉) · C. Sulzbachner · J. Kogler
AIT—Austrian Institute of Technology GmbH, Donau-City-Strasse 1 1220 Vienna, Austria
e-mail: Erwin.schoitsch@ait.ac.at

C. Sulzbachner
e-mail: Christoph.sulzbachner@ait.ac.at

J. Kogler
e-mail: Juergen.kogler@ait.ac.at

J. Fischer-Wolfarth and G. Meyer (eds.), *Advanced Microsystems for Automotive Applications 2013*, Lecture Notes in Mobility, DOI: 10.1007/978-3-319-00476-1_11,

# 1 Introduction

A key objective of the EU Transport Policy is to reduce considerably (50 % and more in the future) casualties due to road accidents. Various programs have been established to investigate new sensor technologies and the resulting applications as basis for advanced driver assistance systems. The project ADOSE (Reliable **A**pplication Specific **D**etection of Road Users with Vehicle **O**n-board **Se**nsors, FP7 ICT 216049) addressed the enhancement of preventive and active safety functions through the development of high performance and low cost sensing technologies suitable for detection and classification of vulnerable road users under harsh environmental, weather and light conditions by combining and evaluating complementary technologies.

ADOSE addressed five breakthrough sensing technologies, with the goal to improve the current state-of-the-art in terms of costs, performance and reliability:

- IR-add-on sensor (FIR), with sufficiently good thermal and spatial resolution at lower cost, to be combined to a high resolution imager for enhanced night vision (BOSCH, IZM, UMICORE).
- Low-cost multi-functional and multi-spectral CMOS vision sensor (MFOS), detecting critical environmental parameters (fog, rain etc.) and providing information on the driving scenario (vehicles, VRUs (vulnerable road users) at nigh etc.) (CRF, STM, IZM, MM).
- High spatial resolution and low-cost 3D range camera (3DCAM), by the integration of 3D packaging, optical CMOS and laser radar technologies for short range ADAS requirements (high-speed object recognition and distance measurement) (IMEC).
- Harmonic radar combined to passive nonlinear reflector and active tags (HR-PTAG and HR-ATAG), enabling easy detection of traffic obstacles and vulnerable road users, even in dark or adverse weather conditions (VTT, TRIAD, UU).
- High temporal resolution and low-cost bio-inspired silicon retina stereo sensor, addressing time critical decision applications (SRS) (AIT).
- Additionally, data (pre-) processing and partial functional system integration for demonstrators have been performed by CRF, AIT, Paragon and MM.

ADOSE was set up in the context of EPoSS, the "European Technology Platform on Smart Systems Integration" by its Automotive Working Group. All partners are members of this EPoSS WG. The project was designed to focus on the development of high performance and low cost sensing technologies for ADAS for accident prevention and pre-warning. Sensor fusion and post-processing was addressed in a complementary project PReVENT and some other related projects (see Fig. 1).

Only pre-processing algorithm have been developed and implemented for each sensor, which are sensor specific, but have not been extended to sensor data fusion. Raw data processing has been implemented on laptop processing hardware to be able to demonstrate the sensing technology and its abilities in a real driving environment (done for the review at a test drive area at CRF near Torino).

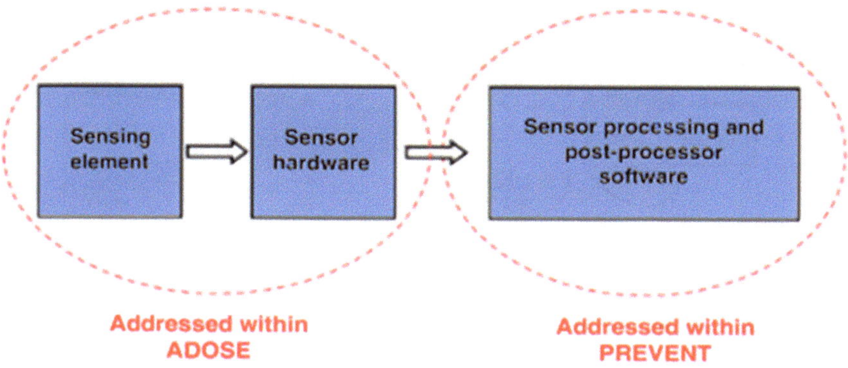


**Fig. 1** ADAS sensors field of interest addressed in ADOSE (as compared to projects like PReVENT)

The sensors have been installed in two demo vehicles, one from CRF and one from AIT, both demonstrating together their "ego-vehicle" capabilities based on the sensors installed. The project focused fully on in-vehicle sensor technology without communication and interaction with sensors outside the vehicle.

## 2 ADOSE Sensors for the In-Vehicle Virtual Safety Belt

The five sensor types listed in "A Warning Algorithm for Intersection Collision Avoidance" have different fields of view and distances where they are optimal for detection of objects, potential collision or side impact. Fig. 2 provides an overview of all sensors evaluated and considered in ADOSE and their potential application area. It can be seen that no single sensor would be able to fulfill all requirements of safe obstacle detection and pre-crash warning functionality under all light and weather conditions. The goal of ADOSE was to improve these sensors, to develop low-cost versions (for mass production) and with extended capabilities (improved parameters), or to provide new implementations like the silicon retina sensor for stereo vision on moving objects (up to now they had been applied statically mounted e.g. to count passing vehicles for traffic monitoring or people, or to detect unusual behavior in elderly care flats etc.).

ADOSE will have impact on the "virtual safety belt" around the vehicle by offering different sensing technologies for a set of complementary safety functions.

## 3 The ADOSE Approach, Selected Driving Scenarios

The ADOSE approach and the selection of the relevant driving scenarios had to take into account the following criteria:

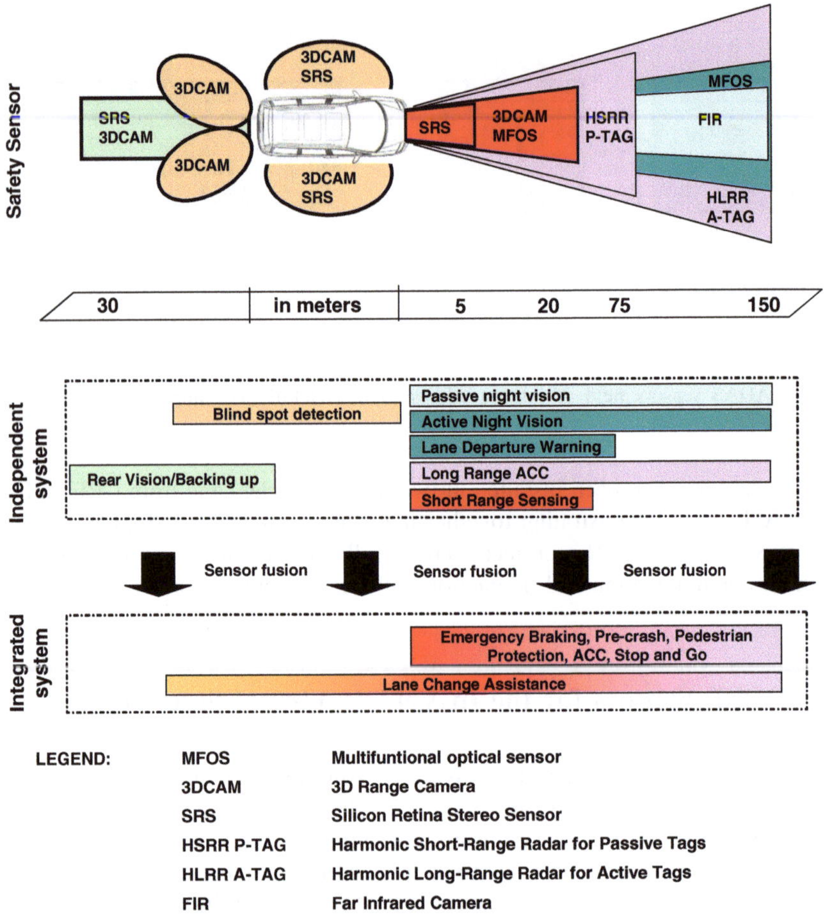


**Fig. 2** ADOSE sensor technologies overview (potential ADAS applications, field of view, operating distance)

- ADOSE aims at enhancing the current ADAS sensors by breakthrough technological developments; the distinctiveness of each sensor (see Fig. 2) limits the number of scenarios.
- ADOSE is a product driven project which means that partners have to select scenarios (preventive and active functions), characterized by higher user demand and acceptance, in order to be competitive on the market.

ADOSE did a state-of-the art analysis of safety functions and systems available on the market (2008), analyzed the scenarios of complementary or related research projects like PReVENT, WATCH-OVER, SAFESPOT or COOPERS, including co-operative smart systems communication with other road users or infrastructure (see also [1]). As a result, ADOSE focused on Ego-vehicle bound scenarios and put these scenarios in relation to ADOSE sensor types (see Fig. 2), identifying their potential and restrictions, and potential measures for pre-crash warning or active safety actions, depending on timing and controllability of the situation.

ADOSE scenarios were because of these criteria limited to two groups:

- Collision Avoidance (urban and extra-urban areas) (1.1, 1.2)
- Pre-Crash Warning/Preparation (front, side and rear impact) (2.1, 2.2, 2.3).

It should be noted, that the side impact (distance 6 m and less over a very short time) allows no reasonable collision-avoiding safety action, only pre-crash preparative actions (e.g. side airbag, warning signal, pre-tensioning of safety belts, stopping flow of fuel). Because of the fast reaction time needed the SRS sensor is best fit for this purpose and was therefore selected for this scenario [2].

The result is best described in a scenario/sensor technology matrix (Table 1):

# 4 Summary of the Results of ADOSE Sensor Development

This text is taken from the ADOSE Final Report, the final publishable summary (which is public by definition), since this is the most concise summary of ADOSE results). This text was prepared by the whole ADOSE project team (partners are listed in the abstract).

## *4.1 FIR Imager (BOSCH, IZM, UMICORE)*

In the last year the FIR imager activities focused on manufacturing of the final integrated FIR array devices targeting the ADOSE specifications (automotive night vision add-on sensor for hot-spot detection and combination with a NIR CMOS camera, [3]). The FIR chip uses micromechanically suspended and thermally insulated multiple thermo-diodes, manufactured simultaneously with the read-out integrated circuit (ROIC). Mono-crystalline silicon as sensor material enables batch wafer-level vacuum packaging.

Although the samples had a flaw within the ROIC and a redesign was impossible within the project deadline, a metal-fix was applied in order to show the basic functionality of the imager. All measures possible within the project constraints were taken and resulted in a demonstrator camera with fair performance to show feasibility of the concept but not fulfilling completely the specification for ADOSE.

Evaluation and testing of the samples and the demonstrator camera was conducted with reasonable results. In parallel steps towards a redesigned and then fully functional ROIC have been taken.

**Table 1** Scenario/sensor technology matrix

| SENSOR scenario | FIR | MFOS | 3DCAM | HR-PTag | HR-A Tag | SRS |
|---|---|---|---|---|---|---|
| 1.1 Collision avoidance extra-urban | Range 120 m max, 160 ms min. reaction, only warm objects, affected by bad weather | Range 150 m max as NIR, data on visibility level, improved sensor performance at night | Low range <25 m (laser power limitation), affected by bad weather (spin drift, fog) | Range 100–150 m, tag needed for harmonic operation, heavy rain reduces performance | Range 100–150 m, active tag needed for harmonic operation, heavy rain reduces performance | No potential because of long distance |
| 1.2 Collision avoidance intra urban | Range 120 m max, 160 ms min. reaction, only warm objects, affected by bad weather, reduced performance in complex scenarios | Range 150 m max as NIR, data on visibility level, improved sensor performance at night | Low range <25 m (laser power limitation), affected by bad weather (spin drift, fog) | Range 100–150 m passive, tag needed for harmonic operation, normal op. for other objects, heavy rain reduces performance | Range 100–150 m active tag, tag needed for harmonic operation, normal op. for other objects, heavy rain reduces performance | No potential because of long distance |
| 2.1 Pre-crash warning/ preparation front impact | No potential with specified FOV, warm objects only | Good lateral resolution, classification of obstacles, lane features, complementary weather data (e.g. rain) | Range <25 m, good lateral resolution, affected by bad weather (spin drift, fog) | Range 25–100 m passive, tag needed for harmonic operation, heavy rain reduces performance | Range 100–150 m active tag, tag needed for harmonic operation, heavy rain reduces performance | Low potential (only short distance) |
| 2.2 Pre-crash warning/ preparation side impact | No potential with specified FOV, warm objects only | No potential: MFOS concept suitable only for front area monitoring | Good potential for limited range, good lateral resolution, affected by bad weather | No potential (HR-TAG is only a forward looking radar) | No potential (forward looking radar) | Very good potential, very fast, reduced performance in bad weather |

(continued)

**Table 1** (continued)

| SENSOR scenario | FIR | MFOS | 3DCAM | HR-PTag | HR-A Tag | SRS |
|---|---|---|---|---|---|---|
| 2.3 Pre-crash warning/ preparation rear impact | No potential with specified FOV, warm objects only | No potential: MFOS concept suitable only for front area monitoring | Good potential for limited range, affected by bad weather (spin drift, fog) | No potential (HR-TAG is only a forward looking radar) | No potential (forward looking radar) | Low potential (only short distance) |

## 4.2 Multifunctional Optical Sensor (CRF, ST, IZM, MM)

Two MFOS camera prototypes, enabling different functional integration, have been developed with optical designs based on light guides concepts: (a) planar optical light guides by IZM, (b) dioptric lenses by CRF.

The plastic optical components have been fabricated respectively with hot-embossing and injection molding technologies. The CMOS imager based on OLGA package has been redesigned and realized by ST to meet the ADOSE specifications. The MFOS easy-to-assemble process has been evaluated with respect to the position tolerances between the imager and the optical light guides affecting the performance reliability.

Fog, twilight, rain and solar functions have been integrated in the two sensor prototypes. The housing of one prototype sensor has been designed and fabricated in order to be integrated in the internal rear view mirror.

Indoor tests have been carried out at component, sensor and functional levels. Outdoor tests have been performed on the twilight, tunnel/bridge and warning night vision functions.

## 4.3 3D Range Camera (IMEC)

Different 3D imaging methods have been analyzed and simulated. A range-imaging hybrid camera concept was completed. The camera consists of a photo sensor and its corresponding readout electronics. These two components are hybridized using existing IMEC capabilities for wafer processing and flip-chip technology.

The photo sensor has been designed and fabricated at IMEC's facilities. It uses back-side illuminated (BSI) technology to increase the quantum efficiency and the fill factor of the detector elements. The readout electronics have been designed to be able to implement different range measurement methods in order to find the optimum one. It was designed at IMEC and fabricated at a commercial CMOS foundry.

The system controlling the camera comprises an FPGA implementing the control of the camera and the communication with the ADOSE central control system and software. It also contains the necessary power supply, bias and clocking sources. The sensor is located in a small PCB which is connected to the main control PCB.

## 4.4 Harmonic Radar and Tags (VTT, TRIAD, UU)

The harmonic radar relies on intermodulation principle, which provides a smaller frequency offset as compared to the traditional harmonic radar. The radar prototype transmits two nearby signals at the 77 GHz band allocated for automotive radars. Harmonic reflectors, or tags have a nonlinear component that create an intermodulation response at a nearby frequency. Because nonlinear reflections are rare in nature, tags can be easily identified from the radar reflection. If the tag is carried by a vulnerable road user, he/she can be detected even in dark or fog.

Several nonlinear elements have been studied: Schottky diodes, ferroelectric varactors and MEMS resonators. For longer detection range, semi-passive or active tags can be used. In these designs, a switch (semi-passive tag) or an amplifier (active tag) is used to achieve modulated backscattering. Tag antennas are developed on a flexible substrate, to make tag integration into textiles and curved surfaces easier. In the last year of ADOSE project, the radar and tag concept has been tested and demonstrated in a laboratory setup.

## *4.5 Silicon Retina Stereo Sensor (AIT)*

The "silicon retina" sensor technology is based on bio-inspired analogue circuits that pre-process the visual information on-chip in parallel for each pixel. These optical sensors provide excellent temporal resolution, a wide dynamic range and have low power consumption [4].

Typical applications include vision systems for roadside traffic data acquisition, real-time stereo vision systems for reliable person counting and in ADOSE high-speed and low-cost ranging sensors for time-critical decision making functions. In a first step the automotive requirements for such a sensor system were collected, and the technical specifications were determined. In ADOSE the stereo high resolution (304 × 240 Pixel) SRSS was used as a pre-crash sensor for side-impact airbags control.

Novel algorithms based on locality and timely correlation of asynchronous data streams have been developed in order to fully exploit the advantages of the silicon retina technology for safety–critical automotive applications. The stereo matching and detection algorithm have been further enhanced and adapted to work with the shorter baseline of 15 cm (45 cm of the laboratory prototype) [5]. The final demonstrator is able to show functionality of novel time–space stereo matching technique.

## *4.6 Data Processing and Functional System Integration (MM, PARAGON, AIT, CRF)*

The ADOSE sensor specifications (related to the communication bus, protocol and data format), the software architectures and the test plans have been defined. Algorithms for pre-processing of raw data have been defined in terms of high level (flow chart) and low level (pseudo-code) design. The development of such algorithms (based on features extraction, timing and spatial common reference between FIR and NIR sensors) has been completed with real data for MFOS and NIR sensors and with data coming from FIR prototype sensor available.

Application oriented algorithm (AdaBoost) for vehicle classification has been defined, developed and trained using different target vehicles in different conditions.

As far as regard the 3DCAM, the SW has been based on simulator of the sensor.

Different working modes have been also provided (Normal, Recording and Testing) in order to give the possibility to manage different situations, allowing calibration procedures, and to reproduce on bench the behavior experienced during the driving sessions for tuning of algorithms and parameters and for the analysis of the results.

The Harmonic Radar (HR-TAG) raw data pre-processing algorithms for detection, early sensor fusion, object tracking and feature extraction have been developed and integrated in the radar ECU software module, and, finally tested successfully using HR-TAG simulated data. Paragon's radar ECU architecture is modular and portable. The module is customizable, may support multiple radar sensors as an early sensor fusion stage, and may retrieve the car dynamics from its CAN-bus.

Classical and advanced stochastic algorithms have been implemented to enhance the characteristics of detected road-users (RU). Objects and features detected by the radar ECU are available to other ECUs, as well as, for further object and situation refinement by any higher ADAS layer, to improve VRU safety.

Test (indoor, outdoor, static and dynamic) procedures, test cases, and dynamic laser testing technology have been defined. Test execution has been performed in both indoor and outdoor environments and in both static and dynamic conditions. Software for the testing technology (LIDAR) has been developed and tuned and synchronization and communication procedures between NI/FIR framework and LIDAR framework have been developed and set up.

## 5 The ADOSE Demonstrators

Two demonstrator vehicles were available for the Final review meeting and a final demonstration was organized at the Centro Sicurezza test track of CRF (Torino):

- MFOS and FIR cameras for Warning Night Vision (CRF vehicle) (Fig. 3),
- SRS sensor for side pre-crash applications (AIT vehicle) (Fig. 3).

The first vehicle was equipped with CMOS NIR multifunctional camera integrated in the internal rear-view mirror, low resolution FIR camera and NIR illuminators integrated in the headlight housings.

The interior of AIT demo vehicle has been modified to allow integration of a fixed rack-mounting system. The camera demonstrator was installed in the inside

**Fig. 3** CRF demo vehicle (FIR, MFOS) (*left*), AIT demo vehicle (SRS, *side* impact test) (C) CRF

of the vehicle at the rear window. At the final demonstration the output of the 3D algorithm was shown in real-time for pre-crash side impact.

**Acknowledgments** The research leading to these results has received funding from the EC Seventh Framework Program (FP7/2007–2013) under grant agreement ICT-216049 (ADOSE). The authors would like to thank all ADOSE participants working on the success of the project. Partners were Fiat CRF (co-ordinator, Italy), BOSCH (Germany), Magneti Marelli (Italy), ST Microelectronics (Italy), TRIAD (Norway), UMICORE (Belgium), Paragon (Greece), IMEC (Belgium), VTT (Finland), Fraunhofer IZM (Germany), Uppsala University (Sweden) and AIT Austrian Institute of Technology.

## References

1. Sun Z, Bebis G, Miller R (2006) On-road vehicle detection: a review. IEEE Trans Pattern Anal Mach Intell 28(5):694–711
2. Kogler J, Sulzbachner C et al (2010) ADOSE—Bio-inspired in-vehicle sensor technology for active safety. In: Meyer G, Valldorf J (eds) Advanced microsystems for automotive applications 2010, Smart systems for green cars and safe mobility, Springer, Berlin
3. Herrmann I, Hattaß M et al (2011) Low-cost platform technology for LWIR sensor arrays for use in automotive night vision and other applications. In: Meyer G, Valldorf J, Advanced Microsystems for Automotive Applications, Springer, Berlin, pp 89–97
4. Lichtsteiner P, Posch C, Delbruck T (2008) A 128 × 128 120 dB 15 s latency asynchronous temporal contrast vision sensor. IEEE J Solid-State Circuits 43(2):566–576
5. Humenberger M, Zinner C, Weber M, Kubinger W, Vinze M (2010) A fast stereo matching algorithm suitable for embedded real-time systems. Comput Vision Image Underst J (Special Issue on Embedded Vision) 114(11):1080–1202

# Smart and Green ACC, Safety and Efficiency for a Longitudinal Driving Assistance

**Sebastien Glaser, Sagar Akhegaonkar, Olivier Orfila, Lydie Nouveliere, Volker Scheuch and Frederic Holzmann**

**Abstract** Driving Assistances aim at enhancing the driver safety and the comfort. Nowadays, the consumption is also a major criterion which must be integrated in the driving assistances. Then, we propose to redefine the behavior of an ACC with energy efficiency consideration to perform a Smart and Green ACC. We apply our development to the specific use case of the electric vehicle that allows regenerative braking. The ACC, once activated, operates under two possible modes (speed control and headway spacing control). We define the behavior of the driving assistance under these both possible modes, focusing on the distance control. We present the efficiency of various strategies without trading off safety. We conclude on the efficiency by presenting several use cases that show the SAGA behavior.

**Keywords** Driving assistance · Longitudinal control · Energy efficiency

---

S. Glaser (✉) · O. Orfila · L. Nouveliere
LIVIC (Laboratory on Interactions Vehicles-Infrastructure-Drivers),
A research unit of IFSTTAR, Batiment 824,
14 Route de la Minière, 78000 Versailles, France
e-mail: Sebastien.glaser@ifsttar.fr

O. Orfila
e-mail: olivier.orfila@ifsttar.fr

L. Nouveliere
e-mail: lydie.nouveliere@ifsttar.fr

S. Akhegaonkar · V. Scheuch · F. Holzmann
INTEDIS GmbH and Co. KG, Max-Mengeringhausen-Strae 5,
97084 Wuerzburg, Germany
e-mail: Sagar.Akhegaonkar@intedis.com

V. Scheuch
e-mail: volker.scheuch@intedis.com

F. Holzmann
e-mail: Frederic.Holzmann@intedis.com

J. Fischer-Wolfarth and G. Meyer (eds.), *Advanced Microsystems for Automotive Applications 2013*, Lecture Notes in Mobility, DOI: 10.1007/978-3-319-00476-1_12, 

# 1 Introduction

Daily traffic congestion or long trip on a highway brings up issues that the driver must face during his driving experience. However, these tasks may generate anger, stress or drowsiness. In a situation where the driver kept a constant clearance for a long time, and suddenly facing braking, his reaction time is higher and may lead to a collision. Automation, and automated driving seem to be one possible answer to these problems, by delegating partly or totally the driving task. Many projects during the 80's and 90's, have proved the feasibility of automated and autonomous (without driver interaction) driving systems. Eureka Prometheus project in Europe, or the US National Automated Highway System consortium conducted experiments on real road of automated driving or platoon. Even if the concepts were not fully adapted by car manufacturers, current vehicles benefit greatly from these research. Since 10 years, the driving assistances are booming. However these driving assistances present two major drawbacks:

- The optimization process behind the driving assistance aims only at maximizing the safety and/or the comfort of the considered vehicle,
- In order to work, the driving assistance relies only on the perception systems embedded in the vehicle: they are autonomous systems.

In today situation, the energy consumption is one of the major topic for the car users: electric vehicles still have a limited range and for conventional vehicles, the oil price has skyrocketed and the greenhouse gas emissions must be reduced. The consumption criterion must be taken into account in the definition of a driving assistance.

Moreover, communication devices and navigation devices become popular. In a vehicle, we can consider that we have access to these systems to exchange data with other vehicles and with the infrastructure. The driving assistance systems are now cooperative and the driving assistance systems can sense the environment behind the vehicle sensors range.

In the eFuture project, we focus on the shared control between the vehicle automated systems and the driver for electric vehicles. The driving assistance we propose, the Smart and Green ACC (SAGA), derives from a standard ACC (Adaptive Cruise Control). It aims at optimizing the common criteria and also the energy consumption. Moreover, the required variables come from the vehicle sensors and from a digital map which includes information on the upcoming road.

With an Adaptive Cruise Control system, the driver delegates the longitudinal control task. When the system is active, the vehicle speed is controlled automatically either to maintain a given clearance to a forward vehicle, or to maintain the driver desired speed, whichever is lower. Since 1997, car manufacturers propose this system on their high-end cars. However, research is still active. Researchers aim at evaluating the impact of the ACC on traffic, under congested situation [1] or with improved strategies [2]. They also extend the range of possible speeds, driver comfort, safety or road capacity [3, 4]. In 2006, the introduction of a vehicle to vehicle communication (cooperation ACC, C-ACC) allows to decrease drastically the clearance to a forward vehicle [5] and also to create stable vehicle platoon. The evaluation of the C-ACC [6, 7] shows promising results on road capacity and safety. An ISO standard now defines the intended performance of the ACC [8].

In the following, we develop the Smart and Green ACC function. In the next section, we define the function, the notations and the consumption model. Section 3 explains the two problems: speed control and distance control. This last point, being the main issue, as it means to handle both consumption and safety criteria, will be developed in Sect. 4. The Sect. 5 presents simulation results. In the last section, we conclude on this work.

## 2 Problem Definition

### *2.1 Adaptive Cruise Control*

When an ACC system operates, the function is either in a speed control or distance control mode (Fig. 1). In the first case, there is no vehicle in front of the considered vehicle or in the distance of sensing. The vehicle aims at reaching a driver's desired speed. In the second case, a vehicle is in front of our vehicle. The system aims at maintaining a clearance defined by the driver, as soon as the lead vehicle speed is lower than the driver's desired speed.

The general variables are represented on the Fig. 1, along with the Table 1, for the description and units.

Moreover, [8] also defines the operating range of the driving assistance. The assistance cannot be activated below a given speed $V_{min}$, which must be higher than 7 m/s. The average automatic deceleration of ACC systems shall not exceed 3.5 m/s$^2$, while the acceleration is limited to 2 m/s$^2$. the average rate of change of an automatic deceleration (jerk) shall not exceed 2.5 m/s$^2$. The ISO standard also defines the minimal performance of the perception system according to the possible value of the speed and of the time gap.

Then, we have to define the behavior of the SAGA function for these two modes: Speed Control and Distance Control.

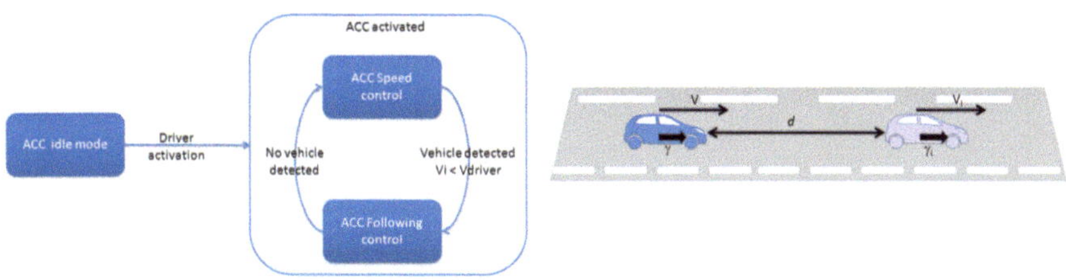


**Fig. 1** ACC architecture and generic use case

**Table 1** ACC related variables

| Variables | Description | Units |
|---|---|---|
| $d$ | Clearances to the lead vehicle | m |
| $V$, $V_d$ | Speed of the ego vehicle, drive desired speed | m/s |
| $\gamma$ | Acceleration of the ego vehicle | m/s$^2$ |
| $T$, $T_d$ | Time Headway ($T = d/V$), driver desired time headway | s |
| $V_i$ | Speed of the lead vehicle | m/s |
| $\gamma_i$ | Acceleration of the lead vehicle | m/s$^2$ |
| $\Delta V$ | Relative speed ($\Delta V = V_i - V$) | m/s |

## 2.2 Consumption and Efficiency Model

The consumption model that we define here, is based on the evaluation of the torque needed, at each of the two motorized wheel, to overcome resisting forces and generate the desired acceleration. It could be defined as:

$$T = \frac{R_w}{2}\left(\frac{1}{2}\rho SC_x V^2 + MgC_{rr} + MG\sin\Phi_r + M\gamma\right) \qquad (1)$$

where $R_w$ is the wheel radius, the air volumetric mass, $SC_x$ the air drag coefficient, $V$ the current vehicle speed, $M$ the vehicle's mass, $C_{rr}$ the rolling resistance coefficient, $\phi_r$ the slope and $\gamma$ the vehicle acceleration. The engine speed, supposing without sliding, is $w_e = {}^{V}\!/_{R_w}$. According with the torque definition, we evaluate the kinetic energy and the electric energy that is either consumed or regenerated during a period of time $dt$, depending on the value of the torque, in Table 2, where $\eta_g, \eta_b, \eta$ are respectively the efficiency of the transmission, the battery and the motor. This last parameter depends on the torque demand and on the motor speed.

Using previous equations, we can then define the specific regenerative braking area for a given electric motor associated with the regenerated power. A real characteristic of regenerative deceleration is presented on the Fig. 2. It is obtained from the first prototype of eFuture project. At low speed, the regenerative braking is not high. The main reasons are technological choices and to avoid that a regenerative braking leads to a blocked wheel. However, at low speed, the energy that could be regenerated is low, because of the low speed of the motor.

The regenerative braking area could be easily approximated by the following function :

$$\gamma_d(V) = \begin{cases} \gamma_{dec_1} : V < V_1 \\ \frac{A}{V} : V \geq V_1 \end{cases} \qquad (2)$$

**Table 2** Energy definition for the consumption model

| Energy | Consumption | Regeneration |
|---|---|---|
| Mechanic | $E_m = 1/\eta_g T_e w_e dt$ | $E_m = \eta_g T_e w_e dt$ |
| Electric | $E_m = 1/\eta_b E_m \eta(T_e, w_e)$ | $E_m = \eta_b E_m \eta(T_e, w_e)$ |

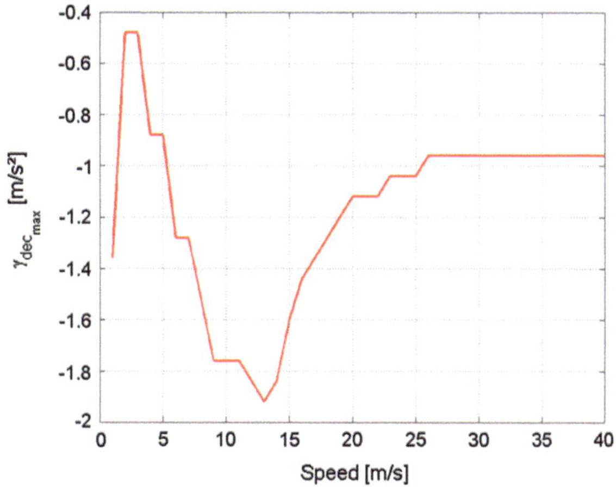


**Fig. 2** Regenerative braking area for an electric motor

where $A$ is a negative constant, $\gamma_{dec_1}$ is the maximal deceleration below a given speed $V_1$.

## 3 SAGA Function

As described in the previous section, our SAGA function must cope with two operating domain. The first one corresponds to the speed control case, where the system must follow a driver's desired speed. The second one deals with the problem of distance control with a lead vehicle. The system must regulate the speed to maintain a constant clearance expressed as a driver desired headway time.

### *3.1 SAGA Speed Control Function*

In this operating mode, there is not many safety related issues considering the interaction with the other road users, as, by definition, SAGA system operates in this mode when no vehicle is detected in front of our vehicle.

The main idea is then to supervise the conventional behavior of the ACC by defining a speed profile that includes the regenerative braking limitation and safety issues using a digital map to provide the needed data: legal speed limit if lower, can override the driver desired speed; approaching an intersection, the system automatically decreases the speed limit; using the curvature and slope information, a speed profile is defined to safely pass the curve. In this case, we extend the speed profile computation defined in [9, 10] with the deceleration limitation of the regenerative braking described previously.

### *3.2 SAGA Distance Control Function*

The distance control function is far more critical as it directly deals with the interaction between vehicles. The objective of the ACC is to regulate the error on the clearance $e_d (e_d = d - T_d V)$ around 0. Depending on the sensors used to measure the distance, the algorithm may be more robust by integrating the error on the relative speed $\Delta V$. The resulting acceleration is a function of this two errors, which is limited by the definition of the ACC. However, the regenerative braking is often below this threshold: with only a regenerative braking, we cannot ensure the same safety level of an ACC if we use the same strategy. Figure 3 shows the safety domain of a conventional ACC and an ACC which uses a regenerative braking. In this figure, we suppose that a front vehicle has a constant speed of 20 m/s and that the driver sets the time headway at 2 s.

The objective is then to regulate the acceleration of our vehicle to reach the point $(40\ \mathrm{m}, 20\ \mathrm{m/s})$. We can define a limit curve as being the points $(d, v)$ which allow to reach this point with a given braking capacity. The points below this curve are in a safe area. As the regenerative braking is lower than the conventional ACC, the safe area is smaller. If we want to keep the performance in term of speed, the sensing range must be greater, at the opposite, if we want to keep the same performance on the sensors, then the maximal speed must be smaller. In the following, we suppose that we keep a constant maximal sensing distance and that we limit the maximal speed.

## 4 SAGA and Distance Control Problem

We must handle two specific situations for distance control:

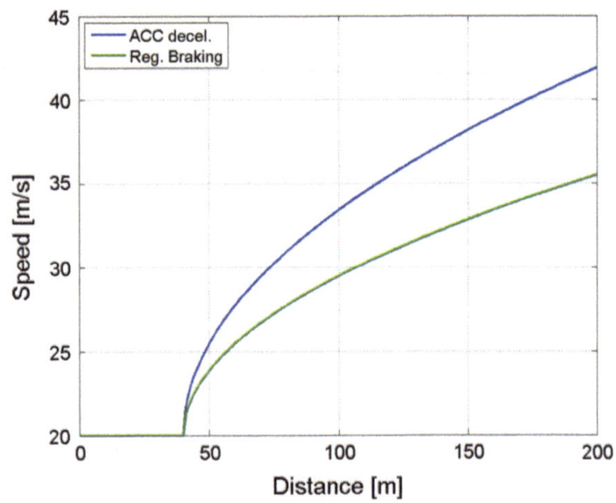


**Fig. 3** Conventional ACC and regenerative braking safety domain

- We approach a slow vehicle, starting at a given distance and speed in the safe domain
- We follow a vehicle with an error $e_d$ at zero. The lead vehicle start to decelerate with a deceleration that is greater than the possible deceleration.

For the first problem (see Fig. 4), we can suppose that the vehicle can either maintain a constant speed until it reaches the limit of the safe area, then decelerate at the regenerative deceleration (red path). Or that the vehicle starts to decelerate early to reach a speed that is slightly higher than the lead vehicle, and then it dece lerates slowly (orange path). Between these two extrema, we can tune the SAGA strategy.

The main drawback of the second strategy is that the gap between the two vehicles diminishes slowly. This situation allows a third vehicle to cut in our lane, leading to possible strong deceleration of our vehicle. If we follow the first strategy, and that the lead vehicle starts to decelerate, we may go out of the safe domain. For instance, if the front vehicle decelerates to 17 m/s (green curve in Fig. 4), our trajectory may quickly go out of the safe domain, losing the regenerative capacity, while the other strategy allows us to remain in the safe domain. Therefore, we need to define a tradeoff between these two strategies.

For the second problem, we need, at least, to achieve at least the same safety than a conventional ACC. For instance, we consider the following use case: a lead vehicle that drives at 30 m/s, our vehicle is at the same speed and the time headway is set at 2 s. The lead vehicle decelerates with a given deceleration $\gamma_i$ up to a speed of 7 m/s. The ACC system can achieve the deceleration without any collision for a large variation of the lead vehicle deceleration, even if the time headway drops to very low value (0.6 s for a considered deceleration of $-8$ m/s$^2$). Figure 5 represents the clearance when the vehicle reaches the same speed than the lead vehicle. Even for a very strong deceleration, the vehicle does not collide.

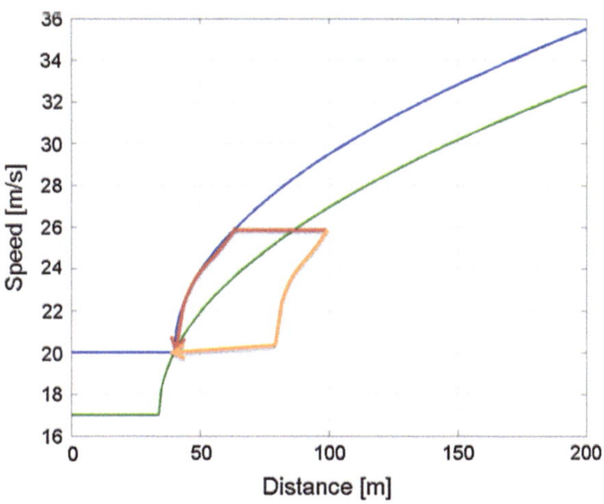


**Fig. 4** Variation of the safe domain with a deceleration of the lead vehicle

Using the regenerative deceleration only, it is not possible to obtain the same safety level than a conventional ACC with the same use case (initial time headway of 2 s, speed of 30 m/s and final speed of the lead vehicle of 7 m/s). The possible solutions are:

- To increase the minimal time headway that is defined by the driver. However, to obtain the same safety, we need to increase the minimal time headway up to 5 s. The resulting distance is hardly achievable as the sensor range is limited and other road user may cut in the space between vehicles.
- To switch to conventional braking if the time headway drops below a given threshold. For instance, we can maintain a collision free system, as for a conventional ACC, with an initial time headway of 3.7 s and an activation of a stronger braking at a threshold on the time headway of 1.5 s.
- To switch to an emergency braking if the deceleration of the lead vehicle and the distance drops below given thresholds. If the emergency braking can generate a deceleration of $-6$ m/s$^2$ when the time to collision (difference of distance divided by the difference of speed) is below 2 s, we can set the minimal time headway at 3 s.

As we want to use only the ACC system, then we choose the second option. However, we do not evaluate the acceptability by the user of the resulting clearance.

## 5 Simulation Results

In the following, we develop two different use cases to present the efficiency of the system. In a first scenario, our vehicle drives at the driver desired speed and it approaches a slow vehicle. The second scenario shows the reaction on a cut

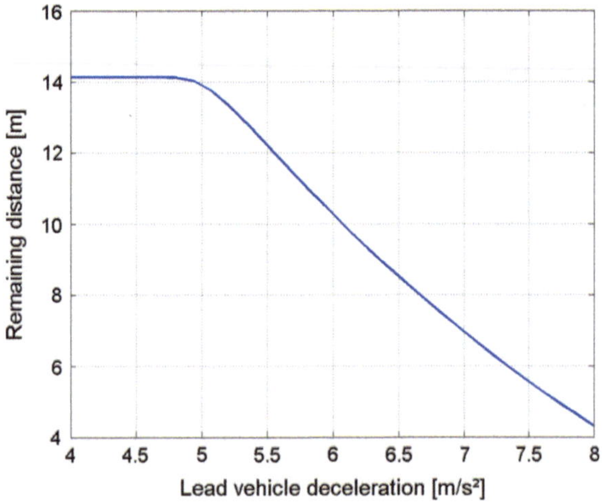


**Fig. 5** Final clearance as a function of the deceleration of the lead vehicle

in situation. The new target first decelerates slowly to increase the clearance with the previous lead vehicle, then accelerates to reach the traffic flow speed.

In the following, the ACC aims to regulate the distance at an headway of 2 s, the SAGA parameter is at 3.5 s. For SAGA, we use the conventional ACC if the headway drops below 1.2 s, and we start to control the vehicle with a 20 % longer distance.

## 5.1 Approaching a Slow Vehicle

In this scenario (see Fig. 6), we suppose that the lead vehicle is at 150 m in front of our vehicle. The lead vehicle's speed is constant and equal to 20 m/s. SAGA takes into account this front vehicle very early and decelerates slowly to reach the same speed and the correct clearance. The clearance decreases slowly and the complete maneuver takes more than 12 s. For the ACC, the same result is achieved in 8 s. During this maneuver, the State of Charge (SoC) increases of 0.07 kw for the SAGA, while it decreases of 0.05 kw for the ACC.

## 5.2 Vehicle Cut In

In this use case (see Fig. 7), we suppose that our vehicle follows a lead vehicle at a speed of 20 m/s with a time headway of 3.5 s. At $t = 1$ s, a vehicle inserts at a distance of 20 m in front of our vehicle. Its speed is 20 m/s and it decreases slowly to 18 m/s in order to increase its gap with the lead vehicle. At $t = 1$ s, our vehicle starts to decelerate, first with a regenerative deceleration, but as the new lead vehicle may be dangerous, the system shift to a conventional ACC, without regenerative capacity. During this deceleration, the vehicle's speed drops to

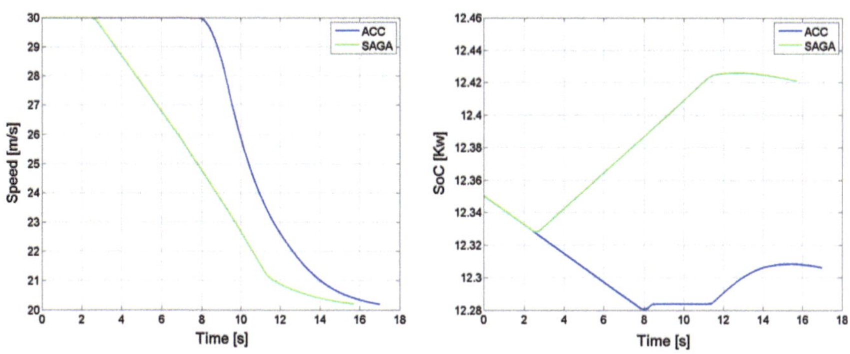


**Fig. 6** Approaching a vehicle, comparison of SAGA and ACC outputs on speed and state of charge

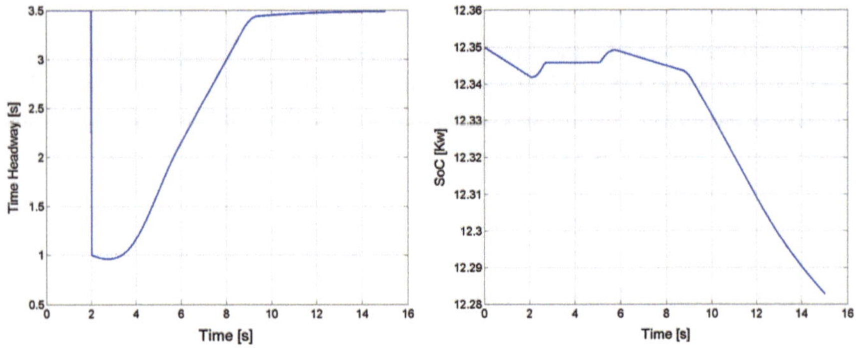


**Fig. 7** Evolution of the time headway and the SoC during a cut in use case

12.7 m/s. In the last phase the vehicle has to accelerate to reach the lead vehicle speed and the driver desired time headway. The system manages this dangerous situation without any collision.

## 6 Conclusion

In this article, we have presented the Smart and Green ACC, namely SAGA. This application enhances the conventional ACC for electric vehicle with regenerative capacity. The main aim of the application is to deliver, at least, the same level of safety than a conventional ACC, and to integrate a consumption criteria. Given the motor specification, and the possible deceleration, the tradeoff is to increase the clearance and to replace the regenerative deceleration with a conventional braking if the clearance drops below a low value. Moreover, the deceleration has to start early to allow a regenerative braking when the vehicle approach a slower vehicle. This behavior could maintain a large clearance for a long time, allowing a third vehicle to cut in the gap. On daily situation, the driving assistance shows to be efficient, increasing or at least maintaining the state of charge of the battery. We demonstrate the efficiency and the safety of our application on two use cases.

However, the behavior of the driving assistance has a huge impact on the acceptability: increased clearance, long deceleration and resulting smaller time headway, which need to be evaluated under the user point of view.

## References

1. Marsden G, McDonald M, Brackstone M, Marsden G (2001) Towards an understanding of adaptive cruise control. Transp Res Part C 9:33–51
2. Kesting A, Trieber M, Schonhof M, Helbing D (2007) Extending adaptive cruise control to adaptive driving strategies. J Transp Res Board N2000:16–24 (Transportation Research Record)

3. Vahidi A, Eskandarian A (2003) Research advances in intelligent collision avoidance and adaptive cruise control, IEEE Trans Intell Transp Syst 4(3):143–153
4. Moon S, Moon I, Yi K (2009) Design, tuning, and evaluation of a fullrange adaptive cruise control system with collision avoidance. Control Eng Pract 442–455
5. Naus G, Vugts R, Ploeg J, van de Molengraft R, Steinbuch M (2009) Cooperative adaptive cruise control, IEEE automotive engineering symposium Eindhoven, The Netherlands, 6 Apr 2009
6. Van Arem B, Van Driel CJG,Visser R (2006) The impact of cooperative adaptive cruise control on traffic flow characteristics, IEEE Trans Intell Transp Syst 7(4):429–436
7. Nowakowski C, Shladover SE, Cody D et al (2010) Cooperative adaptive cruise control: testing drivers choices of following distances. UCB-ITS-PRR-2010-39
8. ISO 15622:2010, Intelligent transport system, adaptive cruise conrol system, performance requirements and test procedures
9. Glaser S, Aguilera V (2003) Vehicle infrastructure driver speed profile : towards the next generation of curve warning systems. In: Proceedings of the 10th ITS world congress, Madrid, Spain, Nov 2003
10. Aguilera V, Glaser S, Von Arnim A (2005) An advanced driver speed assistance in curves: risk function, cooperation modes, system architecture and experimental validation. In: Proceedings of the IEEE intelligent vehicle symposium, Las Vegas, 2005

# Part II
# Networked Vehicles

# An Energy Management System for Light Two-Wheeled Vehicles Based on a Smartphone-in-the-Loop Architecture

**Andrea Dardanelli, Mara Tanelli, Sergio M. Savaresi, Mario Santucci and Onorino di Tanna**

**Abstract** This paper presents an active energy-saving strategy for two-wheeled electric vehicles, based on an innovative Energy Management System (EMS). In particular, a novel model-based EMS is proposed, which aims at reducing the energy consumption on board by actively modifying the driving-style in real-time. The proposed EMS is capable of tracking a desired discharge rate by means of speed and acceleration controllers that alter the vehicle dynamics conforming to the high-level energy control policies. Simulation and experimental results prove the feasibility of the proposed approach.

**Keywords** Electric vehicles · Control systems · Energy saving systems · Mobile devices

---

A. Dardanelli · M. Tanelli · S. M. Savaresi
Dipartimento di Elettronica e Informazione, Politecnico di Milano, Piazza L. da Vinci, 32 20133 Milan, Italy
e-mail: dardanelli@elet.polimi.it

M. Tanelli
e-mail: tanelli@elet.polimi.it

S. M. Savaresi
e-mail: savaresi@elet.polimi.it

M. Santucci · O. d. Tanna (✉)
Piaggio Spa, Viale Rinaldo Piaggio, 25 56025 Pontedera, Pisa, Italy
e-mail: onorino.ditanna@piaggio.com

M. Santucci
e-mail: mario.santucci@piaggio.com

J. Fischer-Wolfarth and G. Meyer (eds.), *Advanced Microsystems for Automotive Applications 2013*, Lecture Notes in Mobility, DOI: 10.1007/978-3-319-00476-1_13,

## 1 Introduction

The increasing cost of fuel, reduced availability of oil and environmental concerns encouraged the spread of fully-electric vehicles (EVs). As EVs do not use oil in the *tank-to-wheel* path, they can be regarded as a solution for the urban mobility of the near future, see e.g., [1–3]. However, range anxiety, i.e., the fear of not having enough battery State of Charge (SoC) to reach the desired destination, constitutes a huge barrier to the spread of EVs. In fact, range anxiety is such that existing EVs capabilities, in terms of driving range, are largely under-exploited. The upcoming increase of recharging points number in urban areas can only solve a part of the problem: charging might become accessible but still be time consuming and thus significantly impacting on the vehicle usability. In view of these issues, it is of the utmost importance to allow the driver feeling safe and confident that the final destination can be in fact reached within the given battery charge.

To this end, EVs must be designed to cover the average driving ranges needed by potential users, while also leaving some margin against variable traffic conditions or more aggressive driving styles. This is quite tough to achieve, as users are used to Internal Combustion Engine (ICE) vehicles, which generally cover longer distances with a full tank, and for which re-charging (i.e., filling the tank when empty) is a matter of a few minutes.

To address this challenging task, this paper proposes a novel Energy Management System (EMS), realized in the form of an active closed-loop state of charge controller for EVs, enforcing limitations on the driving-style. Specifically, once the desired battery discharge rate has been defined according to the specified route to be covered, the corresponding tracking problem determines speed and acceleration bounds that must not be exceeded by the vehicle in order to ensure the desired energy consumption. To this end, low-level motion controllers enforce the vehicle to stay within the prescribed limits by dynamically redefining the relation between the gas handle and the torque generation logic. In this work, we focus on light, two-wheeled EVs, as they represent an intriguing solution for green mobility in urban scenarios. However, it is worth noticing that the proposed approach might be extended to ICE vehicles: in this context, the proposed controller would allow reducing energy consumption (and thus emissions), using exactly the same rationale as that presented herein, with the sole difference that the regulated variable would be fuel consumption in place of the battery discharge rate.

The structure of the paper is as follows: Sect. 2 presents the vehicle setup, while Sect. 3 details the vehicle model. Further, Sect. 4 illustrates the proposed EMS showing also the experimental results.

## 2 Vehicle Setup

The test vehicle is a prototype, two-wheeled EV, equipped with a CAN-bus for signal transmission (see Fig. 1). A Li-Ion battery stores the energy available for the propulsion. An electric motor exchanges power with the rear wheel (the wheel radius is denoted with $r_w$) via a wet transmission with fixed transmission ratio $K_g$. The motor is equipped with an electric motor drive (EMD) which converts the gas handle opening signal $g_e(t)$ into a motor torque set-point $T_{m,ref}(t)$ via a static map, which is an unknown function of $g_e(t)$ and vehicle speed, $v(t)$. Such a set-point is tracked by means of a control strategy which is embedded in the EMD itself and not directly accessible, i.e., $T_m(t) \rightarrow T_{m,ref}(t)$. In the remainder of the chapter, we denote with $T_m(t)$ and $F_w(t)$ the motor torque and the force at the wheel, respectively. The electric motor works both in traction and regenerative braking mode; in particular, traction is achieved when $g_e(t) > 0$, while energy is recovered in the battery by means of regenerative braking policies if and only if $g_e(t) = 0$.

Thus,

$$F_w = f_g(g_e, v) = \begin{cases} \eta_g \kappa_g T_m, & g_e > 0. \\ \frac{1}{\eta_g} \kappa_g T_m, & g_e = 0. \end{cases},$$

where $\eta_g$ is the transmission efficiency and $\kappa_g = K_g r_w$.

The vehicle ECU (the *Gateway* block in Fig. 1) measures the gas handle opening requested by the rider, $g_d(t)$, and processes such signal before sending it to the EMD. Thus, the EMS influences the battery discharge by varying the torque requested to the electric motor for a given value of $g_d(t)$. As a consequence $g_e(t)$ is the control variable. Obviously, the open-loop operation is characterized by the relation $g_e(t) = g_d(t)$. Further, an estimate of the battery State of Charge (SoC) $\xi(t)$ is available for control purposes and the battery current $i_b(t)$ and voltage $v_b(t)$ are measured by the gateway ECU. Finally, the vehicle longitudinal speed $v(t)$ is available on the CAN-bus. All signals are sampled at $f_s = 100$ Hz.

A Bluetooth module embedded on the gateway ECU manages the communication with the smartphone. The smartphone computational capability is exploited

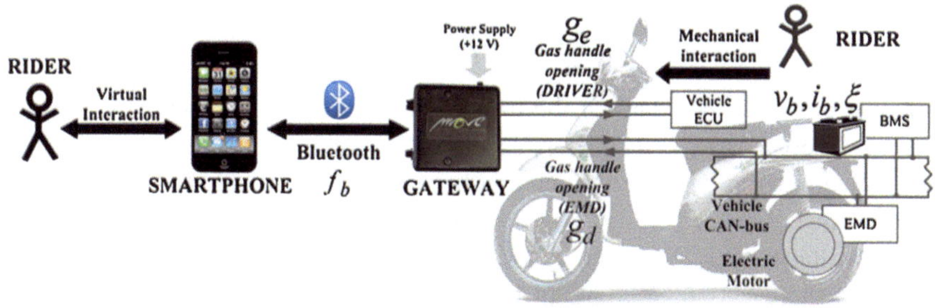


**Fig. 1** Schematic view of the instrumented vehicle

for controller execution, as it runs the high-level control algorithms. Furthermore, by means of a dedicated application (developed on both Android OS and on iOS), the rider is made aware of the vehicle status and can play active role in the vehicle energy management. In fact, the user can modify some parameters of the SoC control loop via the smartphone human–machine-interface (HMI), [4].

## 3 Vehicle Dynamics Model

Consider a vehicle with mass $M$ that proceeds at velocity $v$ on a road with slope $\vartheta$ ($\vartheta > 0$ indicates uphill roads). The vehicle longitudinal acceleration is $\dot{v}$. Denoted with $F_f$ the friction forces, $F_s = Mg \sin \vartheta$ the slope forces (the term $g$ represents the gravitational acceleration) and $F_{br}$ the mechanical braking forces, the model of the longitudinal dynamics has the form (see also [5] and [6])

$$M\dot{v} = F_w - \left(F_f + F_s + F_{br}\right).$$

The friction forces can be described with a lumped-parameter model, i.e.,

$$F_f = \sigma_{f,2}v^2 + \sigma_{f,1}v + Mg\sigma_{f,0}.$$

The quadratic term represents the aerodynamic drag force, the linear term highlights the viscous components and the constant term indicates the rolling resistance. The parameters $\sigma_{f,i}i \in \{1,2,3\}$ were identified as detailed in [7].

To derive a complete energy-oriented model, we express the longitudinal dynamics model in terms of power, namely

$$P_i = P_w - \left(P_f + P_s + P_{br}\right) \rightarrow P_w = P_i + P_f + P_s + P_{br}.$$

Notice that the power at the wheel $P_w = F_w v$ is the sum of the inertial power $P_i = M\dot{v}v$, the friction power $P_f = F_f v$, the slope power $P_s = F_s v$ and the mechanical braking power $P_{br} = F_{br}v$.

Let us now introduce the battery power $P_b = v_b i_b$ and the idle battery power $P_{b,0}$, i.e., the idle constant power. The vehicle efficiency

$$\eta = \begin{cases} \eta_t = \frac{P_w}{P_b - P_{b,0}}, & g_e > 0. \\ \eta_r = \frac{P_b - P_{b,0}}{P_w}, & g_e = 0. \end{cases}$$

links the power at the wheel to the battery power. The battery SoC is defined as $\dot{\xi} = -i_b/Q_0$, where $Q_0$ is the battery capacity. We adopt the *electrical model* of the battery (see [8]) to describe the relation between the electrical quantities and the battery power, i.e., $P_b = v_b i_b = v_{oc} i_b - R_b i_b^2$, where $v_{oc}(\xi)$ indicates the open-circuit voltage, and $R_b$ is the inner battery resistance. Finally, the complete nonlinear energy-oriented model of the vehicle is

$$\dot{\xi} = -\frac{v_{oc} - \sqrt{v_{oc}^2 - 4R_b\left(P_{b,0} + \eta P_w\right)}}{2R_b Q_0},$$

and it outputs the battery SoC given a speed profile.

The unknown parameters were estimated with the following experiments:

- *Coasting down experiments*. The vehicle decelerates subject to the sole friction forces ($F_w = 0$, $F_{br} = 0$), starting from maximum speed.
- *Regenerative braking experiments*. The vehicles decelerates subject to the friction forces + the regenerative braking force ($F_w < 0$, $F_{br} = 0$), starting from maximum speed.
- *Long-run experiments*. The vehicle is ridden along a 13.3 km urban flat path. Several quantities can be derived from these experiments by properly filtering the measurements and approximating nonlinear terms, e.g., the efficiency, using neural networks, see, e.g., [9].

To evaluate the quality of the validation results, we introduce two performance indexes, based on the simulation error $e$, which is defined as the error between the simulated output $\tilde{y}$ and the real measurement $y$.

- *Normalized squared L2-norm of the simulation error*, defined as

$$J_{L2} = \frac{L2(e)^2}{LE(Y)^2},$$

where $L2(y)$ denotes the L2-norm of the signal $y$. This index gives insight into the capability of the model in reproducing the real measurements, i.e., the fitting performance.

- *Normalized mean of the simulation error*, defined as

$$J_m = \frac{m(e)}{m(y)},$$

where $m(y)$ is the sample mean of the signal $y$. This index is suitable for the detection of non-zero mean errors, which may worsen the model prediction capabilities.

**Table 1** Performance measurements evaluated on three validation datasets

| Rider | $J_{L2}$ | $J_m$ |
|---|---|---|
| Rider 1 | 0.06 | 0.007 |
| Rider 2 | 0.053 | 0.055 |
| Rider 3 | 0.045 | 0.039 |

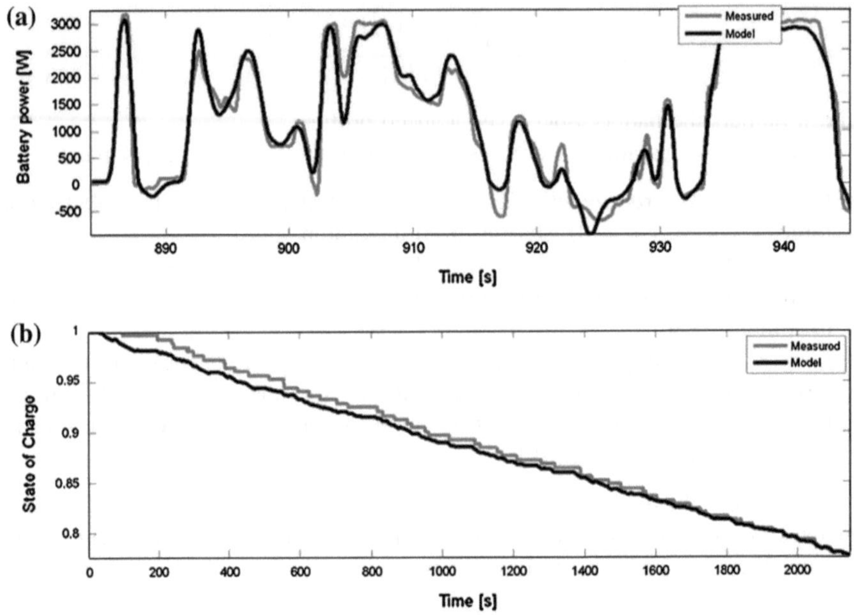


**Fig. 2** Validation of the energetic model. Time histories of the model output (*black solid line*) and measurements (*grey solid line*). **a** Battery power (*zoom*). **b** Battery SoC

The cost functions were evaluated over the battery power data taken from 6 experiments carried out by 3 different riders. Validation results (collected in Table 1) show that the energy-oriented model is suitable for the development of the energy management control system (Fig. 2).

## 4 The Energy Management System

In this section we present the EMS depicted in Fig. 3. The proposed hierarchically distributed architecture is composed of two main elements (refer also to [6]).

- *The Gateway ECU*, which hosts the speed/acceleration controllers, driven by the speed and acceleration bounds provided by the external SoC control-loop. These control systems have been located on the gateway ECU as they actively modify the vehicle behaviour, thus posing safety–critical issues that need a time-triggered control loop with guaranteed execution patterns.
- *The mobile device*. The smartphone runs the high-level control routines and acts as a driver-to-vehicle and vehicle-to-driver interface. The energy manager represents the core of the EMS and it provides the speed/acceleration bounds to the low-level motion controllers.

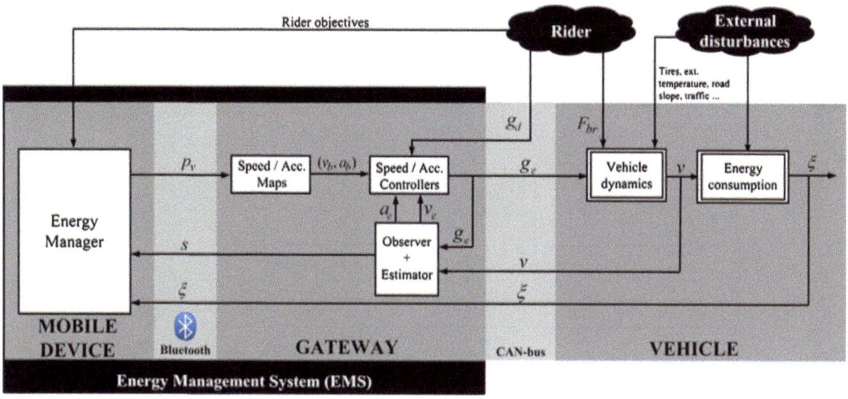


**Fig. 3** Schematic view of the energy management system

Due to the fact that a smartphone is nested in the control loop, we refer to this paradigm as *smartphone-in-the-loop*. The low-level motion controllers are detailed in [6, 9]. In the following, we will focus on the energy manager only.

To better define the working principles of the EMS, let us introduce the *spatially normalized discharge* $n = -d\xi/ds = -(1/v)d\xi/dt$. This quantity is a useful indicator of the energy consumption on-board as it represents the projection of the discharge rate of the battery over the spatial coordinate $s$. Notice that, given a spatial interval $\Delta s = s_{i+1} - s_i$, we can compute the average value of $n$ as

$$n_{avg} = \frac{1}{\Delta s}\int_{s_i}^{s_{i+1}} n\, ds = \frac{1}{Q_0}\frac{i_{b,avg}}{v_{avg}} = \frac{\Delta \xi}{\Delta s} = \frac{\xi_{i+1} - \xi_i}{s_{i+1} - s_i}.$$

Hence, the energy consumption over a flat path with length $\Delta s$ can be fully characterized by looking at the relation between $v_{avg}$ and $n_{avg}$. To study the relation between these two quantities, we focused on a trapezoidal speed profile, whose features (maximum acceleration, coast speed, deceleration, distance travelled) can be regarded as a simple parameterization of a riding-style. Specifically, we studied the impact on $n_{avg}$ of perturbations on the coast speed/maximum acceleration, using the identified vehicle model. The analysis showed that there exists a relation (*consumption map*) between the maximum coast speed $p_v$ and the average spatially normalized discharge, i.e., $n_{avg} = f_c(p_v)$. Furthermore, a feasible value of $n_{avg}$ is always related to one single value of $p_v$, i.e., $p_v = f_c^{-1}(n_{avg})$.

The energy manager structure is depicted in Fig. 4. Note, again, that the inputs of the energy manager are the rider objectives, the battery SoC and the travelled distance $s$, while its outputs are the speed/acceleration bounds.

The user interacts with the system through the SoC reference generation block. Here, the rider specifies the objectives in terms of destination and desired final value of the battery SoC. The smartphone computes the desired route using the Internet connection, querying one of the providers of navigation data. Said $s_f$ the length of the desired route, we denote with $\xi_{ref}(s_f)$ the desired final value of

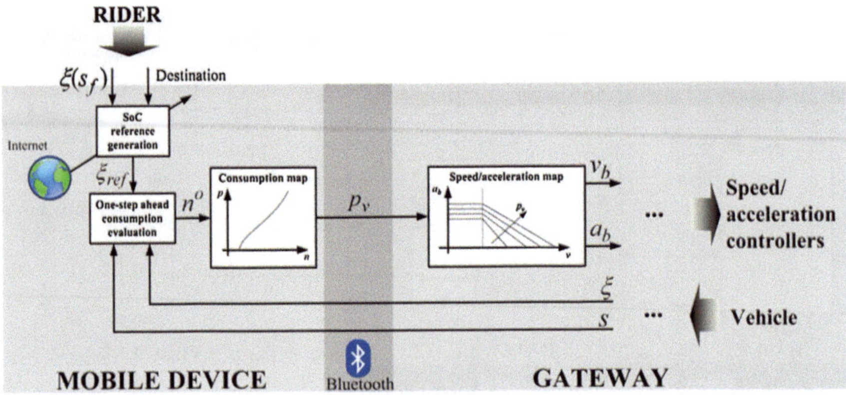


**Fig. 4** Schematic view of the energy manager block

the SoC. On a flat road, an easy way to define a set-point is to track a spatially linear SoC profile, namely

$$\xi_{ref}(s) = \xi(0) - n_{avg,ref} s_f, \quad n_{avg,ref} = \frac{\xi(0) - \xi_{ref}(s_f)}{s_f}.$$

If the road grade is not negligible, the slope power $P_s$ significantly affects the power consumption at the wheel $P_w$ and must be taken into account while defining the control policy, i.e., the reference profile must be modified accordingly. The interested reader may refer to [10] for a thorough analysis.

The energy manager works in a spatial domain. In particular, every $\bar{s}$ meters, it evaluates the one-step ahead consumption, i.e., it defines a value of average spatially-normalized discharge $n^o_{avg}$ that the vehicle must track in the following $\bar{s}$ meters in order for the SoC to track the reference, i.e.,

$$n^o_{avg} : \xi(s + \bar{s}) = \xi_{ref}(s) \rightarrow n^o_{avg} = \frac{\xi(s) - \xi_{ref}(s + \bar{s})}{\bar{s}}.$$

This value is then passed to the consumption map, which provides the maximum coast speed, i.e., $p_v = f_c^{-1}(n_{avg})$. The speed/acceleration maps process the maximum value for the coast speed in order to obtain the driving-style bounds, i.e., $(v_b, a_b)$. These maps are computed offline and have the form

$$a_b(v, p_v) = \begin{cases} a_{b,1} = a_{b,min} + \frac{a_{b,max} - a_{b,min}}{p_{v,max} - p_{v,min}} (p_v - p_{v,min}), & v < v_t \\ a_{b,2} = a_{b,1} - \frac{a_{b,1}}{p_v - v_t} (v - v_t), & v \geq v_t \end{cases}$$

$$v_b = p_v$$

where $v_t$ is a threshold speed (set to 15 km/h) used to guarantee an acceptable level of acceleration at low speeds. The acceleration limitation varies between $a_{b,min}$ and $a_{b,max}$ (maximum vehicle acceleration). Note that when $p_v = p_{v,max}$ (maximum

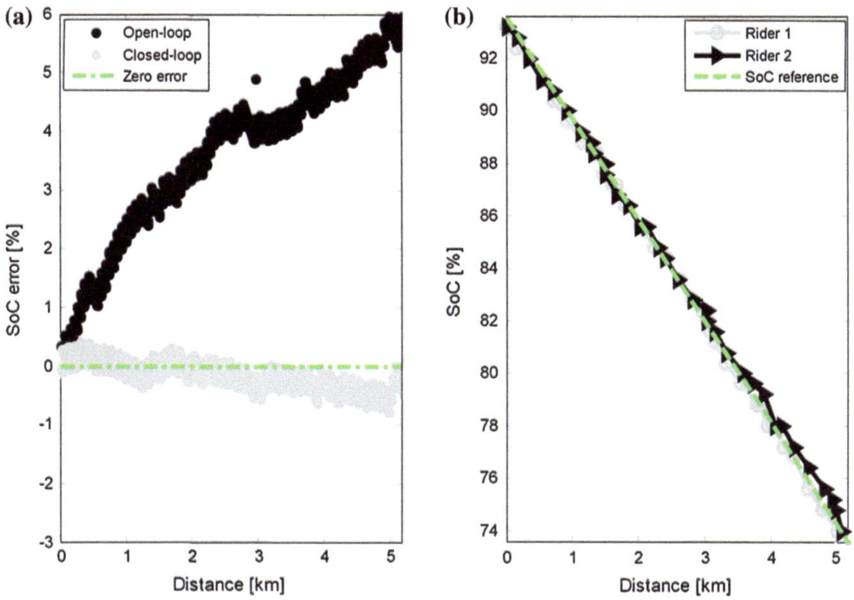


**Fig. 5** Experimental results. SoC error closed-loop vs open-loop as a function of distance in (**a**). Closed-loop SoC as a function of distance for different riders in (**b**)

vehicle speed) the EMS interventions are limited, while $p_v = p_{v,min}$ implies great interventions as the current driving style is not compliant with the desired discharge profile of the rider.

We kept the shape of these maps as simple as possible; furthermore, several riders tested the system with different maps: this shape comes also from their feedbacks about the vehicle dynamics.

Experimental results are shown in Fig. 5. Figure 5a plots a comparison between an open-loop and a closed-loop SoC profile. In particular, we computed the SoC error with respect to a linear reference discharge policy. The percentage refers to the SoC, namely $\xi[\%] = 100\xi$. The vehicle was used in the same external conditions, i.e., rider, path, day hour. Some comments are due: the open-loop SoC profile diverges from the zero-error line, even if the rider tried to adopt a conservative riding style; the closed-loop system keeps the SoC profile as close as possible to the SoC reference [note that the error is kept between $(-0.5\ 0.8)$ %]. This situation is what we call "range assurance": the EMS intervenes so to attain the desired level of energy consumption. In particular, we saved approximately 6 %, which corresponds to the 30 % of the energy depleted in the open-loop test. In Fig. 5b the reader can appreciate the system robustness: two riders used the vehicle over the same route and adopted the same SoC policy, i.e., the same reference profile. Notice that the SoC profiles are quite different: this is mainly due to the fact that the driving styles are rider-dependent. Nevertheless, the closed-loop SoC profile tracks the reference in both cases, thus guaranteeing the correct "range assurance".

## 5 Concluding Remarks

This paper presented a novel model-based energy management system for light-electric two-wheeled vehicles, which allows reducing the energy consumption by actively modifying the driving-style. The proposed hierarchical control architecture is capable of regulating the battery state of charge by imposing a desired discharge rate, which is tracked by means of low-level motion controllers that regulate the vehicle dynamics according to the high-level energy control policies. Extensive experimental results proved the effectiveness of the approach.

## References

1. Hacker F, Harthan R, Matthes F, Zimmer W (2009) Environmental impacts and impact on the electricity market of a large scale introduction of electric cars in Europe, ETC/ACC Technical Paper 2009/4
2. EU Commission (2009) Greening road transport: EU-funded research supports EU's environmental objectives, Memo/09/26, Brussels, 26 Jan 2009
3. Larminie J, Lowry J (2012) Electric vehicle technology explained. Wiley
4. Mangan S, Wang J (2007) Development of a novel sensorless longitudinal road gradient estimation method based on vehicle can bus data. IEEE/ASME Trans Mechatron 12(3):375–386
5. Guzzella L, Sciarretta A (2005) Vehicle propulsion systems: introduction to modeling and optimization. Springer, Berlin
6. Dardanelli A, Tanelli M, Picasso B et al (2011) Speed and acceleration controllers for a light electric two-wheeled vehicle. In: Proceedings of the 50th IEEE conference on decision and control, 2523-2528, Orlando, FL
7. Dardanelli A, Tanelli M, Picasso B, Savaresi S, Di Tanna O, Santucci M (2011) Control-oriented energy-profiling and modelling of urban electric vehicles. In: Proceedings of the 2011 multi conference on systems and control, Denver, CO, pp. 332–337
8. Urbain M, Rael S, Davat B, Desprez P (2007) State estimation of a lithium-ion battery through kalman filter. In: IEEE Power Electronics Specialists conference 2007, pp. 2804–2810
9. Dardanelli A, Tanelli M, Picasso B et al (2012) A smartphone-in-the-loop active state-of-charge manager for electric vehicles. IEEE/ASME Trans Mechatron 17(3):454–463
10. Dardanelli A, Tanelli M, Savaresi S, Santucci M (2012) Active energy management of electric vehicles with cartographic data. In: Proceedings of the 2012 IEEE international electrical vehicle conference (IEVC), pp. 1–6

# Cooperative Systems in Motorway Environment: The Example of Trento Test Site in Italy

**Filippo Visintainer, Leandro D'Orazio, Marco Darin and Luciano Altomare**

**Abstract** Cooperative systems play a major role in the reduction of road accidents. Among several research initiatives, field trials allow to evaluate cooperative system performance as well as their impact on end-users. Field Operational Tests, such as DRIVE C2X EU Project, are being carried out by major players in several sites across Europe. In this context, an Italian Test Site close Trento city addresses free-flow scenarios on a motorway. Applications based on vehicle-to-vehicle and vehicle-to-infrastructure cooperation provide alerts directly on-board. This paper describes the Italian Test Site, including vehicle fleet and infrastructure elements. It outlines a specific application tested and preliminary experimental results.

**Keywords** Test site · Cooperative system · Wrong way driving · Vehicle-to-vehicle · Vehicle-to-infrastructure · ITS-G5 · IEEE802.11p · Communication

## 1 Introduction

Road safety and traffic efficiency still represent major issues to solve for a sustainable transport in future society. Accidents, delays, high fuel consumption as well as discomfort while driving are becoming a less and less acceptable cost of

F. Visintainer (✉) · L. D'Orazio · M. Darin · L. Altomare
Department of Info-Telematic Systems, Centro Ricerche FIAT—Trento Branch,
Via della Stazione, 27 38123 Trento, Italy
e-mail: filippo.visintainer@crf.it

L. D'Orazio
e-mail: leandro.dorazio@crf.it

M. Darin
e-mail: marco.darin@crf.it

L. Altomare
e-mail: luciano.altomare@crf.it

J. Fischer-Wolfarth and G. Meyer (eds.), *Advanced Microsystems for Automotive Applications 2013*, Lecture Notes in Mobility, DOI: 10.1007/978-3-319-00476-1_14,

individual mobility, which sums to the collective cost in terms of environmental impact caused by traffic congestions. Among the approaches for safe and efficient driving, Intelligent Transportation Systems (ITS) based on the cooperation among vehicles and road infrastructure through low-latency wireless communications, have been addressed by several research initiatives. Through vehicle-to-vehicle (V2V) and vehicle-to-infrastructure (V2I) data exchange using ITS-G5/IEEE802.11p, cars, roadside elements and service centres back-ends form a highly responsive cooperative system, capable of rapidly sharing and processing information about the surroundings. In particular, this article will refer to the ETSI standard communication architecture [1], where communication among nodes is primarily based on the transmission of Cooperative Awareness Messages (CAMs) and Decentralized Environmental Notification Messages (DENMs). CAMs [2] are generated by each ITS station providing information related to presence, position and status of itself to other neighbouring nodes. This kind of messages can be transmitted at a frequency varying from 1 to 10 Hz (depending on the vehicle type) and their broadcast diffusion is strictly limited to the radio coverage of the system (i.e., single hop distance). DENMs [3] are event-based messages mainly used by cooperative road hazard warning applications for alerting road users about detected dangerous situations (e.g., collision risk warning, emergency electronic brake light). These messages are repeated at a given frequency by the transmitting ITS station and their diffusion persists for the whole event duration. If critical events are detected, drivers are then provided with early warning whenever a hazard or traffic congestion may affect them, much earlier than their own perception, and even earlier than automotive detection systems. The safety margin is thus increased to the preventive safety region, whereby the user can still react and safely avert dangers, avoid congestions, or behave in a more efficient way in queue-forming conditions. Research projects at national and international level have been going on in Europe, spanning from the enabling technologies and applications [4, 5], and providing guidelines for deployment and standardisation [6]. Lately, cooperative systems are being experimented in large scale Field Operational Tests (FOTs) [7], whereby stakeholders can assess V2V- and V2I-based applications not only from technical perspective, but also from the impact they will have in future society. In DRIVE C2X FOT [8, 9], co-funded by the European Commission DG Connect within the 7th Framework Programme, major OEMs, equipment suppliers, road operators and research institutes are testing a common cooperative system in seven test sites across Europe, covering a whole set of scenarios, environments and local adaptations. One of the test sites is located near the city of Trento, it is led by FIAT Research Centre (CRF) together with the local operator Autostrada del Brennero (A22 motorway), and focuses on free-flow scenarios on a segment of the local motorway. A fleet of equipped vehicles is operational along the motorway, to test V2V–V2I cooperation for on-board warning, e.g. of traffic jams, broken-down cars ahead, speed limits, approaching emergency vehicle. Cooperative systems give the possibility to develop applications that inform incoming drivers more effectively, in terms of timeliness, distance and on-board warning strategies, further reducing collision risks. In this paper the architecture of the cooperative system and the specific Wrong Way Driving function are presented, as deployed at the Italian Test Site. Then, preliminary results of experiments on vehicular ITS-G5 connectivity are reported.

# 2 Test Site Overview

The Italian Test Site is located near the city of Trento in northern Italy, along the A22 motorway. The latter is a 314 km toll road with 2 lanes per direction, connecting Italy with northern Europe, thus having a consistent flow of commercial vehicles, but also tourist flow peaks in specific periods. The motorway is already equipped with ICT infrastructure for mobility. The Test Site is a 9 km long section between "Rovereto Nord" and "Rovereto Sud" toll stations, dedicated to cooperative applications and tested by the FOT fleet.

## *2.1 Vehicle*

The cooperative vehicle prototypes of the Test Site fleet (Fig. 1) are provided by CRF and host the Vehicle ITS Station (VIS) of the DRIVE C2X reference system [1, 10–12].

The system, schematically represented in Fig. 2, is based on two on-board computers: a Communication Control Unit (CCU) and an Application Unit (AU). The AU operating system is Linux-based and runs the cooperative applications via OSGi framework, collecting GPS data for positioning and vehicle data from CAN bus. It is connected to CCU for short range and UMTS for long-range communication. The CCU is mainly dedicated to short range communication using ITS-G5 and it is connected to an omnidirectional antenna on the car roof.

**Fig. 1** CRF fleet: Vehicles aligned at a demonstration (*left*); Lancia Delta prototype (*right*)

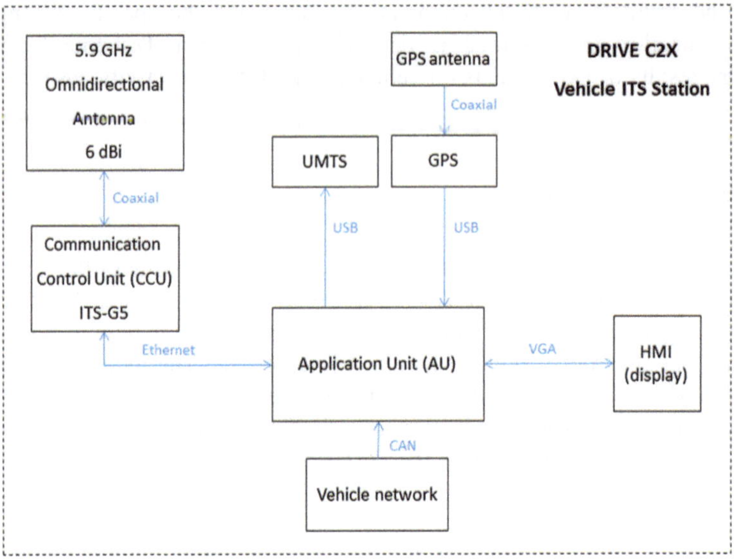


**Fig. 2** Scheme of in-vehicle system demonstrator

One sample vehicle was tested in the System Test Site of DRIVE C2X in Helmond, the Netherlands, to validate the set-up. Then, a fine tuning phase followed in Trento to obtaining the final prototypes.

## *2.2 Road Infrastructure*

Cooperative infrastructure has the primary function to enable roadside-based applications, enhancing the capability of existing equipment, for example Variable Message Signs, through short range communication with incoming vehicles. In turn, existing infrastructure can provide useful data on road conditions to the Test Site. In terms of architecture, Roadside ITS Stations (RIS) like VIS are based on an ITS-G5 Communication Control Unit and an Application Unit. The CCU is connected to reflector antennas placed on roadside poles or gantries. Roadside CCUs are connected to AUs and Service Centre terminals via fibre optics. In parallel to V2I communication, roadside infrastructure is used to measure local conditions. Existing roadside components include traffic and weather sensors, whose interface has been adapted by Autostrada del Brennero to provide data to the Field Operational Test analysts, so that the boundary conditions of each trial can be evaluated. In addition, for V2I, five RIS have been installed. The following picture (Fig. 3) reports their placement along the motorway.

RIS 901 and RIS 905 were placed at the north and south boundary of the Test Site, 902 was placed at the Gas Station, to inform in case of Wrong Way Driving, while RIS 903 and RIS 904 were placed close to each other to test continuous coverage.

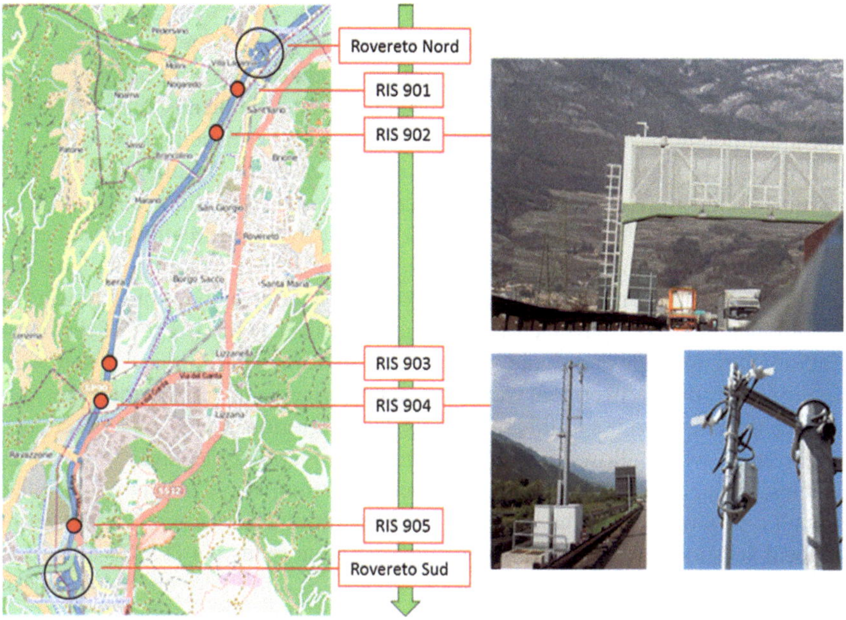


**Fig. 3** Test Site overview

# 3 Functions of the Test Site

The functions addressed by the Italian Test Site are Wrong Way Driving, In Vehicle Signage, Car Breakdown Warning, Road Works Warning, Motorcycle Warning, Traffic Jam Ahead Warning, Slow Vehicle Warning, Approaching Emergency Vehicle. More functions can be implemented and tested in the future. An important aspect of this large scale FOT is the fact that these functions are provided by project OEMs and research centres, e.g. for this Test Site CRF, Opel, PSA, Ford, Honda, Fraunhofer Fokus. Functions are tested and compared in multiple test sites, to account for the distinct geographical and road characteristics as well as for the legacy systems already in place. For instance, In-Vehicle Signage function allows the Service Centre to send messages to the drivers via vehicle-to-infrastructure communication. The driver's behaviour is then monitored with and without the function, to infer the impact of the function on driving behaviour [13]. A specific function of the Italian Test Site is Wrong Way Driving. In motorways, wrong way driving may happen at Gas Stations or rest areas mainly in two kinds of situations. In the first scenario, the driver leaves the station and, due to its own condition or to adverse factors such as lowered visibility, proceeds towards the station entrance instead of the exit. If not informed, the driver will enter the main road in opposite direction. This case is typically unintentional and rare, but when it happens the connected accident risk is very high. In the second scenario, a driver about to enter the gas station suddenly decides to return to the main road. Instead of crossing the station, he/she drives backwards along the entrance lane and resumes driving in the correct direction once on the main road. Some past cases concerned heavy vehicles entering rest areas to stop overnight, but finding no place. They decided to save time moving backwards, instead of crossing the narrow passage among other parked trucks. This scenario happens more frequently than the former one, and although the hazard is limited to the time of reverse maneuver, there is a high risk for vehicles approaching the station in that moment. Wrong Way Driving function should thus guarantee a prompt detection and immediate warning of the approaching cars, as well as the offending car itself, to avoid or limit the consequences of a vehicle proceeding along the highway in opposite direction. An implemented solution is constituted by sensors able to detect the direction of the vehicle passing on the entrance lane. The passage of the vehicle determines the trigger of the alarm condition through light and sound indicators located along the entrance lane to warn the wrong way driver. In critical conditions, the Service Centre is informed. A way to enhance the effectiveness of this solution is the integration of DRIVE C2X cooperative system, which allows in addition to transmit warning messages to incoming vehicles, through Wrong Way Driving function. Wrong Way Driving works either through a command from the Service Centre or through an autonomous trigger by the local sensors. Based on these inputs, the function sends via ITS-G5 a warning message using DENM standard format [3], in order to inform drivers about the risk. The on board application of the Vehicle ITS station receiving the message analyzes its content, checks its validity and finally provides the warning to the driver. Incoming vehicles are thus warned more effectively, both in terms of space–time as well as in terms of warning strategies. Indeed, the alert is displayed directly on board and filtering false alarms and signal redundancies, through algorithms which depend on vehicle direction, approach and geographic position.

## 4 Preliminary Experiments on the Italian Test Site

This chapter illustrates preliminary tests to characterise the message reception by vehicles approaching the RIS used for Wrong Way Driving near the Gas Station. In particular, the communication performance was tested, for the safety margin extension. In addition, another scenario and location was addressed, to provide an example of the dependency of short range communication on the geometrical and geographical characteristics, as also other studies report [14, 15].

### *4.1 Communication Coverage: Experimental Set up and Tests*

The car used was a Lancia Delta demonstrator, equipped with DRIVE C2X Vehicle ITS Station, using NEC Linkbird CCU connected to 5.9 GHz omnidirectional antenna MOBILEMARK ECOM6-5500, with 6 dBi nominal gain. The first Roadside ITS Station addressed, RIS 902, is installed on an overhead gantry 300 m north of the gas station and 11 m above the road surface. It is equipped with a NEC CCU, connected to two corner reflector 5.9 GHz antennas, MOBILEMARK SCR14-5725, each having 14 dBi nominal gain, beam width 30° in azimuth and 60° in elevation. In particular the antenna heading north, informing the vehicles in direction south of possible ghost drivers, was addressed. Both VIS and RIS CCUs have been configured with an output power of 21 dBm. Without considering cable and insertion losses (<2–3 dB), this yields 27 dBm total transmit power on the vehicle, while on the roadside, accounting for power splitter attenuation and antenna gain, 32 dBm total transmit power in each direction of the motorway. First, trials at typical motorway speeds were carried out, namely 10 trials at 90 km/h on the right lane and 10 trials at 130 km/h on the left lane respectively, to evaluate communication in free-flow conditions and to locate the areas of interest for subsequent static accurate measurements. Based on the results, static tests were planned, with the support from A22 operators, for safety reasons. Seven test points were selected (Fig. 4), where the vehicle stopped for the acquisition of RIS 902 CAMs [2]. The test point selection criteria was based on the message reception rate, to spot the area where this rate increased to almost 100 %. The acquisition at each point lasted 10 min, in order to gather a statistically relevant number of samples, considering the 1 Hz frequency of RIS messages. Experiments were repeated on another roadside station, RIS 904, installed on the roadside and 9 m high, for comparison.

The following graph in Fig. 5 reports the measured Packet Delivery Ratio (PDR) versus the distance between Vehicle and Roadside ITS Station. The PDR is intended as the ratio between received and transmitted packets, and is measured in the Vehicle ITS Station collecting the CAMs sent by RIS 902 at 1 Hz. When an incoming message is available for the vehicle application, it is logged with the in-vehicle system timestamp. Both VIS and RIS are synchronised, as the following

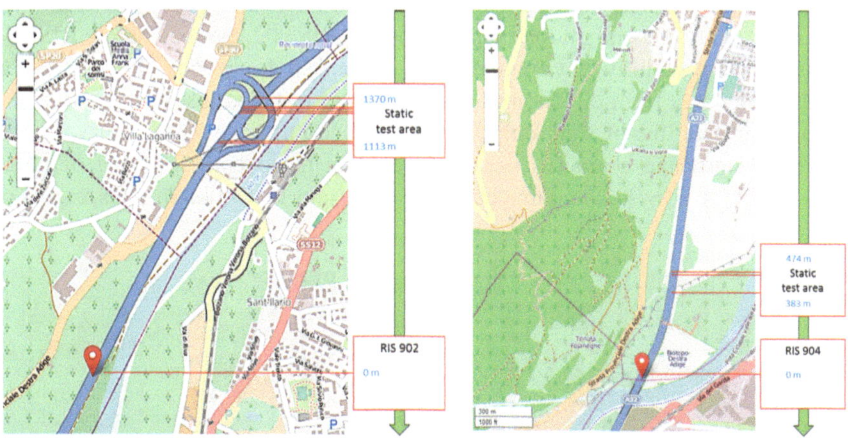


**Fig. 4** RIS 902 and RIS 904 location on OpenStreetMap

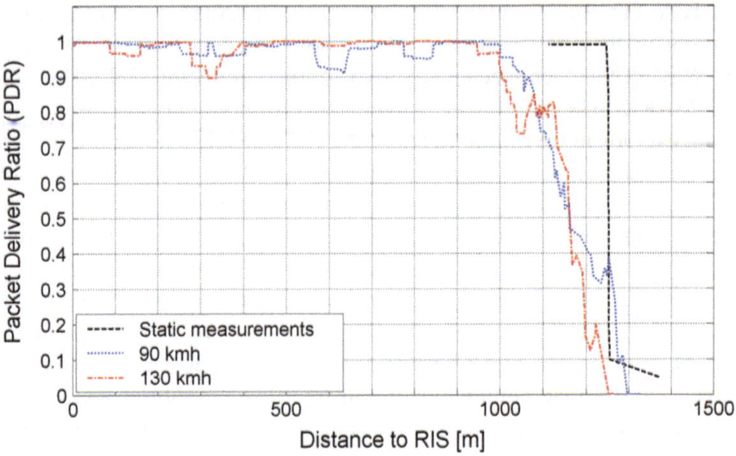


**Fig. 5** RIS 902, PDR versus distance

section reports. In static tests, PDR was computed from a sequence of 500–700 transmitted messages, while in dynamic tests, PDR was inferred from the periodicity of the received messages, averaged over multiple trials across the Test Site.

Test results show that the vehicle started receiving roadside beacons between 1,400 and 1,200 m. A steep increase in the PDR happens between 1,200 and 1,000 m. Static measurements show that a PDR close to 100 % is reached at longer distance, as expected. However, there is a very sudden change of reception rate around 1,250 m (Fig. 6). A possible explanation could be the geographic characteristics. Indeed, RIS902 antenna heading north faces a free and almost flat landscape up to 1 km, followed by a slight bend between 1.1 and 1.5 km.

Moving to test results on RIS 904 (Fig. 7), an analogous behaviour of the PDR was measured, but on a different scale, as can be seen in the following graph. Communication range is around 300 m, much lower than the previous case.

Compared to the straight road segment before RIS 902, the road before RIS 904 passes through a hilly environment and is flanked by woods (Fig. 8). When reaching the area from direction north, RIS 904 is installed just after a curve and becomes in free line of sight only from around 500 m.

These landscape characteristics could explain the overall lowering of range with respect to RIS 902, and maybe the better results of 130 km/h with respect of

**Fig. 6** RIS 902: Static test area (*left*), Static test point (*right*)

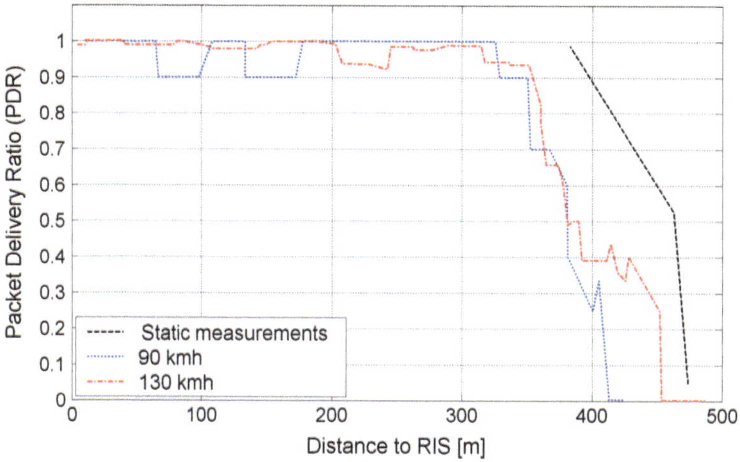


**Fig. 7** RIS 904, PDR versus distance

**Fig. 8** RIS 904: Static test area (*left*), Static test point (*right*)

90 km/h trials, thanks to a longer vehicle-to-infrastructure line of sight in the overtaking lane. Static measurements give the best results, as the good conditions of stopped vehicle might counterbalance the lowered line of sight. Overall, it can thus be expected that a vehicle at 130 km/h with this equipment, receives messages before approaching the area of interest with 27 s advance for RIS 902 and 8 s advance for RIS904, respectively.

## 4.2 Latency: Experimental Set up and Tests

Latency measurements were performed on a warning message from RIS 902. The road side system is time synchronized with a NTP server provided by Autostrada del Brennero, the latter being synchronized with GPS time. The vehicle system is also time synchronized via the specific software module NTPD, used in conjunction with the vehicle GPS sensor and the related software component GPSD in the Application Unit. The car uses a 5 Hz GPS sensor on board (U-BLOX EVK-6H), and the vehicle application unit is running a specific software module NTPD with

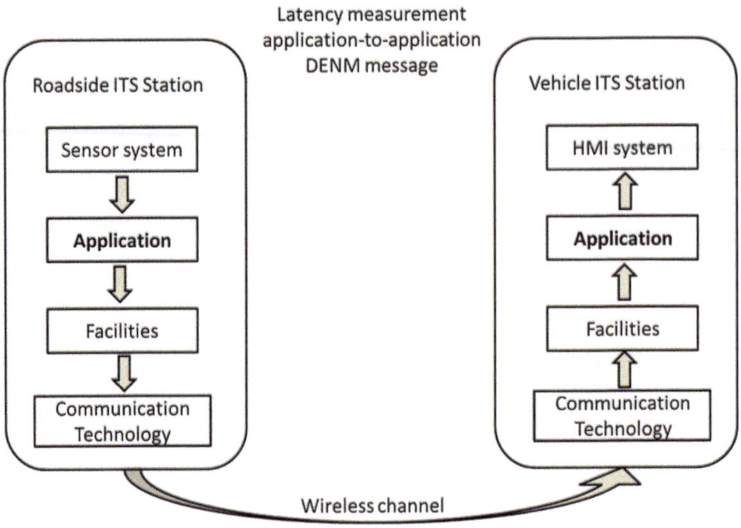


**Fig. 9** Application-to-application latency measurement

GPSD to synchronize the system clock internally and provide time synchronisation to other in-vehicle devices. Measurements were carried out collecting data from the receiving vehicle in static conditions at 350 m distance from RIS 902, for about 30 min. To measure latency, the generation timestamp of the received DENM message was compared with the timestamp corresponding to the instant at which the received message was available for the listener application, since each DENM packet has its generation timestamp stored internally. This means an end-to-end latency between the application layers of emitting RIS and the receiving Vehicle ITS Station (Fig. 9). Measurements over 2,000 received messages gave an average time delay of 42 ms with a standard deviation of 6 ms. This latency was measured to evaluate the time until data are available to the HMI. The measured value is below the requirements of 100 ms set by ETSI for ITS [16].

## 5 Conclusions

Within the worldwide framework of cooperative system, Field Operational Tests allow for a thorough evaluation of C2X systems, the assessment of their impact, and the set-up of test sites as a groundwork of an Interoperable pan-European cooperative system. As a whole, test sites represent an opportunity for both car manufacturers and road providers to assess their applications in real, free-flow scenarios. In the Italian Test Site, cooperative systems have been adapted to the existing motorway infrastructure, with the aim of improving safety thanks to vehicular networking. Several functions are being tested by a fleet of equipped vehicles. During the tests, data are collected of communication, application behaviour, vehicle dynamics and environmental conditions. First steps of the testing were the technical characterisation of the implemented system for the applications addressed, considering the Test Site specific installation. Preliminary tests focussed on the ITS-G5 data reception in the vehicle along the motorway, measured in terms of packet delivery ratio and end-to-end latency of the V2I communication, to evaluate the extension in space–time of the safety margin thanks to the early warning on board the vehicle. The tests showed how the specific environment and installation strongly affect the behaviour of vehicle-to-infrastructure communication, as is evident comparing the figures of two roadside stations versus the local characteristics. As a whole, vehicle connectivity with infrastructure in the motorway environment of the Test Site gave good results up to 1,000 m in free line of sight. These values are confirmed by further tests on other roadside units. Concerning latency of messages, 42 ms was measured as an average end-to-end latency among application layers in these preliminary tests.

## References

1. ETSI EN 302 665, 3 (2010) Intelligent transport systems (ITS), Communications Architecture
2. ETSI TS 102 637-2 (2011) Intelligent transport systems (ITS). Vehicular communications, Basic set of applications, Part 2: specification of cooperative awareness basic service
3. ETSI TS 102 637-3, V1.1.1 (2010), Intelligent transport systems (ITS). Vehicular communications, Basic set of applications, Part 3: specifications of decentralized environmental notification basic service
4. Karagiannis G, Altintas O, Ekici E, Heijenk G, Jarupan B, Lin K, Weil T (2011) Vehicular networking: a survey and tutorial on requirements, architectures, challenges, standards and solutions, communications surveys and tutorials, IEEE, vol 13. pp 584–616
5. Papadimitratos P, de La Fortelle A, Evenssen K, Brignolo R, Cosenza S (2009) Vehicular communication systems: enabling technologies, applications, and future outlook on intelligent transportation. IEEE Commun Mag 47(11):84–95
6. Brakemeier A (2012) Cooperative ITS—Standardization improving road safety and traffic efficiency, European wireless 2012, Poznan, Poland, 18–20 Apr 2012
7. Festag A, Le L, Goleva M (2011) Field operational tests for cooperative systems: a tussle between research, standardization and deployment, VANET'1. In: Proceedings of the Eighth ACM international workshop on vehicular inter-networking, pp 73–78
8. DRIVE C2X EU Project, coordinated by DAIMLER, http://www.drive-c2x.eu/
9. Mäkinen T, Schulze M, Krajzewicz D, Gaugel T, Koskinen S, DRIVE C2X methodology framework, DRIVE C2X public Deliverable D22.1, on http://www.drive-c2x.eu/
10. Stahlmann R, Festag A, Tomatis A, Radusch I, Fischer F (2011) Starting European field tests for Car-2-X communication: the DRIVE C2X framework. In: Proceedings of 18th its world congress and exhibition, Orlando, USA, 2011
11. COMeSafety2 EU Project, coordinated by BMW, http://www.comesafety.org/
12. Kosch T, Kulp I, Bechler M, Strassberger M, Weyl B (2009) Communication architecture for cooperative systems in Europe. IEEE Commun Mag 47(5):116–125
13. Rämä P, Innamaa S, Brignolo R (2012) Creating research hypotheses and defining measurements for FOT evaluation. In: 19th ITS world congress, Vienna, Austria, 2012

14. Gozalvez J, Sepulcre M, Bauza R (2012) IEEE 802.11p vehicle to infrastructure communications in urban environments. IEEE Commun Mag 50(5):176–183
15. Sepulcre M, Gozalvez J, Hernandez J (2013) Cooperative vehicle-to-vehicle active safety testing under challenging conditions. Trans Res C 26:233–255
16. ETSI TR 102 638, V1.1.1 (2006–2009) Intelligent transport systems (ITS), Vehicular communications, Basic set of applications; definitions

# GuideWeb: Information Acquisition Analysis in a Conceptually Infrastructure-Free Vehicle Navigation System

**Bernd X. Weis**

**Abstract** GuideWeb is a support network for vehicle navigation and guidance. It is constituted by the cooperation of a multitude of autonomous MapSynthesiser located in the vehicles. A MapSynthesiser, being the autonomous constitutional core of GuideWeb, receives over radio communication traffic flow information via information-enhanced maps (called map syntheses) from other GuideWeb participants' MapSynthesiser. It creates from received map syntheses and the information of its own travel route a new map synthesis, which is then broadcasted. MapSynthesiser provides timely and accurate information on traffic flow, density and trend on how traffic will develop as well as traversability everywhere within a radius of approx. 100 km to a navigation system for driver assistance. MapSynthesiser cooperation is based on short range radio communication e.g. Wireless Local Area Network (WLAN) according to IEEE 802.11 standards. Further, the information acquisition analysis shows that a MapSynthesiser density of only 2 % in a medium traffic load class enables the acquisition of 500 km of route information in approx. 15 min.

**Keywords** Vehicle navigation · Privacy · Ad-hoc-networks · Autonomous Systems · Data acquisition · Dynamic systems

## 1 Introduction and Background

There is an increasing demand for individual traffic and route information and navigation. The existing systems are typically based either on elaborate infrastructures for traffic measurement or on traffic patterns derived e.g. from individual mobile phone movements. Both concepts have their drawbacks.

B. X. Weis (✉)
BlackForestLightning, Haeberlinstr. 29b, Stuttgart 70563, Germany
e-mail: Bernd.Weis@BlackForestLightning.de

J. Fischer-Wolfarth and G. Meyer (eds.), *Advanced Microsystems for Automotive Applications 2013*, Lecture Notes in Mobility, DOI: 10.1007/978-3-319-00476-1_15,

The first one requires substantial investment into setting up the infrastructure; its maintenance incurs cost and should result in some profits. Therefore, this service for the end user can't be free of charge. Furthermore, the measurement infrastructure is typically only available on motorways.

The second one collects data by tracking individual user-movements e.g. mobile phones. By processing this data and matching it to maps allows the derivation of traffic flow estimations. This process can't ensure privacy as patterns of movement of individuals are recorded. The potential misuse of this data mandates a rigid supervision of the process by independent auditors.

v2v communication allows a variety of applications e.g. communicating to other vehicles dangerous situations [1–3] and/or mobile Internet [4, 5] with their corresponding issues of security [6] etc. A German study [7] indicates that v2v communication networking will not be in place before 2020.

This sets the scene for the concept presented here. The requirements derived from above are:

- to create most up-to-date traffic flow information and navigation support everywhere—city, over-land or motorway (ubiquity),
- to offer most up-to-date information on temporarily traversable/non-traversable routes not provided by the maps of a navigation system,
- to ensure privacy and data security,
- to avoid recurrent service charges.

The implications of these requirements are manifold. The most important one is: no investments into infrastructure and no cost for system maintenance. The direct implication is that no centralized processing, evaluation or distribution unit must be required for the functioning of the process.

The concept GuideWeb presented in the following fulfils these requirements [8–10]. GuideWeb is a systemic process, in which a multitude of autonomous participants are involved. A similar concept has been presented in [11].

This paper describes briefly the functioning and implementation of GuideWeb. First the concept of GuideWeb and the autonomous MapSynthesiser as its constituting element is introduced. The next two sections present the MapSynthesiser and its map processing capabilities. Simulation results that quantify penetration requirements and a conclusion complete the paper.

## 2 GuideWeb and MapSynthesiser

The principle of GuideWeb is intriguingly simple—one gets information, processes it, uses it and distributes the processed information. Information is freely offered and everyone can take it (give and take). GuideWeb is constituted by a multitude of participants where the participants continuously broadcast their information and knowledge about current traffic and environment in the form of map syntheses. A map synthesis is a data compressed form of the synthesized map

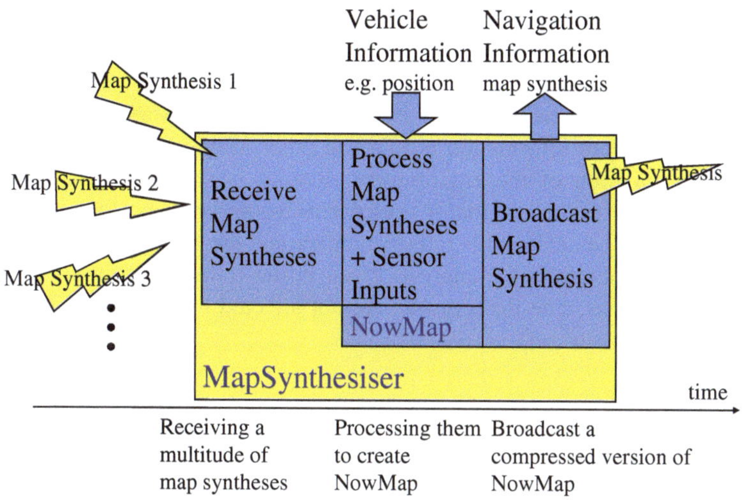


**Fig. 1** Functional diagram of the MapSynthesiser

(NowMap) derived from the aggregation of all routes GuideWeb participants have traversed including averaged speed and traffic density information of route segments. Thus, the individual map synthesis is built up from every participant's best knowledge of traffic flow and environment. In order to facilitate this process each participant is equipped with a device called MapSynthesiser of which the functional diagram is shown in Fig. 1.

The MapSynthesiser is typically a software application (executing together with a navigation application), which autonomously enables the functionalities of the GuideWeb only requiring connectivity from the communication platform and some basic vehicle information. MapSynthesiser provides timely and accurate information on traffic flow and density as well as traversability of roads everywhere within a radius of approx. 100 km. This information can be further processed and displayed to the driver e.g. by a navigation system. The MapSynthesiser generates the NowMap, the representation of the current information and knowledge about the environment, from which the broadcasted map synthesis is derived.

Conceptually GuideWeb is self-organizing and self-contained and independent of any infrastructure. Functionally GuideWeb rests entirely on the cooperation of the multitude of autonomous MapSynthesiser nodes.

## 3 Analysis of Information Acquisition in MapSynthesiser

The following analysis discusses how quantitatively MapSynthesiser acquires route information. Two cases will be considered.

The first case is based on the scenario that all MapSynthesiser in a specific area start collecting information at the same time, i.e. all MapSynthesiser have no information about the environment.

The second case is based on the scenario that only the MapSynthesiser under consideration has no information at all. The other MapSynthesiser have already acquired information of the environment. Reality is somewhere in between—however it is safe to assume that it is closer to the second case than to the first.

The system to be analyzed is rather complex and the full stochastic analysis is rather tedious. Therefore, the analysis given in the following restricts itself to a mean value analysis. The main results are not affected by this limitation.

## 3.1 Definitions and Notations

When describing traffic situations typically the following two parameters are stated:

Traffic flow (or flux) Q Number of vehicles per time unit
Traffic density D Number of vehicles per length unit

Traffic flow and traffic density relate through the average vehicle velocity, i.e. $v_{av} = \frac{dQ}{dD}$ or in the steady state $v_{av\ \mathrm{Steady\ State}} = \frac{Q_{av}}{D_{av}}$

The following table classifies traffic density in traffic load classes with Traffic density = Number of Vehicles per km.

| Traffic density | Traffic load class |
|---|---|
| 0–16 | Low |
| 16–23 | Medium |
| 23–32 | High |
| 32–45 | Very high |
| 45– | Overload |

It is called "encounter" when in traffic two vehicles both equipped with a device executing MapSynthesiser exchange route information. For further explanation of the notations see Fig. 2.

$t_{En}$ Point in time of encounter n

$\tau_n$ Time period between encounters n and n + 1

$K_{in}(t)$ Route information in km stored in the map of MapSynthesiser i after encounter n at time t

$S_i(\tau_n)$ Distance in km driven of MapSynthesiser i between encounters n and n + 1

$K_{max}$ Maximum route information in km that can be stored

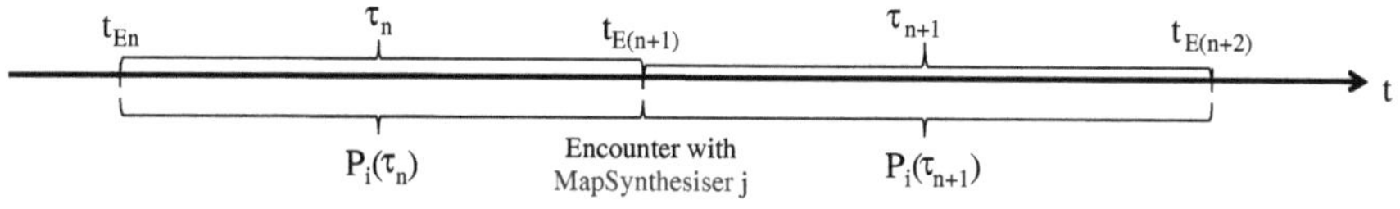


**Fig. 2** Definition of notations

$P_i(\tau_n)$ Maximum route information in km that can be processed in time $\tau_n$ between encounters n and n + 1 (determined by processing power of MapSynthesiser computing platform)

$\alpha$ Parameter to account for maps not completely matching in their respective map areas.

## *3.2 Route Information in MapSynthesiser*

The route information available in MapSynthesiser i at time t is the information at the last encounter $K_{in}(t_{En})$ and the information from its own travel experience $S_i(t - t_{En})$ adjusted by the information load already in place, which accounts for multiple traversed route segments, i.e.

Information load adjustment: $\left(1 - \frac{K_{in}(t_{En})}{K_{max}}\right)$.

During the encounter n + 1 the route information $K_{jm}(t_{En})$ received in encounter n from MapSynthesiser j is processed into the route information $K_{in}(t_{En})$ creating $K_{i(n+1)}(t_{E(n+1)})$. However, since the processing power of the platform on which MapSynthesiser is executing is limited only that information that is actually processed contributes to $K_{i(n+1)}(t_{E(n+1)})$. This accounts for the minimization term. Further, since the some of the route information may not be of relevance the minimization term is adjusted by the parameter $\alpha$ as defined above. Hence,

$$\begin{aligned} K_{i(n+1)}(t_{E(n+1)}) &= K_{in}(t_{En}) + (S_i(\tau_n) \\ &\quad + \alpha \, \mathrm{Min}\{K_{jm}(t_{En}), \; P_i(\tau_n)\})\left(1 - \frac{K_{in}(t_{En})}{K_{max}}\right). \end{aligned}$$

**Case 1: GuideWeb in formation state**

For a more detailed analysis of this formula assume that all MapSynthesiser start at the same time t = 0, i.e. GuideWeb is in the process of being formed. Then, stochastically there is no difference between $K_{in}$ and $K_{jm}$ with respect to the size of the route information available, thus

$$\begin{aligned} K_{i(n+1)}(t_{E(n+1)}) &= K_{in}(t_{En}) + (S_i(\tau_n) \\ &\quad + \alpha \, \mathrm{Min}\{K_{in}(t_{En}), \; P_i(\tau_n)\})\left(1 - \frac{K_{in}(t_{En})}{K_{max}}\right). \end{aligned}$$

**Case 2: GuideWeb in steady state**

In this case the MapSynthesiser enters the system when it is already in steady state. The information acquisition is only limited by the processing power of the computing platform, thus

$$K_{i(n+1)}(t_{E(n+1)}) = K_{in}(t_{En}) + (S_i(\tau_n) + \alpha\, P_i(\tau_n))\left(1 - \frac{K_{in}(t_{En})}{K_{max}}\right).$$

## 3.3 *Average Case Analysis*

In order to evaluate this formula the average case is analyzed. For that assume that the encounters in which map information is exchanged between vehicles in traffic is with respect to time equidistant, i.e. $\tau_n = \tau$ for all n. Thus, $t_{E(n+1)} = t_{En} + \tau$.

The size of the map information available in MapSynthesiser i at the (n + 1)-st encounter is given by

**Case 1:**

$$K_{i(n+1)}(t_{E(n+1)}) = K_{in}(t_{En}) + (S_i(\tau) + \alpha\, \mathrm{Min}\{K_{in}(t_{En}),\ P_i(\tau)\})\left(1 - \frac{K_{in}(t_{En})}{K_{max}}\right).$$

**Case 2:**

$$K_{i(n+1)}(t_{E(n+1)}) = K_{in}(t_{En}) + (S_i(\tau) + \alpha\, P_i(\tau))\left(1 - \frac{K_{in}(t_{En})}{K_{max}}\right).$$

## 3.4 *Average Case Results*

For the following it is assumed that the route information corresponding to 3,000 km road can be handled in MapSynthesiser which corresponds to a payload size for transmission of approx. 100 kB. When the transmission is based on WLAN using the UDP protocol the payload is limited to 1,522 Byte. Thus, the information to be transmitted is broken down in suitable chunks of payload whose sizes don't exceed 1,522 Byte resulting in approx. 75 messages each representing approx. 40 km of road information. A received message ($\sim$1,350 Byte payload) needs less than 1 s processing time. This processing time has been determined on an ASUS 636 N with Windows Mobile 5. But MapSynthesiser is not the only application executing on the

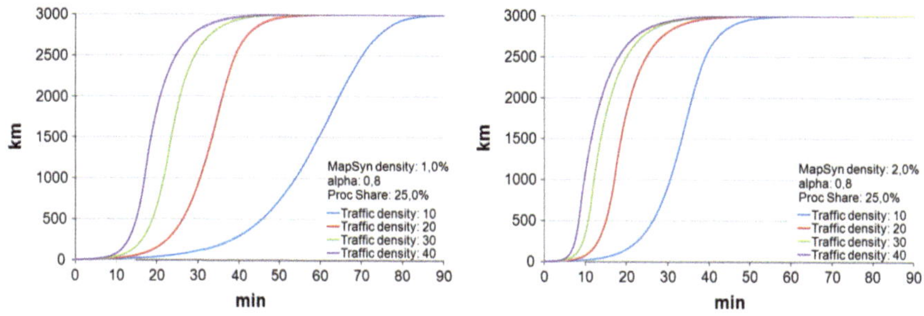


**Fig. 3** Road information in MapSynthesiser in km vs. acquisition time with traffic density as varying parameter

platform, at least a navigation application runs in parallel; therefore only a certain percentage is available for MapSynthesiser execution.

**In summary**

| Maximum length of 1 map ($K_{max}$) | | 3,000 km (<100 kByte) |
|---|---|---|
| 75 messages (per 1 map): | ⇒ | 40 km per message |
| Processing time for 1 message = 1 s | ⇒ | $\mu_{max} = 60$/min |
| Processing time for MapSynthesiser 25 % | ⇒ | $\mu = 15$/min |
| Traffic density: 30 vehicle/km @ 50 km/h (both directions) | ⇒ | Traffic flow: 3,000 vehicle/h |

## Case 1: Evaluation With Respect to Traffic Density

Figure 3 presents the results for the following parameter settings:

| Parameter | Value | Traffic load class |
|---|---|---|
| $\alpha$ = | 0.8 | |
| Traffic density = | 10–40 veh./km | From low to very high |
| Traffic flow = | 1,000–4,000 veh./h | |
| MapSynthesiser density = | 1 and 2 %, resp. | |
| Share of processing power = | 25 % | |

Acquisition time in minutes of 500 km road information

| | Traffic density | | | |
|---|---|---|---|---|
| MapSynthesiser density (%) | 10 | 20 | 30 | 40 |
| 1.0 | 46 | 26 | 21 | 15 |
| 2.0 | 26 | 15 | 11 | 9 |

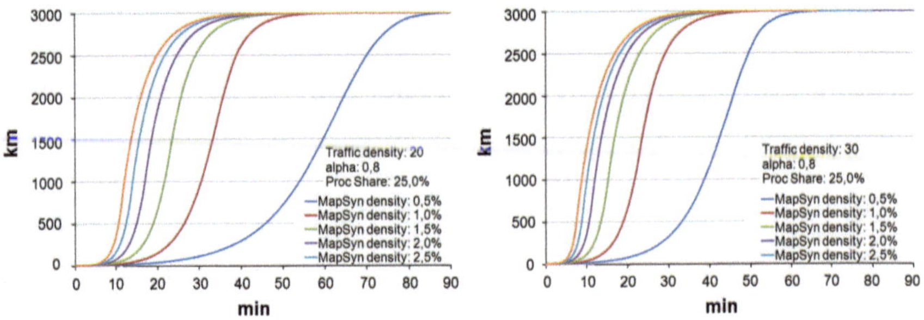


**Fig. 4** Road information in MapSynthesiser in km vs. acquisition time with as MapSynthesiser density varying parameter

## Case 1: Evaluation With Respect to MapSynthesiser Density

Figure 4 presents the results for the following parameter settings:

| Parameter | Value | Traffic load class |
|---|---|---|
| $\alpha =$ | 0.8 | |
| Traffic density = | 20, 30 vehicle/km | Medium and high |
| Traffic flow = | 2,000, 3,000 vehicle/h | |
| MapSynthesiser Density = | 0.5–3.0 % | |
| Share of processing power = | 25 % | |

Acquisition time in minutes of 500 km road information

| | MapSynthesiser density | | | | | |
|---|---|---|---|---|---|---|
| Traffic density | 0.5 % | 1.0 % | 1.5 % | 2.0 % | 2.5 % | 3.0 % |
| 20 | 47 | 26 | 19 | 15 | 13 | 11 |
| 30 | 33 | 19 | 14 | 11 | 9 | 8 |

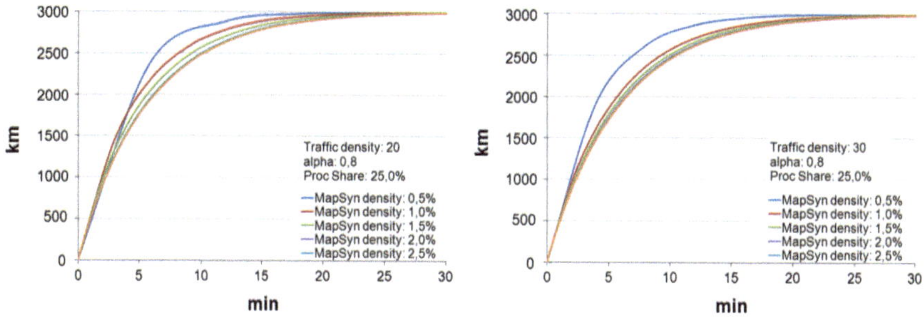


**Fig. 5** Road information in MapSynthesiser in km vs. acquisition time with MapSynthesiser density as varying parameter entering GuideWeb in steady state

### Case 1: Evaluation With Respect to Processing Power Share

The acquisition times in min of 500 km road information for the following parameter settings

| Parameter | Value | Traffic load class |
|---|---|---|
| $\alpha$ = | 0.8 | |
| Traffic density = | 20, 30 vehicle/km | Medium and high |
| Traffic flow = | 2000, 3000 vehicle/h | |
| MapSynthesiser Density = | 2,0 % | |
| Share of processing power = | 25–10 % | |

show only minor variances with respect to the share of processing power.

| | Processing power share | | | |
|---|---|---|---|---|
| Traffic density | 25 % | 20 % | 15 % | 10 % |
| 20 | 15 | 15 | 15 | 15 |
| 30 | 11 | 11 | 11 | 11 |

### Case 2: Evaluation With Respect to MapSynthesiser Density

In this case the MapSynthesiser enters the system when it is already in steady state. The processing of the data exchanged is then only limited by the processing power of the computing platform.

Figure 5 presents the results for the following parameter settings:

| Parameter | Value | Traffic load class |
|---|---|---|
| $\alpha$ = | 0.8 | |
| Traffic density = | 20, 30 vehicle/km | Medium and high |
| Traffic flow = | 2000, 3000 vehicle/h | |
| MapSynthesiser density = | 0.5–3.0 % | |
| Share of processing power = | 25 % | |

The time to acquire 500 km road information is approx. 1 min in both cases.

### 3.5 Summary

**Case 1: A MapSynthesiser enters GuideWeb in formation state**

A MapSynthesiser enters a GuideWeb that is just forming, i.e. the road information exchanged in encounters is likely to be of the same size. From the results presented above for a medium traffic load class and a MapSynthesiser density of 2 % the acquisition time for 500 km road information is **approx. 15 min**.

**Case 2: A MapSynthesiser enters GuideWeb in already steady state**

A MapSynthesiser enters a GuideWeb that is already in a steady state, i.e. the road information received is of maximum size. In this case the acquisition time of 500 km road information is **approx. 1 min**.

## 4 Conclusion

In this paper the concept GuideWeb for vehicle navigation support based on v2v communica-tion has been presented. It has been shown that GuideWeb is a very suitable candidate for a commercially viable introduction of v2v communication imposing a minimum requirement on networking. By using a broadcast communication concept, important difficulties in v2v communi-cation are overcome or circumvented, and due to its coordinate-based exchange format it is independent of map suppliers and easily integrated in any driver assistance system. GuideWeb finds its application window beginning now until all the challenges of the v2v networking capabilities are resolved.

Further, the insights derived from GuideWeb deployment and its behavior allows learning about v2v communication system performance. Keeping historical data of the route segment attributes adds a new level of complexity—the systemic memory. However, the advantages of historical data (detection of implausible attributes or malicious users, computing trends...) trade off favorably with the implications of the systemic memory as shown by simulation. Furthermore, by concept the penetration level required for GuideWeb to function is much lower than in other v2v communication systems. The processing methodology allows that information about specific vehicles can neither be extracted nor traced, i.e. that privacy is ensured.

## References


1. Schnaufer S et al (2006) Vehicular ad-hoc networks: single-hop broadcast is not enough. In: Proceedings of 3rd international workshop on intelligent transportation (WIT), Hamburg, pp 49–54
2. Torrent-Moreno M (2007) Inter-vehicle communications: assessing information dissemination under safety constraints. In: 4th annual IEEE/IFIP conference on wireless on demand network systems and services (WONS), Obergurgl, Austria
3. Baldessari R et al (2007) NEMO meets VANET: a deployability analysis of network mobility in vehicular communication. In: Proceedings of 7th international conference on ITS telecommunications (ITST 2007), Sophia Antipolis, France, pp 375–380
4. Bechler M et al (2003) Mobile internet access in FleetNet. In: Proceedings of KiVS 2003, Leipzig, Germany
5. Baldessari R et al (2006) Flexible connectivity management in vehicular communication networks. In: Proceedings of 3rd international work-shop on intelligent transportation (WIT), Hamburg, Germany, pp 211–216
6. Harsch C et al (2008) Secure position-based routing for VANETs. In: Proceedings of IEEE 66th vehicular technology conference (VTC Fall), Baltimore, MD
7. Zukunft und Zukunftsfähigkeit der Informations- und Kommunikationstechnologien und Medien, Internationale Delphi-Studie 2030, Münchner Kreis e.V., EICT GmbH, Deutsche Telekom AG, TNS Infratest GmbH, 2009
8. Weis B, Sandweg A (2010) GuideWeb: a new paradigm for navigation support based on v2v communication. In: Meyer G (ed.) Advanced Microsystems in Automotive Applications, Springer, Berlin
9. Weis B, Sandweg A (2010) GuideWeb: an introductory solution for V2V communication. In: Proceedings of ITS world congress, Busan, Korea, SP23-1
10. Weis B, Sandweg A (2012) GuideWeb: a conceptually infrastructure-free vehicle navigation system. IET Intell Transp Syst 6(2): 139–144
11. Wischhof L et al (2003) SOTIS—a self-organizing traffic information system. In: Proceedings of 57th IEEE semiannual vehicular technology conference VTC 2003-Spring, Jeju, South Korea

memory. However, the advantages of historical data detection of [illegible] the attributes of malicious users, computing trends [illegible] trade off favorably with the implications of the system memory as shown by simulation. Furthermore, by concept, the penetration level required for GuideWeb to function is much lower [illegible]

[illegible]

## References

[illegible]

# Mobile Probe for Green Traffic Management in the INTEGREEN Project Considering Both Traffic and Environmental Information

**Reinhard Kloibhofer, Franco Fresolone and Roberto Cavaliere**

**Abstract** The main objective of the EU LIFE + project INTEGREEN is to introduce a demonstrative system for the municipal mobility management centre of the city of Bolzano that aims at providing the local public authorities with distributed correlated traffic and environmental information for the adoption of eco-friendly real-time traffic management policies. To achieve this goal a new innovative mobile probe with both traffic and environmental monitoring units will be developed. The mobile probe transmits the sensor data to the traffic management centre and receives information for the driver to offer him/her the possibility to more efficiently optimize his/her trip while en-route. The proposed system architecture is designed in a way, such that it can be easily exploited for other similar urban areas.

**Keywords** Traffic policies · Greener driving · Eco-friendly · Stakeholders · Commuters · Quantitative impact · FRAME Architecture · Wireless communications · CAN-Bus · Stop-and-go · Test vehicle · Pre-trip · On-trip · Urban areas · Acceleration sensor · Pollution sensor · Traffic monitoring unit · Environmental monitoring unit · Traffic management centre · Pollutants · ppb concentration

R. Kloibhofer (✉) · F. Fresolone
Safety and Security Department, AIT Austrian Institute of Technology GmbH, Donau-City-Straße 1, Vienna 1220, Austria
e-mail: Reinhard.Kloibhofer@ait.ac.at

F. Fresolone
e-mail: Franco.Fresolone@ait.ac.at

R. Cavaliere
TIS Innovation PARK, Free Software and Open Technologies, Via Siemens 19, Bolzano 39100, Italy
e-mail: Roberto.Cavaliere@tis.bz.it

J. Fischer-Wolfarth and G. Meyer (eds.), *Advanced Microsystems for Automotive Applications 2013*, Lecture Notes in Mobility, DOI: 10.1007/978-3-319-00476-1_16,

# 1 Introduction and Overview

The project INTEGREEN (INTEgration of traffic and environmental data for improving GREEN policies in the city of Bolzano) is an EU LIFE + project to improve traffic and environmental issuers with new IT technologies. The main objective is the design and implementation of a demonstrative system for a municipal mobility management centre that aims to provide the public authorities with distributed correlated traffic and environmental information for the adoption of eco-friendly traffic management policies.

To achieve this objective the INTEGREEN system will integrate traffic and environmental data from driving vehicles as well as from static traffic and environmental sensors. Moreover, on the base of the validated INTEGREEN framework, the project aims at studying and demonstrating the quantitative impact of specific traffic policies on the urban environment, including effects of novel strategies. This study will include a quantitative assessment of the importance of providing processed information to the whole population through the use of different communication means, including existing Variable Message Signs (VMS) and through new web-based tools.

The demonstration of the system will take place in the city of Bolzano in South Tyrol. South Tyrol is an Italian bordering region in the north of Italy fully collocated in the alpine chain. The main centre of the region is Bolzano, a medium city of about 100,000 inhabitants, which hosts about one-fifth of the entire South Tyrolean population. Bolzano is located in a bowl at the confluence of three natural valleys (Adige valley, Isarco valley and Sarentino valley) and represents the main connection point of the local arterial roads [1, 2].

Mobility in the area of Bolzano is affected by several complementary aspects:

- Being Bolzano the main centre of the Province, where most of the central Public Administration services, private companies, schools (including the university) are located, the town is the daily destination for a large amount of commuters who every day leave their residence placed in a remote town in one of the surrounding valleys in order to reach their working or studying place.
- Bolzano is a node connecting the Province to the Corridor 1 (Berlin-Palermo motorway line), and therefore acts as a transit point for the relevant people and goods traffic flows which every day travel along the Brenner Corridor through the A22 motorway.
- The town is a well-known Point-of-Interest for mass tourism, due to the beauty of the surrounding natural landscapes (e.g. Dolomites), skiing facilities, summer resorts as well as local events such as the Christmas market. Consequently, traffic peaks of relevant intensity in conjunction to particular circumstances periodically appear.

In order to achieve the objectives of the project, a set of technological actions and a set of social actions are planned.

From a technological point of view, the project consists in analysing, designing, implementing and testing new functional modules to be integrated in the existing mobility centre of the city of Bolzano. The novel components that will be introduced during the project are mainly a mobile probe which is equipped with a traffic monitoring unit and an environmental monitoring unit. Also the number of the stationary traffic sensors (mainly video cameras and vehicle counting inductive loops) which currently is the main input for the traffic management centre will be enlarged. Neither environmental stations nor meteorological stations are currently available to the mobility centre for the gathering of environmental (and weather) information and for the correlation of such data with traffic conditions.

From a social point of view, the objective is to perform an active involvement of the local mobility participants (including tourists), for an improvement of drivers' behaviour and habits in relation to the environment (optimisation of displacements, eco-driving, co-modality, etc.). Secondly, local stakeholders will be involved to promote a more integrated approach in the management of the urban environment and for the exploitation of the INTEGREEN system. Lastly, EU actors will also be involved to disseminate the active sharing of the mission of this initiative through best-practice exchange and active involvement in EU networks.

## 2 Architecture, the FRAME Concept

In the INTEGREEN project the objective is to combine data from mobile and roadside units in order to proactively manage and improve the traffic and air-quality of the city. The introduction of this new subsystem in the existing traffic management system is very challenging task.

For creating Intelligent Transport Systems (ITS) the EC funds projects that help to create better ITS infrastructure. Following the recommendation of the High Level Group on Telematics, and a resolution of the Transport Council, the European ITS Framework Architecture, colloquially known as "The FRAME Architecture", was introduced by the EC funded project KAREN (1998–2000).

It has been maintained and enhanced continuously since then—with cooperative systems being added by the current project E-FRAME (2008–2011) [3] (Fig 1).

From the INTEGREEN project "Stakeholders Aspiration" which are the high level requirements the formal "User Needs" will be extracted. After this FRAME offers a tool to generate the Functional View which are the functions or processes needed in order to satisfy the User Needs. Here are also considered the data stores and data exchange between the functions and processes.

The next step is the development of the physical architecture with the FRAME-tool. This shows where the processes are physically located.

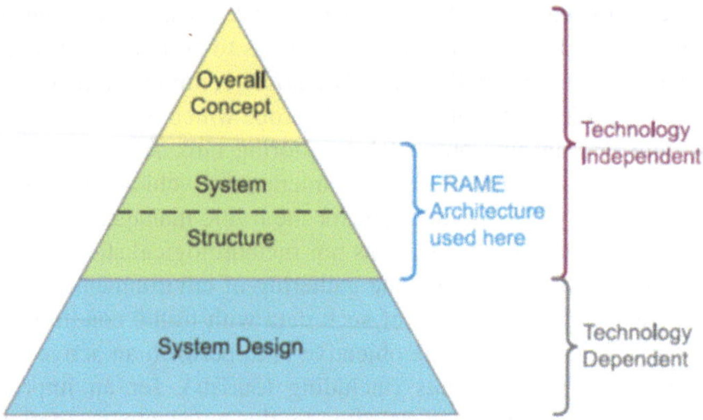


**Fig. 1** ITS architecture showing the FRAME components [3]

The last step is the definition of the communication Layer. Here all data rates between the physical locations will be defined. This part is not included in the FRAME-tool (Fig 2).

The result is a standardized view of the whole architecture. The tool helps with a the FRAME databank not to forget some functions and data and it gives a good

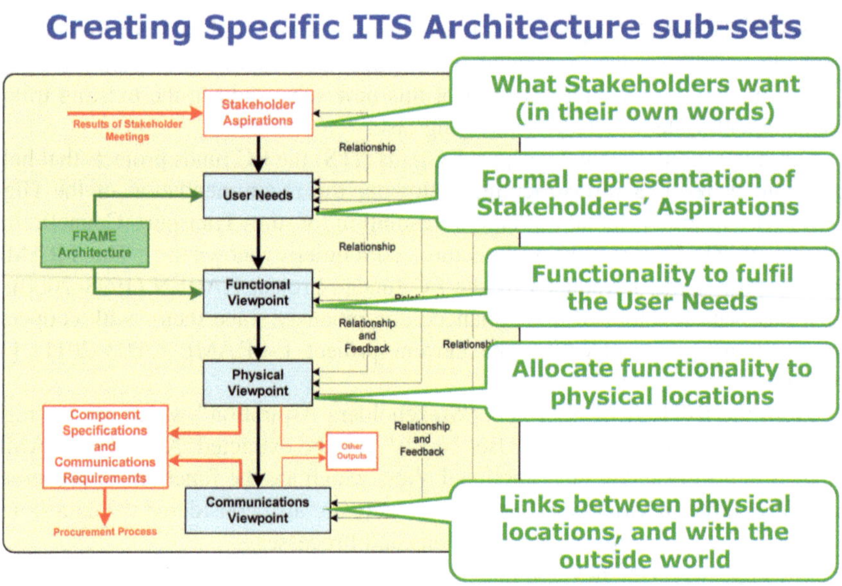


**Fig. 2** View of ITS architecture sub-set creation [3]

basis for expansions or changes in the future. With the FRAME approach it is also much easier to adapt the INTEGREEN reference architecture for the specific needs of other locations or cities.

## 3 Use Cases

The use cases of the INTEGREEN system are various and they cover the thematic field of traffic and environmental planning, navigation, traffic control, environmental control and evaluation of traffic management policies.

In the context of traffic issues we can distinguish three categories of requirements:

- Pre-trip use cases
- On-trip use cases
- Use cases for traffic management.

In the pre-trip use cases a local traveller can get information about the traffic situation in the area where he will be driving with the vehicle. This information will be made available on public channels like a web page or a smartphone application. He gets also information about the environmental state in the city. If the driver will know there are areas with higher pollution, then he can choose a different route. Perhaps it is possible to do the trip at a different time where the status will be better. Not only private users can profit from the system but also local fleet manager can plan their trips in a way to avoid peak periods and zones with high traffic and pollution load.

In the on-trip use cases if the driver has the mobile INTEGREEN system installed on his vehicle he will get information about the actual state of his route with messages on a small display and he will be able to choose a different route with less pollution or less traffic. Also if the driver doesn't have the INTEGREEN mobile system on board he will still get actual and dynamic information through the Variable Message Sign (VMS) which are controlled from the traffic management centre.

The traffic management centre has also different use cases. It gathers all the traffic and environmental data of the different mobile probes moving in the city as well as the data of the stationary sensors and evaluates carefully the status in the city. If there are critical situations (traffic and or pollution) it can react in real time and manage the traffic according to predefined policies. Some possibilities are short-time and some are long-time, such as:

- giving information on a dedicated web page
- providing messages on the VMS
- opening or closing lanes
- changing driving direction of one-way streets
- changing the cycles of the traffic lights
- changing the speed limits in specific areas of the city.

# 4 The Car as a Sensor

The mobile probe of the INTEGREEN system should be able to be integrated into commercial vehicles like city buses, delivery vans or Taxis as well as in private cars which are traveling frequently in the city.

Modern vehicles are already equipped with a large number of sensors for the safety and comfort of the driver. Therefore it makes sense to use these sensors also for the INTEGREEN mobile system.

The sensors of interest for the mobile probe are in particular the vehicle speed sensors and the acceleration sensors. Additionally also the climatic sensors for the external temperature and humidity are of interest. The sensor values can easily be read from the CAN-Bus in commercial vehicles.

# 5 The Mobile Probe

The mobile probe is a device which can be mounted in private cars or in commercial vehicles. It is composed of different modules. So it is very flexible for upgrades (i.e. new sensors) and or functional extensions.

In Fig. 3 the main units are illustrated:

- Telematic Unit: control and data processing. Directly connected there are a communication unit for the data-exchange with the traffic management centre and an Human Machine Interface (HMI) for the vehicle driver
- Environmental Monitoring Unit: sensor unit for environmental measurements
- Traffic Monitoring Unit: sensor unit for traffic data collection.

## *5.1 Traffic Monitoring Unit*

The Traffic Monitoring Unit is equipped with kinematic sensors and a CAN-Bus Interface for traffic state estimation.

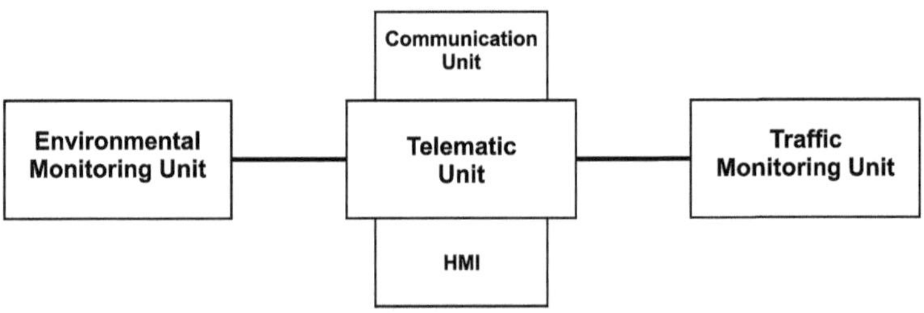


**Fig. 3** Architecture of the on-board system

To estimate traffic status from the own driving vehicle is very challenging task, and different projects have addressed this argument. The most important inputs are the vehicle speed and the characteristics of acceleration and deceleration in the direction of driving. When this information together with a GPS sensor data and the local map data is processed, a typical local traffic status can be derived (i.e. fluid traffic, stop-and-go).

Additionally, the actuation of the brake-pedal can be monitored from the CAN-Bus in order to further improve the derived results. If no CAN-Bus access is possible the speed and acceleration of the vehicle can be derived with separate acceleration sensors.

## 5.2 Environmental Monitoring Unit

The Environmental Monitoring Unit (EMU) will have the function of measuring the air quality. To perform accurate measurements with moving vehicles to determine the air quality and at an affordable price is a very challenging task. Short sample periods, high sensitivity (ppb concentration) and small dimensions are the main requirements,

The parameters of interest for traffic environment are the following gases and pollutants:

- $NO_x$ (nitrogen oxides)
- $O_3$ (ozone)
- CO (carbon monoxide)
- $PM_{10}$ (particulate matter).

In addition also the climatic status is of interest because the environmental sensors typically vary with these parameters:

- temperature
- humidity.

In INTEGREEN the goal is to develop an EMU based on existing sensors rather than designing new ad-hoc sensors.

For measuring the most important pollutants a very large number of off-the-shelf sensors and measurement devices are available on the market. The characteristics of these devices differ strongly in accuracy, linearity, measurement range, stability, cross sensibility to other pollutants, measurement update rate, power-on time, dimension, weight and price.

The most common principles for electrical measurement of gas concentrations are:

- Chemo resistive, Metal Oxide Semiconductors (MOX)
- Capacitive

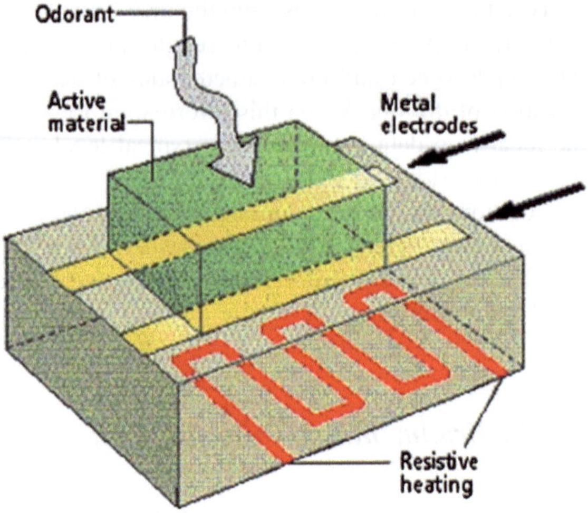


**Fig. 4** View of a MOX-sensor [4]

- Thermic (Pellistor)
- Optical, based on the Beer—Lambert law.

The most accurate sensors can be found in the stationary measurement devices and are typically based on optical gas sensors.

These stations are designed only for stationary mode of operation for high precision measurements and medium to long sample periods. The size and weight are normally large and the price is quite high.

In addition to fixed stations there are also a few measurement devices available with fast sampling time characteristics based on **D**ifferential **U**ltra**V**iolet **A**bsorption **S**pectroscopy (DUVAS) from the start-up company Duvas Technologies Ltd. Although it's a smaller and transportable solution it is still too large for integration in normal vehicles.

In the INTEGREEN project an environmental monitoring unit with physically small and economic sensors will be developed. Therefor the MOX type sensors with a good sensitivity are suitable candidate for INTEGREEN.

The working principle is based on the variation of conductivity in the presence of oxidising and reducing gases. The electron density changes with the adsorption and desorption of oxygen (O, $O_2$). Adsorbed oxygen gives rise to potential barriers at grain boundaries and thus increases the resistance of the sensor surface (Fig 4).

## 5.3 Communication

The communication with the traffic management centre is based on existing wireless communication standards. The mobile probe transmits sensor data to the traffic management centre and receives messages for the driver to be displayed on the HMI.

Therefore a bidirectional communication link is necessary. The following communication techniques can be used:

- GPRS
- UMTS
- W-LAN
- Bluetooth.

As the on-board system is built on a modular architecture, the communication unit can be easily upgraded to meet future ITS communication standards like CALM (Communications Access for Land Mobiles).

## 6 Conclusions

In the project INTEGREEN from the EU LIFE + Programme, an improved sensor system for the traffic management centre of the city of Bolzano will be developed. The novelty and innovation in this project is that static and mobile data sources are combined together in order to achieve a more efficient and environmental-aware traffic management, and all this on a real-time basis. In order to accomplish this, a certain number of vehicles driving frequently in the city will be equipped with a flexible on-board unit. The vehicles will then act as a mobile sensor offering a much wider geographical coverage than fixed monitoring stations. The data is transmitted in real time to the traffic management centre. The driver gets information back about traffic and environmental state on his route and can proactively react by adopting a greener driving. Given these preconditions, the INTEGREEN system paves the way for a near future introduction of cooperative schemes for further increasing the efficiency of how urban traffic is controlled.

## References

1. "Calcolo e valutazione delle emissioni di $CO_2$ e definizione di scenari di riduzione per la città di Bolzano" (2010) A study produced by EURAC research for the Municipality of Bolzano, http://www.comune.bolzano.it/UploadDocs/7528_BolzanoCO2_Report_Ita_100201.pdf
2. "Valutazione dell'impatto sulla qualità dell'aria nella città di Bolzano" (2011) A study produced by CISMA for the Municipality of Bolzano, http://www.comune.bolzano.it/UploadDocs/9152_Qualita_aria2_27_01_2011.pdf
3. FRAME architecture (2011) http://www.frame-online.net
4. Soure: Università degli Studi di Siena http://www.dii.unisi.it/~electron/lab/Ricerca/MOX%20Sensors.htm

# Part III
# Electrified Vehicles

# Multi-Voltage Domain Communication in Electric Vehicles and Consequences for E/E Architectures

**Frank Schade and Steffen Müller**

**Abstract** The electrical and electronic (E/E) architecture of electrical vehicles (EV) consists of at least two voltage domains where vehicle modules are placed in relation to their application. Some communicate across voltage domains. There is a need to protect humans from dangerous voltages and electronics from damage. Limiting the high-voltage domain to 48 V has safety advantages and avoids the need for galvanic isolation but is of limited use for EVs that require higher supply voltages. Even within 48 V systems, unwanted effects may occur when bridging the 12 and 48 V domains. We show exemplary communication architectures and explain effects between voltage domains that may result in damaged electronic sub-systems. EMC effects are investigated in the context of isolation and study results on noise propagation across voltage domains are discussed.

**Keywords** CAN transceiver · Galvanic isolator · Isolator techniques · CAN isolator · 48 V · High-voltage E/E architecture · TJA1052i

## 1 Introduction of Multi-Voltage Domains in E/E Architecture

In electric and hybrid vehicles, a further high-voltage onboard power supply expands the 12 V power supply. Figure 1 shows an exemplary E/E architecture with one motor per wheel. Some control units connect to both voltage domains.

F. Schade · S. Müller (✉)
NXP Semiconductors Germany GmbH, Stresemannallee 101,
22529 Hamburg, Germany
e-mail: st.mueller@nxp.com

F. Schade
e-mail: frank.schade@nxp.com

J. Fischer-Wolfarth and G. Meyer (eds.), *Advanced Microsystems for Automotive Applications 2013*, Lecture Notes in Mobility, DOI: 10.1007/978-3-319-00476-1_17,

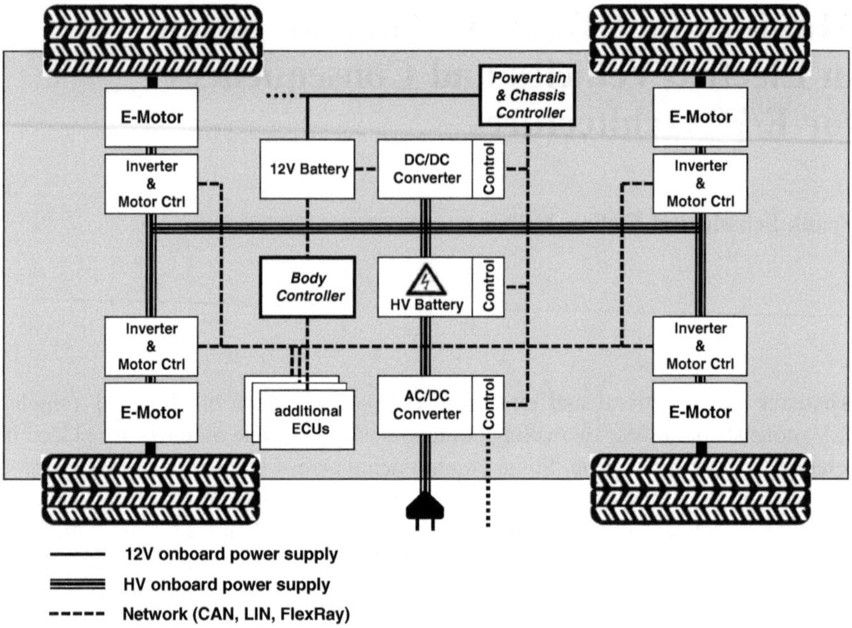


**Fig. 1** Exemplary EV architecture; some components connect to multiple voltage domains

The complexity of cross-voltage domain communication depends on the EV type and number of channels that share control data during operation. For instance, a micro-hybrid with 48 V power supply for a start/stop system requires less cross-domain interaction than the EV that powers its drive with for example 600 V.

The Heating, vent, and air conditioning system exemplifies a vehicle sub-system that is typically located in the high-voltage domain due to its power consumption; in Fig. 1 it relates to the "additional ECUs" box. Control and sensor data exchange with the climate control unit (12 V domain) requires galvanic isolation of the path.

Technical aspects are touch safety and protection of electronics against high voltages. The a.m. micro-hybrid solves touch safety by staying below 60 V (DC), but the destruction of electronics remains a possibility in the event of failure.

## 2 Effects in Cross-Voltage Domain Communication Systems

In general, the power supply voltage greater than 12 or 24 V creates a challenge for communication networks. Most bus transceivers are designed for interferences in 12 or 24 V systems but not above. The high-voltage vehicle supply causes EMC-related interferences. Root causes include ground shift, loss of ground, reverse currents, transient pulses, and voltage surges [2]. In high-voltage systems, the phenomena place significant challenges on bus transceivers and may destroy most of those that are available today. With voltages above 60 V, protection against human touch is an important requirement. Thus, it is vital to shield and isolate voltages that are dangerous for humans in all places that are accessible during vehicle operation and maintenance.

## 2.1 Electrical Networking Effects in Multi-Voltage Architectures

A potential problem may occur in the event of loss of ground of a control unit. As known from 12 V systems, the electronic sub-system of this control unit will drift towards battery voltage if the ground is disconnected. The drift speed towards the battery supply voltage depends on the buffer capacities. For bus transceivers in modules with disconnected ground, bus lines appear to have a negative potential difference up to battery voltage while other control units are properly supplied and keep the bus lines at a normal voltage level.

However, if the ungrounded control unit is supplied with 48 V, the voltage potential compared to the bus potential will be accordingly higher than is known from a supply of 12 V. The ungrounded transceiver (internal resistor network) creates a voltage divider in conjunction with the remaining network of other bus transceivers that still have a ground connection (Fig. 2).

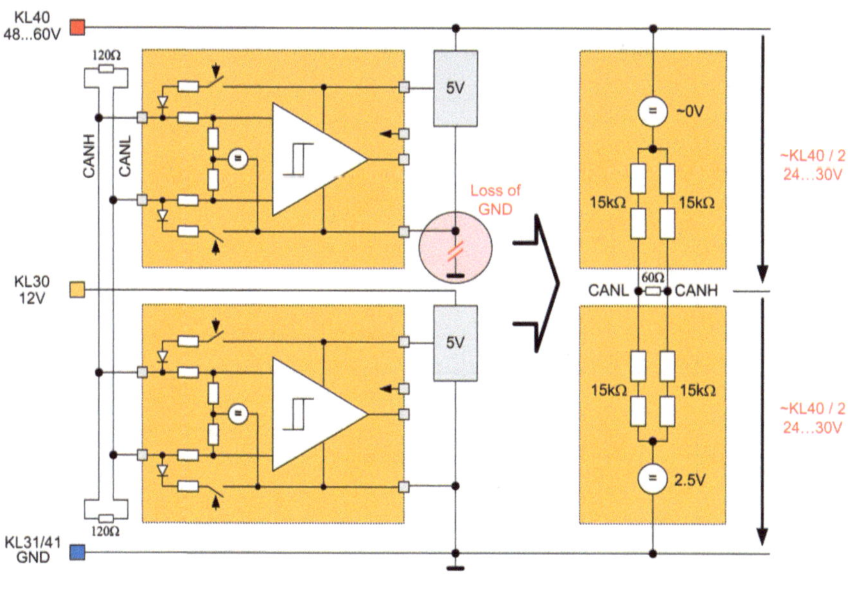


**Fig. 2** Potential situation with CAN transceiver with loss of ground of 48 V control unit

The described voltage divider settles in the recessive bus state. Assuming a CAN bus system, the first active "start of frame" edge of the still supplied network is a timing-critical synchronizing reference. Delaying the first edge will result in a major synchronization deviation between network nodes. In the worst-case scenario, this leads to a stop of communication (Fig. 3). In contrast to a FlexRay bus transceiver, a CAN bus transceiver does not actively drive the so-called recessive bits.

This will result in each recessive bit being repeatedly exposed to the voltage divider by the non-supplied network node and influences the subsequent dominant bus timing. This effect also exists in 12 V systems with reduced potential shift. A loss of ground (GND) impacts the 'bit timing' (signal RxD), as shown in Fig. 3.

A further aspect is the possible damage of the bus transceiver caused by a "loss of ground", e.g. in 48 V systems. That may cause a voltage between the faulty connected module and the remaining network of up to 48 V. Consequently, bus transceivers are supposed to have bus voltage robustness of 48 V or galvanic isolation with separated ground supplies for both voltage domains.

## 2.2 *Electrical Effects in Control Units in Multi-Voltage Architectures*

Galvanic isolation is beneficial for 12/48 V systems with separated ground connections (KL31, KL41 in Fig. 4) when the electronic control unit uses both voltage levels on one PCB. As both grounds are already connected at vehicle level, ground pins connected a second time at a different location create ground loops.

When ground connection is lost on one side of the control unit, uncontrolled cross-currents may occur that may cause unforeseen damage. Galvanic isolation is an ideal

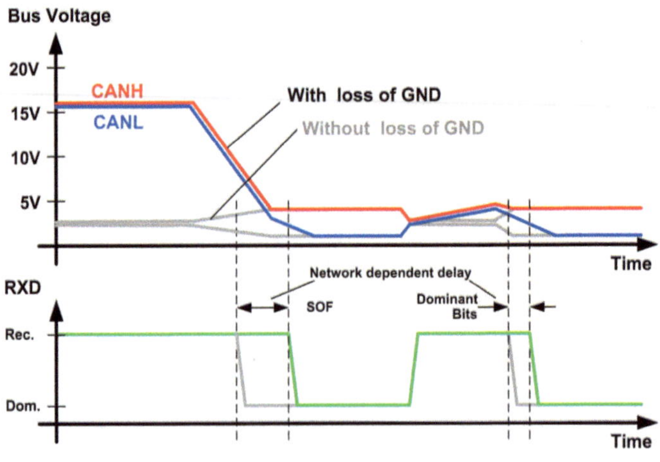


**Fig. 3** CAN network with mixed 12/48 V control units in case of loss of ground

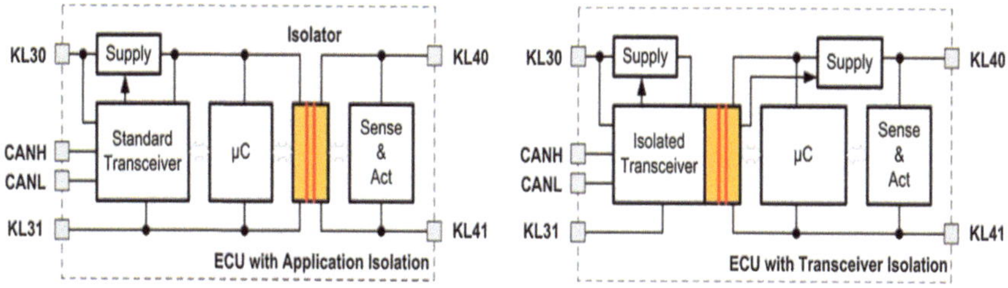


**Fig. 4** Variants for location of galvanic isolation in electronic control units

means to prevent this. Two schematics are possible, shown in Fig. 4 with different locations of the isolator. The location has to be chosen on a case-by-case basis.

### *2.3 EMC Effects in Multi-Voltage Architectures and EMC Limits*

The main sources of interference in hybrid and electric vehicles are electric motors, converters (three-phase alternators), and DC/DC converters (instead of alternators) [4]. With the number of functions, the number of potential disturbers increases in sensitive frequency bands where electronic systems must not be disturbed.

This imposes sharper EM emission limits in general. VHF radio is a prominent example but other sensitive frequency ranges may play a more important role in vehicles in the future. Relaxed limits for emissions might be permitted in frequency ranges with less significance. Figure 5 shows such frequency ranges marked in different colors to separate applications. The lines refer to limit specifications by German OEMs [3].

## 3 Galvanic Isolated CAN Bus Transceiver and Results

Integrated galvanic isolated bus transceivers provide a solution for communication across voltage domains: optimized bus transceivers and galvanic isolation without compromising on functionality, safety, and EMC.

### *3.1 Basics of Galvanic Isolation and Technology Selection*

Table 1 compares galvanic isolation techniques for bus transceivers. After the assessment of different galvanic isolation principles, NXP Semiconductors implemented capacitive isolation in a CAN transceiver. This solution combines the

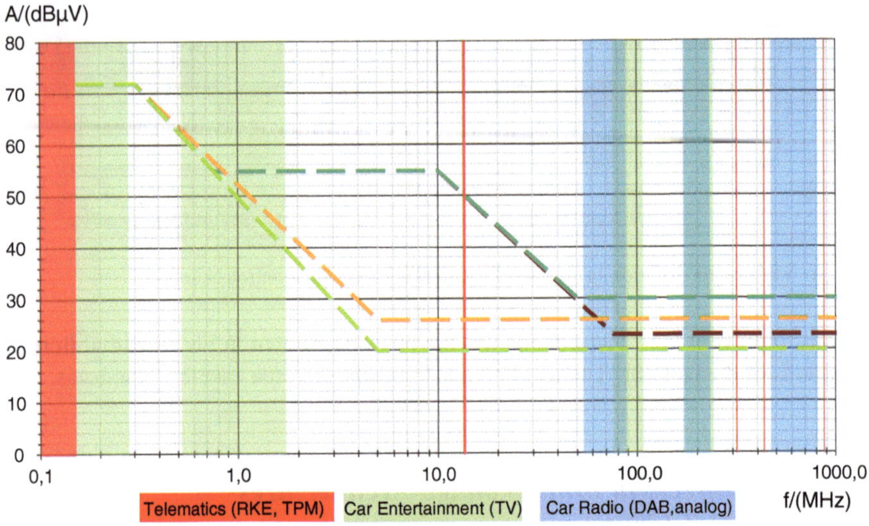


**Fig. 5** Sensitive frequency ranges including limits from German OEMs [3]

advantages of low current consumption, short processing times, and stable pulse widths. By good implementation, the sensitivity to HF fields of capacitive isolators can be limited to a non-application-relevant degree.

TJA1052i [1] combines the advantages of galvanic isolation of up to 5,000 $V_{RMS}$ with excellent EMC characteristics of CAN bus transceivers from NXP Semiconductors. Isolator (two smaller dies) and bus transceiver die are shown in Fig. 6.

## *3.2 Compliancy versus Standard*

The NXP TJA1052i complies with ISO 11898-2 and has a loop delay of 220 ns including isolation that enables a baud rate of up to 1 MBaud in networks.

## *3.3 Compared EMC Performance*

An important EMC parameter for isolated bus transceivers is immunity against conducted interferences. Immunity describes the ability of a device, piece of equipment, or system to perform without degradation in the presence of EM disturbance. Measurement results show minor differences compared to a conventional transceiver in the observed frequency range. Detailed results are shown in [5].

**Table 1** Overview of the most popular galvanic isolation techniques

<table>
<tr><td>Optical isolators</td><td>Inductive isolators</td></tr>
<tr><td>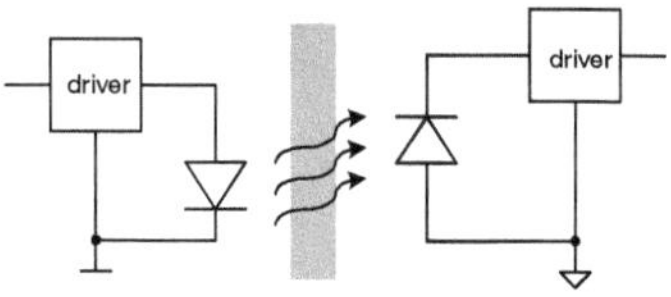
</td><td>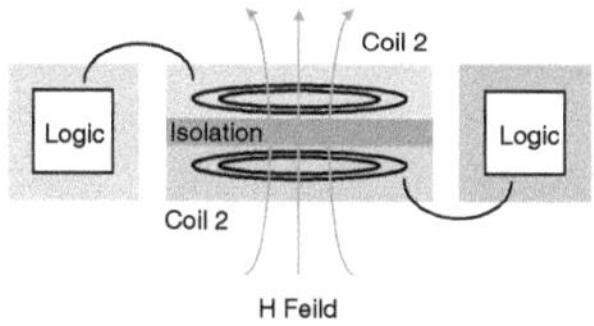
</td></tr>
<tr><td>• Quasi-standard galvanic isolation<br>• Direct & slow alternating signals<br>• Opto-couplers up to 50 MHz<br>• Sensitive to parasitic capacitive coupling paths and crosstalk<br>• Not sensitive to E/M fields<br>• Long signal throughput (60–100 ns)<br>• Aging of optical isolation layer<br>• High power consumption</td><td>• Current and voltage transducer for energy transmission<br>• Modulated direct and slow alternating voltages <150 MHz<br>• Sensitive to variable M fields<br>• Impulse compliance difficult (inductivity, crosstalk of inductors)<br>• Immune common-mode interfer<br>• Average signal throughput times<br>• Low power consumption</td></tr>
<tr><td>Capacitive isolators</td><td>Giant magneto-resistive (GMR) isolators</td></tr>
<tr><td>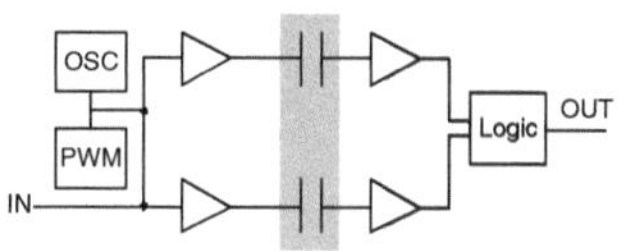
</td><td>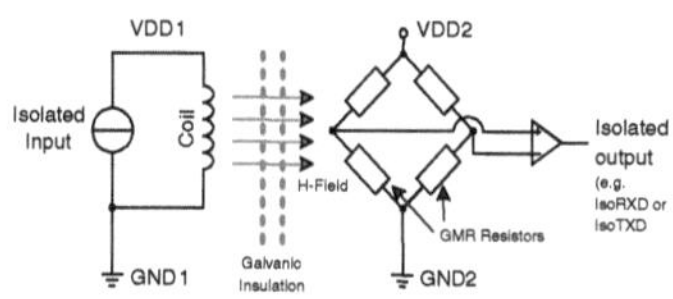
</td></tr>
<tr><td>• Feasible with standard technology<br>• Modulated direct and slow alternating < 150 MHz<br>• Sensitive to external HF fields<br>• Short signal throughput times<br>• Low power consumption</td><td>• Exciter coil and GMR detector<br>• M-field proportional. to current<br>• Direct and alternating signals<br>• Short signal throughput times<br>• Undefined output signals when connecting to power supply</td></tr>
</table>

## 3.4 Voltage Domain Crosstalk

Bus transceivers transfer high-frequency interferences on battery or ground pin to the communication network of the vehicle. Galvanically isolated bus transceivers offer the advantage of extra attenuation between isolated voltage domains. Emissions on one power supply/ground (led from control units to bus and vice versa) are reduced and interferences between voltage domains are suppressed.

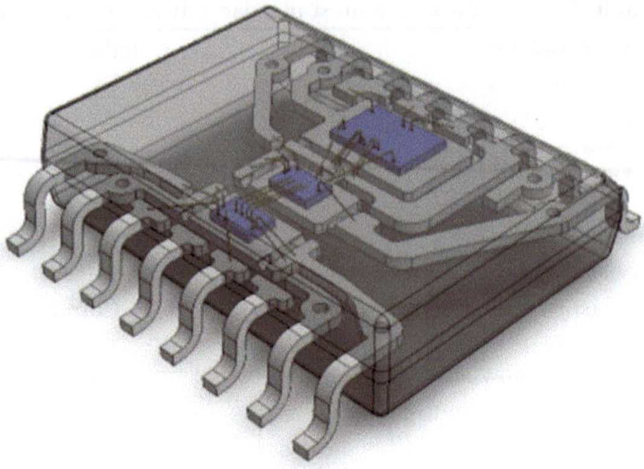

**Fig. 6** Mechanical build of the TJA1052i

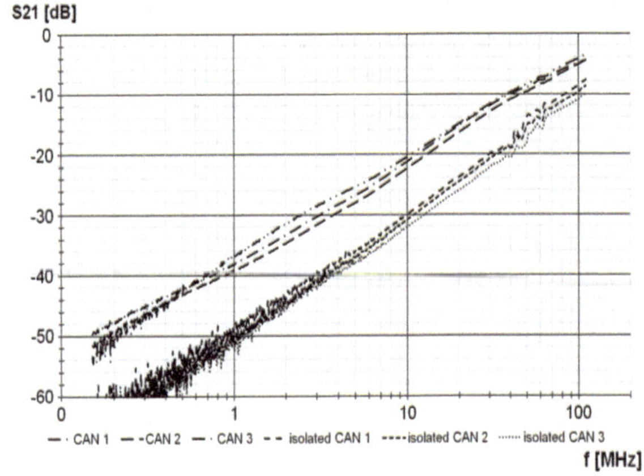


**Fig. 7** Comparison of pin transfer impedance S21 from VCC to CANH for different conventional CAN transceivers (*upper string*) and isolated CAN transceivers (*lower string*)

Figure 7 displays the S21 transfer impedance measurements of conventional and galvanically isolated CAN transceivers that are available on the market. The tests were directly carried out on the respective pin and thus without voltage supply. The lower string of curves shows isolated transceivers and an attenuation of 10 dB between power supply and bus pins in comparison with non-isolated transceivers.

## 4 Summary

Galvanically isolation is a must in multi-voltage domain data communication above 60 V; aspects are touch safety and protection of the electronic systems.

48 V architectures make isolation obsolete with the correct CAN transceiver selection (right level of bus pin voltage robustness) and a carefully designed

communication network taking control unit effects when "loss of ground" occurs into consideration.

Galvanic isolated transceivers based on capacitive coupling technologies like TJA1052i have significant advantages versus opto-couplers, e.g. the aging effect with opto-couplers and the higher loop delay.

EMC performance remains at the same level when comparing galvanically isolated and conventional transceivers; an advantage of 10 dB was measured for domain crosstalk.

## References

1. TJA1052i (2012) Galvanically isolated high-speed CAN-transceiver. Objective data sheet; Rev. 0.04, 20 Jan 2012; NXP B.V
2. ISO 7637-2 (2004) Road vehicles—electrical disturbances from conduction and coupling, part 2: electrical transient conduction along supply lines. Kunz
3. Hardware Requirements for LIN, CAN and FlexRay interfaces in automotive application—AUDI, BMW, Daimler, Porsche, Volkswagen—revision 1.2/2011-03-25
4. EMV von Elektro-Hybridfahrzeugen [EMC of electric hybrid vehicles]; Hillmer, Volkswagen AG; GMM symposium munich, 2009
5. Challenges for future vehicle networks using different voltage domains with regard to electromagnetic compatibility (EMC); Gunnar Schulz, Adrien Schoof, Frank Schade, NXP semiconductors, HDT 2012

# Development of a Scalable Multi-Controller ECU for a Smart, Safe and Efficient Battery Electric Vehicle

**Ulrich Köhler, Nikolaus Decius, Christopher Masjosthusmann and Ulrich Büker**

**Abstract** The goal of the public funded project eFuture is to develop a safe and efficient battery electric vehicle to widen the BEV's bottleneck that is the still expensive battery, limited in capacity and lifetime. A novel E/E architecture designed to merge different the function domains HMI, ADAS, Driving Dynamics, and Energy uses a central decision unit approach based on a central control unit. An ECU developed as the core of the new architecture—yet providing carry-over functionality—is a scalable multi-controller unit with four independent 32-bit microcontroller systems. The viability of this ECU concept is successfully demonstrated on the basis of a Tata Indica Vista EV.

**Keywords** Electric vehicle · Electronic control unit

## 1 Introduction

Increasing the degree of electrification of the drivetrain, up to and including purely electric vehicles, constitutes an important part of today's efforts in greening mobility. Ultimately, the scarcity of fossile resources brings about the need to use other energy sources. Moving to electric energy both as an energy source and storage enables using sustainable sources, be it solar or water power or wind energy.

U. Köhler (✉) · N. Decius · C. Masjosthusmann · U. Büker
HELLA KGaA Hueck & Co, Beckumer Straße 130, 59552 Lippstadt, Germany
e-mail: ulrich.koehler@hella.com

N. Decius
e-mail: nikolaus.decius@hella.com

C. Masjosthusmann
e-mail: c.masjosthusmann@hella.com

U. Büker
e-mail: ulrich.bueker@hella.com

J. Fischer-Wolfarth and G. Meyer (eds.), *Advanced Microsystems for Automotive Applications 2013*, Lecture Notes in Mobility, DOI: 10.1007/978-3-319-00476-1_18,

Moreover, moving to a more electrified vehicle power system enables a higher degree of control of the different power flows between the individual components. Energy management on vehicle level is employed to control and prioretize the energy budget of different functions. In this regard it is obviously very important to either keep the driver closely informed, so as to ensure his compliance, or other to keep him completely unaware of possible limitations. The latter is of course limited to functions unrelated to immediate safety issues.

The work presented here is part of the research project eFuture. Started in 2010 as part of the European Commission's Green Car Initiative, its goal lies in investigating new avenues towards an efficient and safe electric vehicle [1, 2]. These new ways are being looked at from different perspectives: the user experience or the vehicle-driver interface, as mentioned above, predictive driver support through camera or navigation information, improvements in high voltage battery and drive train such as torque vectoring; however the E/E architecture offers a range of possibilities to improve on [3]. Within the frame of this paper, only a certain aspect of the different angles investigated in the course of the project will be presented, the development of a central control unit as the heart of a novel E/E architecture. As such it is housing new functions, and incorporating a new functional architecture [3, 4]. This functional architecture comprises a classical layer structure with command and execution layer. Each of the layers incorporates a so-called Decision Unit (DU). In case of the DU 1 of the command layer determines the proper prioretization and magnitude of the various control inputs, for example the driver's input, the vehicle energy management or ADAS functions, and generating the appropriate control vector. The DU 2 in turn contains algorithms which transpose the control vector into actuator actions. This operating principle is shown in Fig. 1.

From Fig. 1 another important aspect can be identified: altogether there are four levels that make up the functional architecture: the aforementioned command and execution layers, plus perception and energy layer, respectively. Within the eFuture project, the responsibilities for the different layers are split between the partners into the areas Driving Dynamics, Energy, Guiding ADAS and HMI.

## 2 ECU Realization

Based on the deliberations of the preceding section, a concept of a new electronic control unit (ECU) was developed and realized. This concept would make use of the function distribution discussed, at the same time incorporating standardization and modularization.

A key aspect lies in assigning and partitioning the computing resources. In the project presented here, four Freescale Bolero micro controllers are used. This basic aspect is clearly oriented towards the fact that within the eFuture project four partners contribute software (SW) to run on the ECU discussed here. In fact, with a function distribution onto four layers, using four controllers, and having four

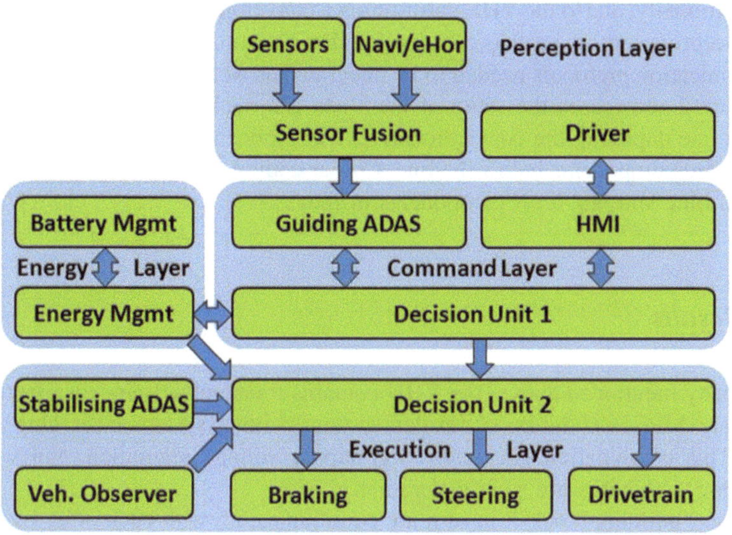


**Fig. 1** Layer-based functional architecture using a decision unit approach

partners supplying SW, it is relatively straightforward to really decouple the different software functions.

## *2.1 ECU Concept and Boundary Conditions*

As this paper details results achieved within the European project eFuture, it is necessary to outline the boundary conditions met, which of course influence the choice of concept.

Most importantly, the project demonstration vehicle is a Tata Indica Vista EV. Being an adaption of a conventionally ICE driven compact car, the base EV is further adapted. Therefore care had to be taken in order to preserve some carry over functionality contained in an ECU which was substituted by the so-called Vehicle Head Unit (VHU) presented here. An important difference between the Vista EV and the modified eFuture vehicle lies in the use of a different drivetrain, enabling torque vectoring by employing two motors, one for each front wheel. The torque vectoring algorithm runs on the VHU, and a safe and reliable communication between VHU and the motors/inverters is imperative [5, 6].

Therefore, special care was taken on designing a robust system. Robustness in turn requires efforts with regard of both hardware and software. Concerning the latter, an AUTOSAR oriented approach towards base and application software was chosen. This enables a reuse of base software for all controllers, assures a free function distribution on the application SW level, and as a result facilitates modularization and transfer to other products.

Additionally, due to the VHU substituting a conventional ECU and incorporating exterioceptive sensors such as camera or radar, a number of input/output lines and communication protocols needed to be integrated as well.

Figure 2 showcases the setup of the Vista EV, with the schematic above the dashed line depicting the conventional 12 V components, and the project specific HV related components below the line. Colour coding shows components on 12 V level in blue and HV level components in red.

## 2.2 Results

As already mentioned above, the VHU contains four mutually independent 32 bit Freescale MPC5607 micro controller systems with a system basis chip (SBC) each. This combination guarantees e.g. supply voltage regulation, and window watchdog with hardware reset generation.

These "building blocks" in turn are communicating with each other by means of a private asynchronous CAN bus.

In terms of software, the use of an AUTOSAR structure with base software layer, virtual function bus, and application layer enables a high degree of modularity on one side, and on the other side provides a well-defined method of

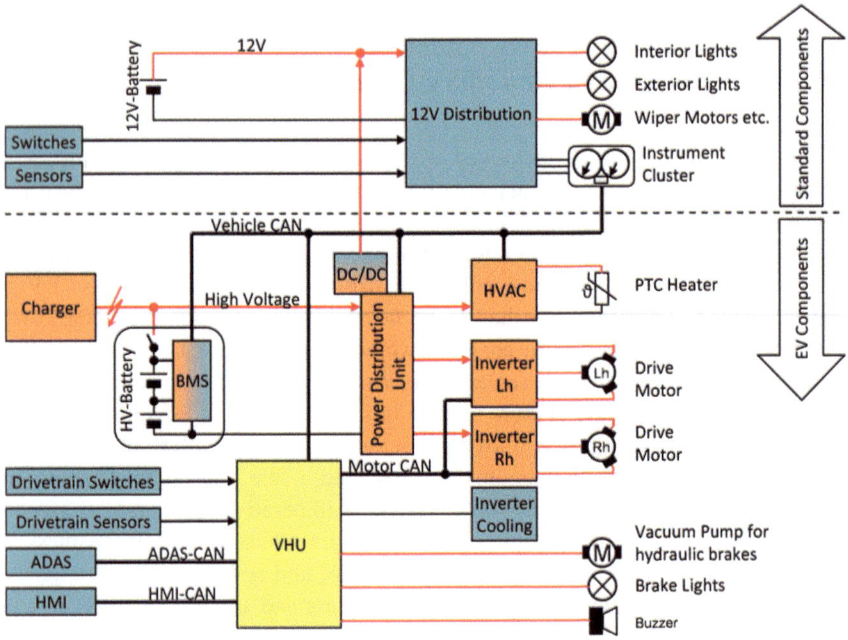


**Fig. 2** Block schematic showing the principal layout of EV components, conventional components, and the relation to eFuture's VHU control unit

communication between different application modules, i.e. functions, keeping a high degree of independence at the same time.

This unique combination served well within the project consortium, since the modularity and relative independence of the software application enabled the distribution of the individual software development efforts between the partners as well. As a consequence, responsibility for the four function levels shown in Fig. 1 is distributed too.

Using this approach an easy scalability of this method is apparent.

Another important aspect in developing this ECU was a high degree of robustness. This level could be achieved by a number of different measures:

Two micro controllers are mutually monitoring their status. In case of a detected malfunction may one controller take over duties of the other one, e.g. providing a limp home mode with limited functionality, however preventing total shutdown or even catastrophic failure. As a necessary prerequisite to this capability, important communication lines are fed to a pair of micro controllers.

As can be easily deduced from Fig. 3, this robustness is also demonstrated by the choice of connectors. Essentially, the use of two individual connectors guarantees a limited functionality even in case of one connector failing completely. Of course, the lines are not simply duplicated, however vital connections as e.g. power feed are redundant. At the same time, the wiring harness may be split in under bonnet and cabin wiring too.

Figure 4 depicts the engine compartment of the Vista EV with a mounted VHU in the centre. The vehicle was successfully integrated.

During the last project year, all software functionalities will be integrated onto the ECU. Drive tests will be performed to validate the implemented functions. Moreover the test results will be used for validating and parametrizing vehicle simulation models, as well as optimizing the control algorithm parameters underlying the various functions.

**Fig. 3** Realized VHU with outer package, connectors, and four micro controllers clearly visible

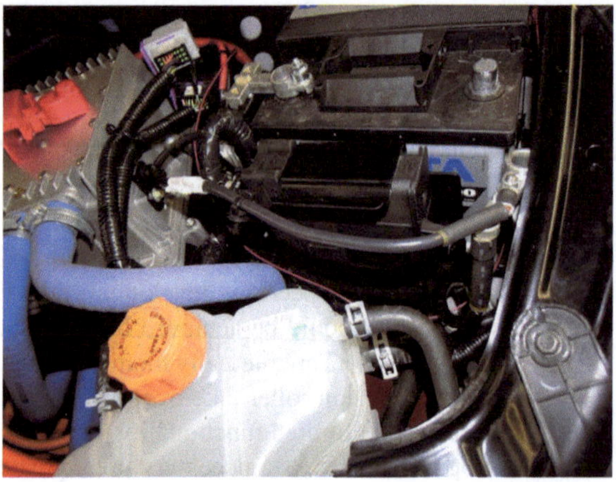

**Fig. 4** VHU in mounting position within the engine compartment

# 3 Conclusions

In this paper, a scalable ECU with a modular approach to both hardware and software was introduced. The ECU was developed within the frame of the European research project eFuture. It is part of a new E/E architecture designed to enable a safer and more efficient electric vehicle. While the vehicle has already been integrated, and base driving functionality could be successfully demonstrated, the full functionality still needs to be implemented. Validation tests are ongoing.

The authors gratefully acknowledge the financial support under the FP 7 Green Cars Initiative, grant number 258133.

# References

1. www.efuture-eu.org
2. Scheuch V (2011) E/E-Architektur für Batterie-Elektrische Fahrzeuge, ATZ elektronik, 2011
3. Scheuch V, Kaiser G, Holzmann F, Glaser S (2012) Simplified architecture by the use of decision units. In: Proceedings of AMAA 2012, Berlin
4. Scheuch V, Kaiser G, Holzmann F (2012) Eine innovative Funktionsarchitektur für das Elektrofahrzeug. Essen
5. Kaiser G, Holzmann F, Chretien B, Korte M, Werner H (2011) Torque vectoring with a feedback and feed forward controller—applied to a through the road hybrid electric vehicle. In: Proceedings of IEEE intelligent vehicle conference, Baden Baden, Germany
6. Glaser S, Orfila O, Schmäche M (2012) Safety of a torque vectoring LKAS on an in-wheel motor electric vehicle. In: Proceedings of EEVC 2012, Bruxelles, Belgium, 2012

# Wireless Charging: The Future of Electric Vehicles

**Anthony Thomson**

**Abstract** Bringing any new and complex technology to a global market requires specific industrial capabilities and key business acumen. Wireless Electric Vehicle Charging (WEVC) is now viewed as a go-to-market solution for the EV industry due to the technology's simple and effortless ability to make EV charging hassle free. WEVC also expands the EV use cases into city taxis and car share schemes that can struggle with conductive, plug-in charging solutions. While the use cases for WEVC deliver user simplicity, the commercial deployment into the complex automotive industry is challenging. This presentation will consider some of the commercial, technological and business issues that need to be addressed in order to make WEVC viable in the near term and to deliver long-term automotive and environmental advantages.

**Keywords** Electric vehicle · Wireless charging

## 1 Introduction

Bringing any new and complex technology to a global market requires specific industrial capabilities and key business acumen. Wireless Electric Vehicle Charging (WEVC) is now viewed as a go-to-market solution for the EV industry due to the technology's simple and effortless ability to make EV charging hassle free. WEVC also expands the EV use cases into city taxis and car share schemes that can struggle with conductive, plug-in charging solutions.

A. Thomson (✉)
Qualcomm Europe, Chiswick Park, 566 Chiswick High Rd, Chiswick W4 5YE, London
e-mail: athomson@qti.qualcomm.com

J. Fischer-Wolfarth and G. Meyer (eds.), *Advanced Microsystems for Automotive Applications 2013*, Lecture Notes in Mobility, DOI: 10.1007/978-3-319-00476-1_19,

While the use cases for WEVC deliver user simplicity, the commercial deployment into the complex automotive industry is challenging. This presentation will consider some of the commercial, technological and business issues that need to be addressed in order to make WEVC viable in the near term and to deliver long-term automotive and environmental advantages.

## 2 Technology Challenges

Delivering a commercially compliant WEVC system into the automotive market will require a deep understanding of the technology and its performance characteristics. Compliance requirements are also vital for engineers to meet in order for car manufacturers to have confidence in wireless EV charging. None interference with on-board electronics and keyless entry systems must be considered, for example.

At a technology level, there are a number of design options that could be considered. Pad and coil architectures as well as operating frequency are key to how the system performs across a wide spectrum of criteria.

Coil design decision is predominantly a choice between double pole and variations of single pole structures. While double pole wireless charging systems have been prevalent in industrial installations including LCD clean rooms for a number of years, automotive requirements—in particular EMC regulations—have resulted in the design and construction of more efficient and compliant circular single pole pad configurations. Single and multiple coil designs can now deliver end to end power transfer efficiencies of 90 % and above while tolerating high levels of pad offset, both laterally, (x, y) and vertically, (z). Tolerance to offset of the pads during charging is considered vital to ensure simplicity of use by drivers. The complexity of automatic guidance and parking systems while offering a good alignment solution for double pole based systems add cost and increase the time to park—both of which will stifle user adoption if implemented.

The operating frequency of the resonant magnetic induction system has also been discussed at length in open forums and in standards groups such as the Society of Automotive Engineers (SAE). Early systems have tended to focus around the sub 40 kHz band due to the cost and availability of high power electronic components at high frequencies. There is however a direct link to power transfer efficiency and frequency and therefore the industry is discussing alternative options below 150 kHz, generally regarded as the upper limit due to a broadcast license being needed in many regions above this frequency. Companies, including Qualcomm are designing and testing systems at frequencies above 70 kHz and early results will be published in time for the EVS27 congress.

# 3 Regulatory Compliance and Electromagnetic Compatibility

Compliance to requirements for non-interference with implantable medical devices has been a key focus of Qualcomm's compliance engineering team based in San Diego. The team benefits from more than 20 years of EMC and regulatory engineering experience and has been a key function in ensuring the Qualcomm Halo WEVC system meets the delivery criteria of the partners participating in the London wireless EV charging trial.

Figure 1 shows a typical example of the work carried out by the compliance engineering team on an existing electric vehicle. Bodywork structure, layout and materials can be included in modelling to analyse the effects on the wireless charging magnetic fields to determine compliance with Pacemaker immunity. This assessment is key when considering the integration of WEVC into the vehicle.

RF (radio frequency) exposure must also be assessed when considering operation of the WEVC system to ensure compliance to regulated exposure limits such as those based on the International Commission for Non Ionizing Radiation Protection (ICNIRP). Again, working directly with OEMs on their EV design enables simulations to be carried out to ascertain the exposure levels that would be expected with different car materials and charging pad locations as well as offset positions of the base-charging and vehicle-charging pads.

Figure 2 shows an EV saloon with considerations for assessing induced effects into bystanders outside the vehicle as well as humans inside the vehicle.

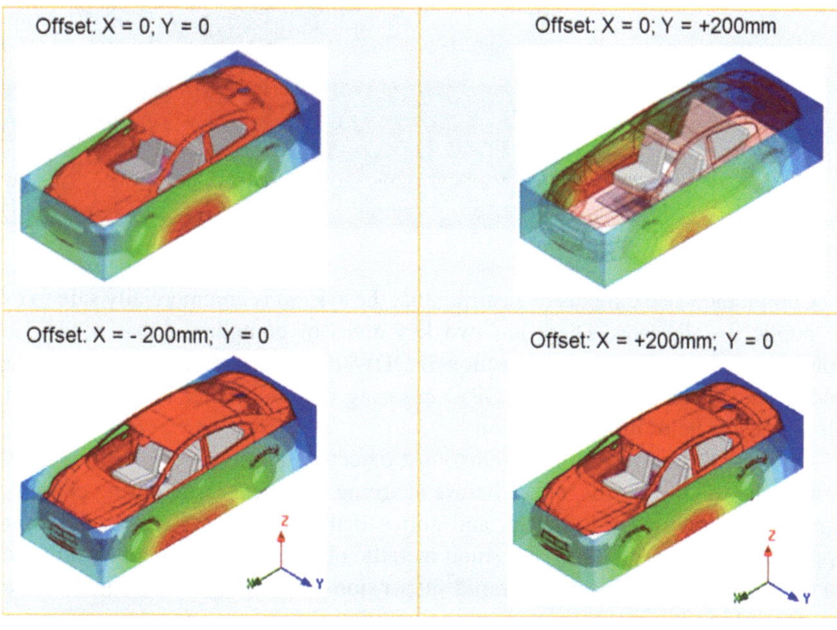


**Fig. 1** Pacemaker EMI analysis (PC69-2007 limit-bystander)

**Fig. 2** RF exposure assessments

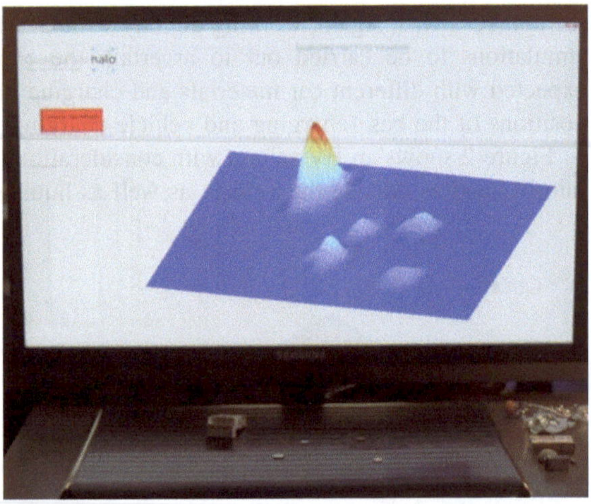

**Fig. 3** Foreign object detection

Compliance also extends to ensuring that the system is commercially safe to use in public installations. There are two key areas of compliance that need to be considered; Foreign Object Detection (FOD)—detection of all ferrous material; and Living Object Protection (LOP)—ensuring organic matter is not exposed to the magnetic field during operation.

Foreign Object Detection systems must detect all instances of a ferrous object coming between the pads, either before charging commences, or during charging. Large objects are easy to detect, and notice if they are on the pad as the driver approaches the charging bay, but small metallic objects can be more difficult to see so the FOD system must ensure rapid suspension of charging the instant detection occurs in Fig. 3.

Living Object Protection can be enabled using various technologies; in particular, off the shelf 60 GHz radar sensors typically used for reverse parking indicators could pick up any movement under the vehicle and automatically turn off the system.

## 4 Business Model

Technology and compliance to industry standards is only one half of the WEVC commercial proposition. Any company that is bringing a new complex technology to market must be able to invest in both engineering resources to support its automotive customers and in on going research and development to ensure a strong forward roadmap as the technology develops.

WEVC is evolving from a strong foundation of 3.3 kW for home use to 7 kW for home and public use, as well as 20 kW fast charge solutions for commercial and public site installations. This development requires companies to be able and willing to invest long term in WEVC technology and to support a range of diverse customers with varying needs. Future deployments may also incorporate semi-dynamic and full-dynamic charge-on-the-move capabilities, which will require substantial R&D investments to make this version of WEVC a commercial reality. There are a limited number of companies that can deliver on the WEVC vision and Qualcomm is one of those that has the knowledge, history and experience to do just that.

# With System Integration and Lightweight Design to Highest Energy Densities

**Klaus Höhne and Eva Hirtz**

**Abstract** Electric Vehicles will play an important role in future mobility. Despite the high efficiency of the electric motor electric vehicles have only a limited range due to the low energy density of battery systems. The progress on cell level will be only evolutionary, but there is a high potential for weight reduction of other components of traction batteries. Within the FP7 project Smart and Safe Integration of Batteries in Electric Vehicles (SmartBatt) a consortium of nine partners from five European countries pushed up the energy density on system level from 80 to 148 Wh/kg by a smart lightweight design.

**Keywords** Battery system · Lightweight design · Electrical vehicle · Energy density

## 1 Introduction

Climate change is a global issue with a high impact on environment and society as it can change dramatically the conditions of life in many areas worldwide. The Intergovernmental Panel on Climate Change (IPCC) determined a very high confidence, i.e. a probability higher than 90 %, that human activities have been the source of global warming. The global increases in concentration of the major greenhouse gas (GHG) carbon dioxide ($CO_2$) are due primarily to fossil fuel use and land use change [1]. The Stern Review showed clearly that climate change has as well an enormous economic impact [2].

K. Höhne · E. Hirtz (✉)
Fraunhofer-Institute for Structural Durability and System Reliability LBF, Bartningstraße 47, 64287 Darmstadt, Germany
e-mail: eva.maria.hirtz@lbf.fraunhofer.de

K. Höhne
e-mail: hoehne.klaus@web.de

J. Fischer-Wolfarth and G. Meyer (eds.), *Advanced Microsystems for Automotive Applications 2013*, Lecture Notes in Mobility, DOI: 10.1007/978-3-319-00476-1_20,

Transport is a source of prosperity and economic growth as well as pollution and illness. Transport is responsible for 32 % of EU's final energy use and 21 % of GHG emissions. The transformation towards a green transport is of utmost importance. Electric mobility will be one of the future modes of transport. It offers several advantages: (1) No local emissions of GHG, nitrogen oxides ($NO_x$), and particulate matter (PM). (2) In combination with renewable energies carbon neutral transportation and (3) reduced dependency on imported fossil fuels.

The European Union set ambitious targets to decarbonise the European road transport. The new car emissions of $CO_2$ have to be below 95 g/km in 2020. In order to reach this ambitious target the affiliation of electric vehicles (EV) to the portfolios of the vehicle manufacturers is without alternative. But a successful introduction of EVs to the market is only possible, if these vehicles meet the expectations of the users in terms of safety, purpose and costs. The main advantage of EVs lies in the high efficiency of electric motors whereas the main handicaps are the low energy density of batteries as well as the high price of batteries.

The FP7 project Smart and Safe Integration of Batteries in Electric Vehicles (SmartBatt) addressed the important topics: energy density, safety, and costs of batteries for electric vehicles. Nine partners from academia and industry of five EU member states contribute with their competences. This paper shows how the energy density of battery systems can be significantly increased by consequent lightweight design.

## 2 Scientific Targets

Experts are convinced that electric mobility will develop first in urban areas. Therefore the SmartBatt battery system was designed to meet the requirements of townspeople and commuters, i.e. a segment C vehicle with a range of at least 100 km in the New European Driving Cycle (NEDC). This results in a battery system that must be able to store more than 20 kWh of energy and has a nominal voltage around 300 V [3]. As a reference vehicle the FP6 result SuperLIGHT-CAR (SLC) was used [4]. The goal was a smart integration of the battery system into the SLC structure without any loss of performance, i.e. keep or even increase the crash worthiness and the torsional stiffness of the vehicle body.

Chemists all over the world are working hard to increase the energy density of lithium ion cells. Despite this there will be only evolutionary progress. New technologies like lithium sulphur or lithium air promise big progress, but an introduction into the market will not be before 2025 [5]. An improvement of the energy density of lithium ion cells was not in the scope of the SmartBatt project. But there is a high potential for weight reduction of other components of traction batteries. In actual EVs the cells contribute to 60 % of the total mass of the battery system. The target of the SmartBatt project was to increase this value up to 75 % by smart integration and lightweight design.

The key element to reach the target is the functional integration of the battery system to the floor structure of the EV. Since the project focused on battery integration and not on the development of a pure EV, the consortium followed a conversion design approach. Using this approach two figures are the cornerstones of the concept: the available design space and the cell selection.

## 3 Design Space

Up to date the expectations of European consumers on EVs are similar as for conventional cars. Therefore no or maximal minor reductions of the safety level or the volume of the passenger/luggage compartment are acceptable. In order to meet this demand the car structure of the SLC was analysed in detail and possible positions for the SmartBatt battery were defined. The available design space must not be deformed during any possible crash event and has to offer a sufficient volume to integrate the whole battery system. The elementary structural analysis of the SLC body-in-white revealed the necessity of a redesign of the original SLC floor structure in order to integrate the battery system and to meet the strength and crashworthiness criteria. Further structural analysis of the SLC/EV concept showed a considerable potential of the new configuration to increase both structural stiffness as well as crashworthiness of the electric vehicle, if battery housing is integrated into the EV's load carrying structure. New design ideas were assessed with regard to crashworthiness by using the macro element (ME) approach [6, 7]. The modelling and CPU time in ME approach is up to 3 orders of magnitude faster compared to FE analysis. Over 2,000 variants were evaluated. For the most promising variants a detailed FE analysis was conducted. The investigation pointed out, that the space below the rear seats and in the tunnel are safe positions for the battery system. Furthermore it was shown, that with some modifications of the floor structure a battery pack below the front seats can enhance the crash safety during a side pole crash for both battery and passengers [8].

## 4 Cell Selection

As denoted above the cell selection is crucial for several attributes of the EV. Among those are design space, performance, costs, reliability and safety. Lithium ion technology offers high energy density, high power density, and high cycle life time. These properties make these batteries attractive for the use in fully EVs. Efficient production processes and a mass market promise to reduce the costs for automotive cells by approximately two third until 2020 [5]. Nevertheless there is still a lot of research and development (R&D) and standardization necessary to get cells, which comply with automotive requirements on performance, safety and costs.

As the project focused on a smart and safe integration of batteries into EVs *of the shelf* cells were used. There is a wide range of different lithium ion cells available on the market. They differ in chemistry, shape, capacity, and safety. The shape can be cylindrical, prismatic or pouch cells. Chemistries are based on electro-active oxide or spinel, such as the three element lithium nickel cobalt manganese oxide (NCM) cells or lithium cobalt manganese spinel (LCMO), or on lithiated metal phosphate, for example lithium iron phosphate (LFP). A detailed analysis of the market with respect the requirements of the SmartBatt battery system led to four different cells, LCMO prismatic, NCM pouch, LFP pouch and LFP prismatic, which were further investigated on multiple attributes.

## 5 Concepts and Their Assessment

The used cell type influences the design on all levels: module and pack housing, cooling requirements, electrical design, sensors, etc. The previous results led to five different design ideas for the SmartBatt battery system: NCM pouch cells below the luggage compartment and the rear seats, NCM pouch cells as well as LCMO prismatic cells below the rear and the front seats, LFP pouch cells as well as LFP prismatic cells below rear and front seats and in the tunnel. A successive assessment on multiple attributes lead to the final SmartBatt concept.

Mass and mass distribution differ for each concept. Numerical simulations on the vehicle stability were performed to ensure driving safety. The impact of the different approaches was investigated by simulating several driving manoeuvres: brake test, steady-state skid-pad test, slalom test and lane change test. No significant deviation on the vehicle stability compared to the SLC was found in any concept.

A thermal analysis was performed for different concepts. For this analysis a software tool was used. Cell and thermal model were verified by experiments before implementing them. Cell parameters were determined by measurements. The simulation uses the ARTEMIS driving cycle at different ambient temperatures from −30 °C up to +40 °C. Four different types of cooling (none, air convection, forced air, and water) were investigated for each concept. The LCMO prismatic cells are the only cells which do not need an active thermal management, if one accepts that they cannot quite meet the target power at −20 °C. The NCM pouch cells require active cooling and heating, while the LFP pouch cells just require active cooling.

Preliminary and therefore rough weight and cost estimations for the whole battery system (including battery management system (BMS), housing, etc.) showed that concept 1, LCMO prismatic cells below the rear and the front seats, has the lowest mass and the lowest price. As it is expected that future EVs will be powered by large format pouch cells, a second concept with NCM pouch cells below rear and front seats was kept. In order to meet the voltage and energy requirements the battery topology is 84 cells in series for concept 2 and an 88s16p-configuration, i.e. 88 cells in series 16 in parallel, for concept 1. The result is a

22.92 kWh battery system with a nominal voltage of 325.6 V for concept 1 and 27.04 kWh at 310.80 V for concept 2. Both are well over the above defined targets for a segment C electric vehicle.

## 6 Modules

The cells chosen for the two concepts are representatives for two classes of cell types and differ considerably in their layout and their dimensions. Therefore different constructions for the module housing were needed. Several design versions of the modules were built up in CAD for packaging studies. Due to the dimensions of the large format pouch cells it is not possible to fill the available design space by only one type of modules. So for the second concept two different versions of modules were built up. The front pack version with six cells and an eleven cells version for the rear pack. The smaller modules of concept 1 are fitting well in the design space.

The SmartBatt solution for concept 1 has only four different parts, is easy to assemble, uses standard welding technologies, and was optimised on short electrical connections. The design for concept 2 is completely different. For active thermal management of the NCM pouch cells the concept 2 module structure must include cooling plates. Furthermore pouch cells need an additional supporting, since they have no hard casing. Therefore the module housing needs parts for cell supporting. Hence 19 different parts are needed and the handling in the assembling process is quite complex. One of the main targets of the SmartBatt project is to increase the mass ratio cells/system up to 75 %. To reach this target the mass of the module housing has to be as low as possible. The concept 2 module housings contribute with 18 % for the 11 cells module respectively with 20 % for the 6 cells module. In the case of the concept 1 module housing it is only 12 %.

Several tests on cell level were performed to assess the safety of the cells. The mechanical shock testing, which was based on the results achieved from crash simulations, and the thermal shock test according to SAE J2464 showed no safety issues on both cell types. Overcharge and nail penetration test, both according to SAE J2464, showed significant differences in the behaviour of the two candidate cells. The LCMO prismatic cells passed this abuse tests. In contrast the NCM pouch cells showed a severe reaction in both tests. A brief summary of the assessment results can be found in Table 1.

## 7 Housing/System

The concept 1 with the LCMO prismatic cells showed the greatest promise concerning weight, costs, safety and design. So this concept was carried out for final design of the SmartBatt battery system. There are different approaches for

**Table 1** Comparison of the two different concepts

| | | Concept 1 | Concept 2 | |
|---|---|---|---|---|
| Cell | Cell type | Prismatic | Pouch | |
| | Chemistry | LCMO | NCM | |
| | Energy density | 181 Wh/kg | 180 Wh/kg | |
| | Mechanical shock | Passed | Passed | |
| | Thermal shock | Passed | Passed | |
| | Overcharge | Passed | Severe reaction | |
| | Nail penetration | Passed | Severe reaction | |
| Module | Module variants | 1 | 2 | |
| | Topology | 1s16p | 11s1p | 6s1p |
| | Energy | 0.26 kWh | 3.54 kWh | 1.93 kWh |
| | Mass | 1.628 kg | 24.686 kg | 14.000 kg |
| | Energy density | 160 Wh/kg | 143 Wh/kg | 138 Wh/kg |
| | Different parts | 4 | 19 | |
| | Assembly | Simple | Complex | |
| | Number of modules | 88 | 6 | 3 |
| | Thermal management | No | Active | |
| | Vehicle dynamics | No change | No change | |

lightweight design, e.g. using lightweight materials, structural durability assessment or functional integration. The SmartBatt consortium used a mix of approaches to reach the ambitious lightweight design targets.

## *7.1 Functional Integration*

The basic idea of the SmartBatt project is a high functional integration. The functional integration of the battery housing into the floor structure, i.e. using the floor structure as top of the battery housing, offer both, a significant weight reduction of the battery system as well as an enhanced crash safety of the EV.

The SmartBatt battery system consists of three parts. The modules are partitioned in a front pack with 16 modules and a rear pack with 72 modules. Both packs are connected via a tunnel in which most parts of the BMS and the service disconnect are located. The connectors to the on-board electrical system and the communication bus are placed on a small balcony of the front pack. The SmartBatt battery system is shown in Fig. 1.

Considering the design space restriction to keep the passenger compartment as big as possible the floor structure of the SLC was adapted to the SmartBatt concept. Major changes are the *side pockets* which redirect the load path in a side pole crash and influence the crash behaviour in this scenario positively.

The battery housing consists of a bottom cover shaped like a flat pan which carries the vertical load and closes the SmartBatt battery system against the environment. Frames around front and rear pack carry the longitudinal and

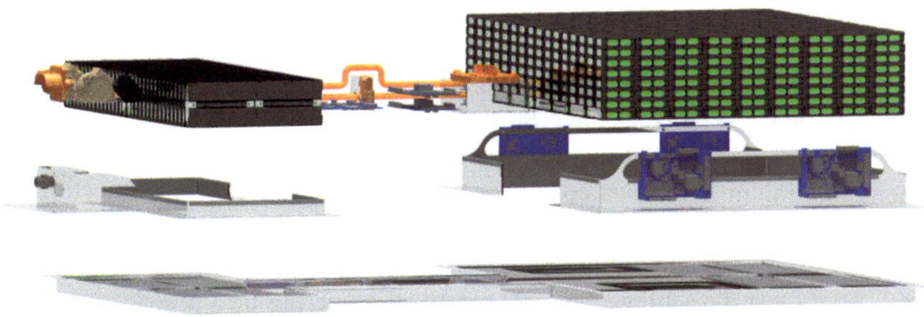

**Fig. 1** Final design of the SmartBatt battery system

transversal forces resulting from the inertia of the modules. The tunnel enables the battery housing to stiffen the whole car structure. With the integration of this mechanical functionality secondary weight reduction effects could be achieved for the vehicle structure.

## *7.2 Innovative Materials*

In addition to the functional integration the use of innovative materials boost the mass reduction of the SmartBatt housing. A benefit analysis for different materials, sandwich material as well as sheet metal, results in the decision for aluminium foam sandwich material. The use of sandwich material for the battery cover is a novel approach. Top and bottom layer are aluminium sheet metal of a thickness of 0.5 mm. Spheres of aluminium foam with a diameter of 4 mm are used as core material and a two component epoxy as adhesive [9]. The overall density of the sandwich material is 0.94 $g/cm^3$, which is a third of the density of aluminium (2.7 $g/cm^3$). This material offers a high bending stiffness at low weight. A 5 mm aluminium foam sandwich of length 1 m has a bending stiffness of $3.54 \cdot 10^8$ $Nmm^2$ and a mass of 4.72 kg. The bending stiffness of a 3.3 mm monolithic aluminium sheet is only $2.1 \cdot 10^8$ $Nmm^2$ (57 % less) and the mass 8.91 kg (89 % more). Due to the thickness and good energy absorbing properties of the aluminium foam sandwich material an enhanced intrusion protection was realized. In addition the thermal properties of this material are another benefit. Measurements of the coefficient of thermal conductivity $\lambda_{\text{them}}$ of the aluminium foam sandwich material result in a value of $0.4\,\text{W}/(\text{m} \cdot \text{K})$, which is over 500 times below the coefficient of thermal conductivity of aluminium $[220\,\text{W}/(\text{m} \cdot \text{K})]$.

The tunnel fulfils several functions. It physically connects front and rear part, stiffens the whole vehicle structure, and hosts many electronic components. To integrate different components and functionalities complex shapes have to be realised. Assuming mass production die casting is a cost efficient process to cope with this challenge. The use of cast aluminium for vehicle floor structure is again a novel and promising approach. Especially high pressure die casting offers the possibility for a cost efficient production of optimised complex shapes.

Using small format LCMO prismatic cells led to an 88s16p configuration and 1,408 cells have to be connected. So the electrical connectors can have a significant influence on the total mass of the battery system. In the SmartBatt battery system aluminium is used as conductor material for the bus bars connecting the cells in parallel in the modules and for serial connection of the modules. The specific resistance of aluminium is $2.65 \cdot 10^{-8}\,\Omega\text{m}$, which is higher than for copper ($1.68 \cdot 10^{-8}\,\Omega\text{m}$). To achieve the same electrical properties the cross-sectional area of the electrical connector has to be 1.6 times larger. Nevertheless the density of aluminium is less than a third of copper (2.7 g/cm$^3$ vs. 8.9 g/cm$^3$). Therefore electrical connectors out of aluminium have only half the mass at equivalent resistance.

## 7.3 Smart Design

The use of lightweight materials and high functional integration may not result in lightweight products. A weight optimised solution can only be achieved if a smart design combines the advantages of the different lightweight design approaches.

For example it was shown that the use of aluminium for the electrical inter module bus bars can save half the mass compared to copper. The inter module bus bars are designed to have a second functionality. They fix the modules mechanically to each other. Depending on the position inside the pack one of four different types of bus bars connects the modules. So there is a single solution for the electrical and mechanical connection. But the best lightweight design is to use as less material as possible. Different topologies for the positioning of the modules were analysed with respect to the total length of the inter module bus bars. The final wiring concept is symmetric and minimizes the material usage for the electric and mechanical connection of the modules. The total mass of the electrical and mechanical inter module connection is only 579 g. In contrast the copper cable used to connect electrically front and rear pack has a mass of 600 g.

The BMS is modular. It consists of a central battery control module and six voltage and temperature balance modules (VTBM). Each VTBM controls up to 16 modules. To keep the mass of the sensor harness as low as possible the VTBMs were placed close to the modules, i.e. four on the frame of the rear pack and two in the tunnel next to the front pack. Despite this measure the mass of the wires related to the BMS and their fixation is with 1,193 g similar to the complete high voltage net of the battery system. Taking into account the increasing amount of electronics in modern cars especially in EV wires with small diameter made out of aluminium can contribute to an overall lightweight design.

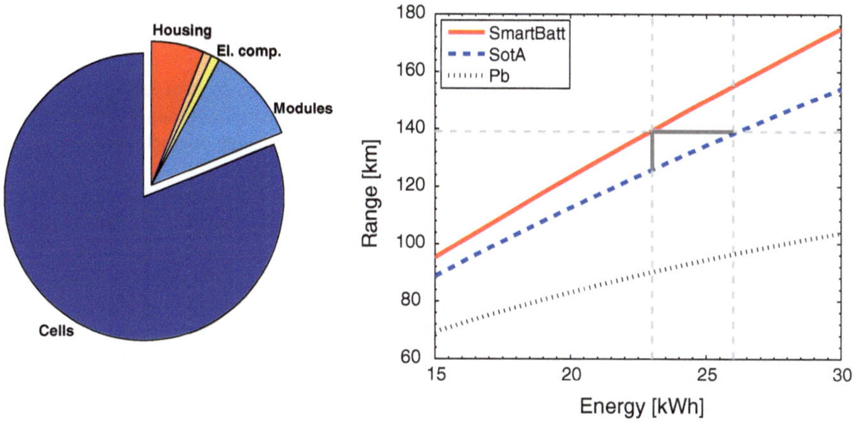


**Fig. 2** *Left* Mass fraction of the components of the SmartBatt battery system: cells (*dark blue*, 81 %), modules without cells (*light blue*, 11 %), electrical components (*yellow*, 1 %), electrical connections (*orange*, 1 %), battery housing (*red*, 6 %). *Right* EV range as a function of energy for lead acid, state of the art, and the SmartBatt battery systems

# 8 Conclusions

Electric vehicles are promising candidates to decarbonise European road transport. Despite the high efficiency of electric motors EVs have only a limited range due to the low energy density of state of the art battery systems. There will be only evolutionary progress on cell level until the market introduction of new technologies like lithium sulphur or lithium air, which are not expected before 2025. Today there is a high potential for the weight reduction of the other components of battery systems. The SmartBatt consortium used a mix of different approaches to lightweight design (innovative materials, functional integration, and smart design) to push up the energy density on system level from 80 to 148 Wh/kg. The effect is shown in Fig. 2 (right). For the SLC/EV the range as a function of the energy for different battery systems was calculated. The increased energy density of the SmartBatt solution results in a 10 % higher range or a 3 kWh (−12 %) smaller battery system to realize the same range as with state of the art battery systems. The mass ratio cell/system increased from 60 % up to 81 % (c.f. Fig. 2, left). Lightweight design results in lower mass and longer range with the same energy – i.e. more efficient electric vehicles.

**Acknowledgments** The work presented in this paper was done cooperatively by all partners of the SmartBatt consortium. The partners are AIT Mobility, Axeon Technologies Ltd, Fraunhofer IFAM, Fraunhofer LBF, Impact Design Europe, Leichtmetallkompetenzzentrum Ranshofen, Ricardo, SP Technical Research Institute of Sweden, Graz University of Technology, and Volkswagen. The paper is dedicated to Włodek Abramowicz, who died during the project term. His work is an essential result of the project. He inspired the project by introducing novel ideas, with useful comments and enriched the consortium with his personality. The SmartBatt project was funded by the European Union within the 7th Framework Programme (Grand Agreement No 266074). Additional information can be found under www.smartbatt.eu.

## References

1. Solomon S et al (2007) IPCC, 2007: summary for policymakers. In Proceedings of climate change 2007: the physical science basis. Contribution of working group I to the fourth assessment report of the intergovernmental panel on climate change, Camebridge University Press, Cambridge
2. Stern N (2007) The economics of climate change. Cambridge University Press, Cambridge
3. Broussely M (2010) Battery requirements for HEVs, PEVs, and EVs: an overview, in electric and hybrid vehicles. Elsevier B.V, Amsterdam
4. Goede M, Stehlin M (2009) SuperLIGHT-car project—an integrated research approach for lightweight car body innovations. In: Proceedings of the international SLC conference: innovative developments for lightweight vehicle structures, Wolfsburg
5. National Platform for Electric Mobility (2010) Interim report of the national platform for electric mobility, GGEMO, Berlin
6. Abramowicz W (2001) Macro element method in crashworthiness of vehicles. In: Ambrosio J (ed) Crashworthiness—energy management and occupant protection. Springer, Vienna
7. Abramowicz W (2004) An alternative formulation of the FE method for arbitrary discrete/ continuous models. Int J Impact Eng 30:1081–1098
8. Luttenberger P et al (2012) Structural analysis of a body in white for battery integration using finite element and macro element with focus on pole crash optimization. In: Proceedings of European congress on computational methods in applied science and engineering, Vienna
9. Jörg Weise et al (2012) Epoxy aluminum hybrid foam—an innovative sandwich core material with improved energy absorption characteristics. In: Proceedings of 10th international conference on sandwich structures, Nantes

# Rotor Position Sensor for Hybrid and Electric Drives

**Olivier Brunel and Rainer Moller**

**Abstract** Electrification of the power train will play a key role in the struggle for higher energy efficiencies and reduced emissions of vehicles. Optimized control of modern electric motors requires precise measurement of the rotor position. Electricfil has developed an Electric Motor Position Sensor (EMPOS) that provides a number of advantages over traditional sensors, including very high robustness to mechanical tolerances, a digital interface and low production costs [1]. After having validated the concept on prototypes, EFI Automotive is now working on the industrialisation of the product that will be ready for mass production scheduled to begin in early 2015.

**Keywords** Electric motor position sensor · Permanent magnet synchronous · Alterating current · Application specific integrated circuit · Printed circuit board · Electric vehicles · Hybrid electric vehicle · Electro magnetic force · Electro magnetic interference · Surface mount device · Field programmable gate array

## 1 Background and Challenges

### *1.1 Introduction*

Greenhouse effect and emission pollutant reduction is within the main objective of the world community. Transportation represents 20 % of greenhouse gas effect, passenger cars contribute to half of this emission. All countries have committed to

O. Brunel (✉) · R. Moller
Electricfil Automotive, 77, allée des Grandes Combes, Z.I. Ouest Beynost 01708, Miribel cedex, France
e-mail: Olivier.brunel@electricfil.com

R. Moller
e-mail: rainer.moller@electricfil.com

J. Fischer-Wolfarth and G. Meyer (eds.), *Advanced Microsystems for Automotive Applications 2013*, Lecture Notes in Mobility, DOI: 10.1007/978-3-319-00476-1_21,

challenging $CO_2$ emission targets that will require extensive use of electric drives. Marketing studies forecast that annual production of electric drives vehicles will reach more than 10 million units per year. This is more than 5 times actual production quantities. Thanks to their high efficiency and high power densities, permanent magnet synchronous motor (PMSM) are actually the preferred choice of OEMs building electric drives vehicles, hybrids or full electric. The electric motor control generate optimum phase shift between the stator coil induced magnetic field and the rotor permanent magnet field. This phase shift is 90° within low to medium speed and is then gradually increased as a defluxing strategy to avoid overvoltage. The optimum angle is controlled with space vector processing in which the position measured by the sensor is a critical variable for the Park transformation. Sensor errors or delay will lead to torque ripple, reduced efficiency and reduced maximum torque capability.

### *1.2 Challenges*

Electric motors for automotive applications mostly use a resolver, the only now available robust solution. Resolvers are basically transformers with one primary and two secondary windings, where the coupling between primary and secondary windings depends on the angular position of the rotor [2]. The construction of a resolver is similar to that of an electric motor, with a primary winding on the rotor and the secondary windings on the stator. An external conditioning circuit provides the excitation signal and processes the two secondary signals.

This technology is well proven and very robust to harsh environment. But the drawbacks are high sensitivity to external magnetic field requiring shieldings, accurate positioning of sensor and target and high cost for large diameter.

The goal of Electricfil was to develop a compact and flexible technology with high immunity to external magnet field, low cost and with optional digital output with below specification: (Table 1).

## 2 EMPOS Technology

Electric **M**otor **Po**sition **S**ensor called **EMPOS** is based on Eddy current technology. The EMPOS working principal is quite similar to that of a resolver. However, the excitation frequency is much higher, typically a few MHz, making it possible to use printed circuit air coils instead of winded coils on ferromagnetic sheet metals.

A primary winding and two secondary windings are printed on the same PCB which is fixed to the stator. A trigger wheel made of aluminium or steel with N teeth is fixed on the rotor. When a high frequency AC current is injected into the

**Table 1** EMPOS specification

| Functional | |
|---|---|
| Range | 0–360° (electrical) |
| Angular speed | 0–200,000 RPM |
| Refresh time | 0.25 μs |
| Accuracy | ±1° |
| Resolution | 12 bit |
| Noise | <0.1° rms |
| Airgap range | 0.5–3 mm |
| *Electrical* | |
| Power supply voltage | 5 ± 0.5 V |
| Power supply current | <15 mA |
| Output protocol | SIN/COS |
| | SIN/COS modulated |
| | DIGITAL SSI |

primary winding, a magnetic field is generated. This high frequency magnetic field causes Eddy currents to circulate in the teeth of the trigger wheel.

The eddy currents in return generate a magnetic field that is opposed to the original field, thereby modifying the coupling between primary and secondary windings. The sensor coils are designed to provide a coupling factor between primary and secondary windings that depends on the position of the trigger wheel.

We obtain two secondary voltages given by

$$\mathrm{V1} = \mathrm{A} \times \sin(\mathrm{N} \times \mathrm{q}) \times \sin(\omega t) \text{ and}$$
$$\mathrm{V2} = \mathrm{A} \times \cos(\mathrm{N} \times \mathrm{q}) \times \sin(\omega t)$$

where **A** is a constant that depends on sensor parameters, $\omega$ is the excitation signal frequency, **t** is the time, $\theta$ is the angular position of the trigger wheel, and N is number of pole pairs ($\mathbf{N} \times \boldsymbol{\theta}$ is the electrical angle).

We can then determine the angular position of the rotor as follows:

$$\frac{V1}{V2} = \frac{\sin(N \times \theta)}{\cos(N \times \theta)} = \tan(N \times \theta)$$
$$\Leftrightarrow N \times \theta = \arctan\left(\frac{V1}{V2}\right)$$

An application specific integrated circuit (ASIC) generates the excitation signal and assures the whole signal treatment chain, including the analogue part (amplification, demodulation, filtering, sampling…) and the digital part (calculation of ratio and arctan function, digital filtering…). The ASIC also provides diagnostic functions in order to avoid false position information in case of any single failure.

The ASIC is soldered directly on the Sensor PCB assuring short signal paths for high frequency analog signals. The position signal and a diagnostic signal are

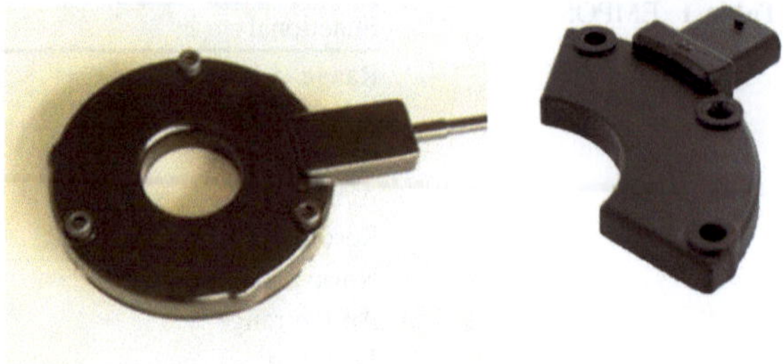

**Fig. 1** First generation "ring design"—second generation "moon design"

transmitted to the motor controller via either an analog or a digital serial interface, the later assures no loss of accuracy due to ADC accuracy, noise and EMI.

The sensor consists of a multi layer PCB with printed coils. The trigger wheel is an aluminium or steel part. The sensor is very light, because it doesn't need any ferromagnetic materials for flux concentration.

All required electronic components including an ASIC and some passive SMD components are directly soldered onto the sensor PCB. The sensor PCB is packaged in a sealed plastic housing with either a cable output or an integrated connector. The total thickness of the packaged sensor can be less than 10 mm. The trigger wheel teeth should be about 2 mm thick to guarantee mechanical robustness.

The first generation of EMPOS was with a ring PCB and package. A second generation with a moon shape has been developed in order to improve integration, serviceability and cost (Fig. 1).

# 3 Integration

## 3.1 Mechanical Interface

EMPOS is a design that adapts easily to electric motor topology, size, and number of pole pairs thanks to a flexible printed circuit board technology.

The PCB size is adjusted according to the inner diameter of the shaft, and the layout period is suited to the number of pole pair. The sensor is bolted on the motor stator while the trigger wheel is attached to the stator with press fit, bolt (Fig. 2).

## 3.2 Electrical Interface

The ASIC has been design in order to be compatible with 3 electrical interfaces.

In term of system performance to cost ratio we would recommend the digital interface but analog interface are proposed to be compatible with available Motor Control Unit (MCU) (Fig. 3) (Table 2).

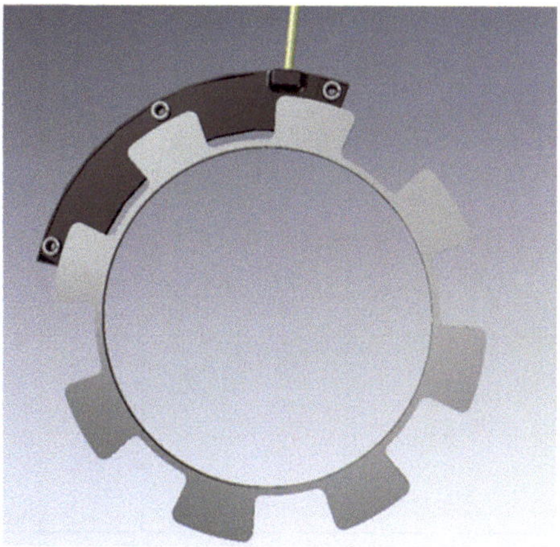

**Fig. 2** Mechanical interface

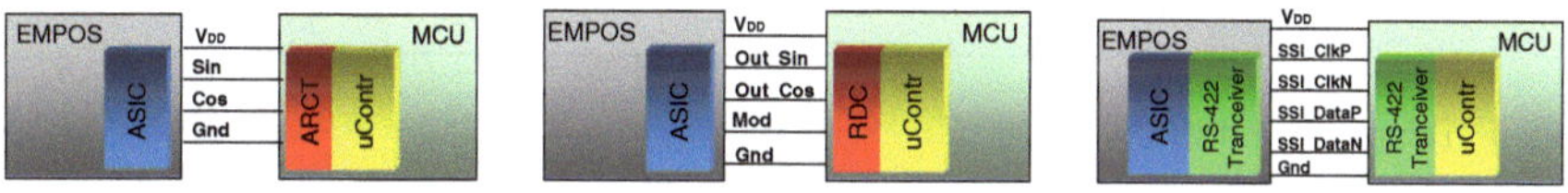


**Fig. 3** Electrical interface options

**Table 2** Electrical interface comparison

| | ANALOG non modulated | ANALOG modulated | DIGITAL SSI |
|---|---|---|---|
| EMC immunity | – | – | + |
| Offset immunity | – | + | + |
| Signal delay control | – | – | + |
| Transmission Diagnostic | Not applicable | | Yes |
| Electronic processing complexity | 0 | – | + |

## 4 Performances

The sensor has inherent errors due to magnetic coupling, non-linearity, and electronic processing errors (offset, gain, ADC…). The sensor performance has minor dependence on the motor shaft diameter (within range of 30 mm to 300 mm) and number of pole pairs (3–16 pole pairs) when we consider electrical angle (related to an electric period). In nominal condition the sensor accuracy is ±0.6°.

The external parameters, such as operating temperature, motor speed, mechanical tolerances (axial, radial, tilt) will create additional errors.

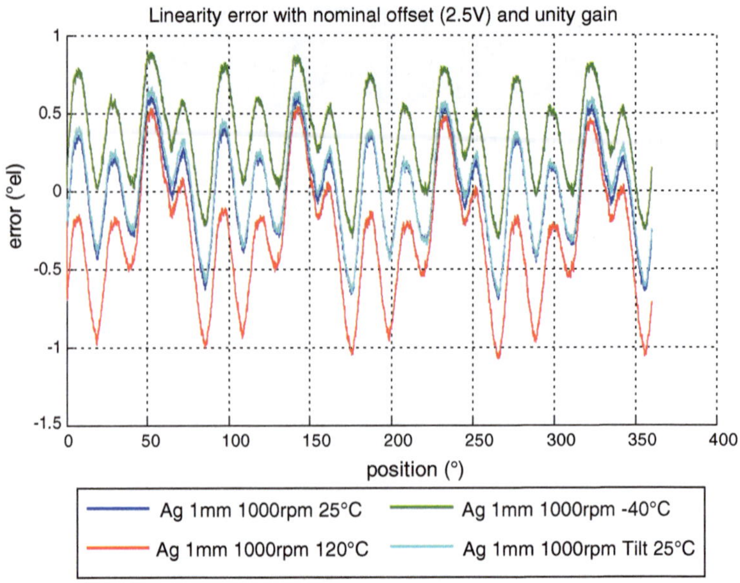


**Fig. 4** Temperature drift (−40/120° C)

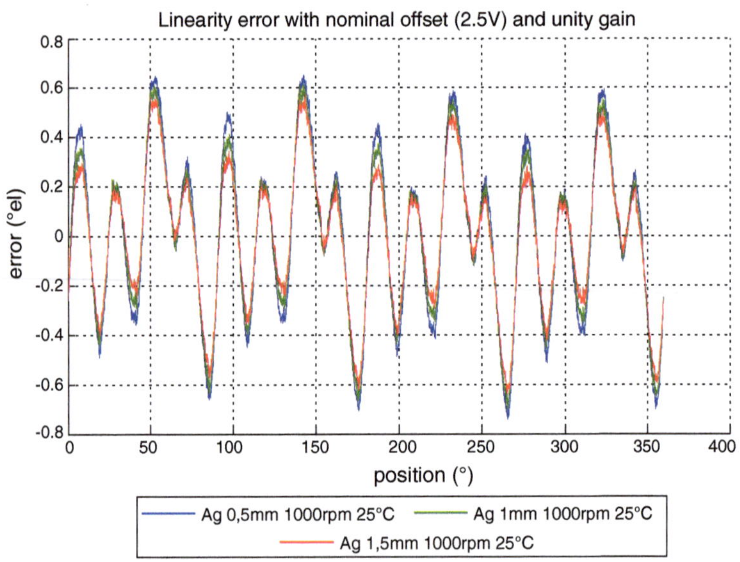


**Fig. 5** Axial tolerance effect (±0.5 mm)

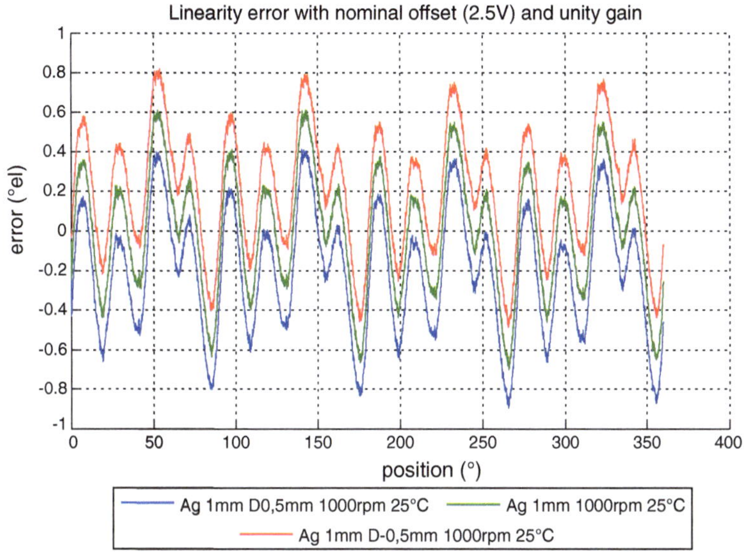


**Fig. 6** Radial tolerance effect (±0.5 mm)

Total errors for the "ring design" are within ±1° for all operating conditions and mechanical tolerances less than:

- ±1 mm axial
- ±0.5 mm radial

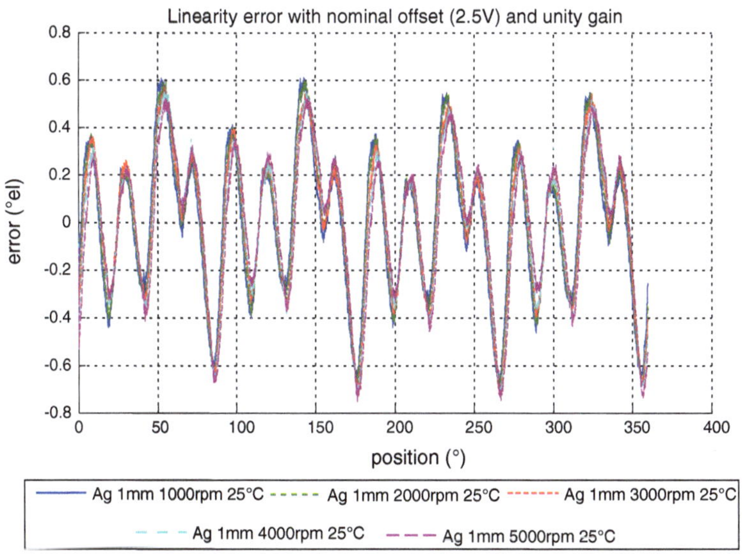


**Fig. 7** Speed effect (5,000 RPM)

Integrated signal processing provides error compensation for electrical speeds above 12,000 RPM.

The following graphs shows the influence of above parameters (Figs. 4, 5, 6, and 7).

## 5 Conclusion

Electrification of the power train will play a key role in the struggle for higher energy efficiencies and reduced emissions of vehicles. The drawbacks of current sensor technologies like high costs or sensibility to EMI will not be acceptable any more when electric power trains go into mass production.

EFI Automotive has developed EMPOS (an Electric Motor Position Sensor) based on eddy-current technology. Less than 10 mm thick and weighing under 100 g, the sensor can be easily integrated in HEV and EV applications. Its sealed package is designed to keep water, oil and pollution out. With no ferromagnetic parts, it is insensitive to EMI, vibrations and positioning errors. In addition, the trigger wheel has been designed to keep production costs down.

The integrated ASIC handles the entire signal-treatment chain, including analog and digital processing. It includes diagnostic functions to avoid false position information in the event of any single failure.

After having validated the concept on prototypes, EFI Automotive is now working on the industrialisation of the product that will be ready for mass production scheduled to begin in early 2015.

## References

1. Brunel O, Moller R (2011) Rotor position sensor for hybrid drives and electric drives, new generation eddy current position sensor. SAE Detroit, 2011
2. Ocket T (2012) Accurate and cost effective resolvers for electrical vehicles, EEVC Belgium, 2012

# Advanced Modular Drive Train Concepts for Electric Vehicles

**Tobias Lange, Hauke van Hoek, Christoph Schäper and Rik W. De Doncker**

**Abstract** The paper gives a detailed overview on modular drive train concepts and how significant benefits can be obtained with different levels of modularity. The advantages are assessed with respect to the additional effort. Modularity offers an inherent scalability of power and driving range, which enables car concepts to be composed with building blocks. This lifts the boundaries between different car types, as all consist of the same blocks. Obviously, a high level of modularity comes along with standardization of components as well as interfaces and offers a substantially high impact on cost reduction due to the economy of scale. Other important aspects are component integration, increased reliability due to redundancy and impacts of the system voltage levels on the safety effort.

**Keywords** Modular drive train · Low-voltage (LV) battery · Fast high-power dc-charging · Electric vehicle

T. Lange (✉) · H. van Hoek · C. Schäper · R. W. De Doncker
Institute for Power Electronics and Electrical Driver (ISEA), RWTH Aachen University, Jaegerstr. 17-19 52066 Aachen, Germany
e-mail: tobias.lange@isea.rwth-aachen.de

H. van Hoek
e-mail: hauke.vanhoek@isea.rwth-aachen.de

C. Schäper
e-mail: christoph.schaeper@isea.rwth-aachen.de

R. W. De Doncker
e-mail: post@isea.rwth-aachen.de

J. Fischer-Wolfarth and G. Meyer (eds.), *Advanced Microsystems for Automotive Applications 2013*, Lecture Notes in Mobility, DOI: 10.1007/978-3-319-00476-1_22, 

# 1 Introduction

The perspective for a successful market introduction of the electric car has never been that promising. Upcoming resource shortage, urban pollution, global warming, as well as environmental disasters are all arguments which emphasize the need for sustainable energy generation and consumption concepts. The electric mobility in combination with renewable energies forms the most promising solution. As a result, numerous governments, for example Germany, recently subsidize a lot of projects to encourage car manufacturers to add electric cars to their portfolio. However, currently, the high prices prevent people from buying electric cars.

With the combustion engine car as competitor, the electric car has to compete with a century of innovation and research, invested in a product with a production volume of above 80 Million pieces in 2011. It is comprehensible that it takes time for the individual parts of the electric car to be able to catch up.

So far, many electric car concepts are based on a combustion engine car platform where the drive train is replaced. Figure 1 depicts the standard solution, which is commonly used in these cars. It consists of three basic components: the battery pack, the inverter and the electrical machine that is connected to a motor shaft. For varied car types, these components have to be adjusted individually.

Although technology can be adapted and innovation synergies achieved, this production approach restricts the possibility to decrease costs with high production volumes. Within this paper, an overview will be given how standardization and modularity can lift these restrictions.

# 2 Economy of Scale

For a market breakthrough of electric vehicles the mass production of components is a key aspect. To reach the aim of the German government to establish a million electric cars in Germany by 2020, it is necessary to achieve the cost goals

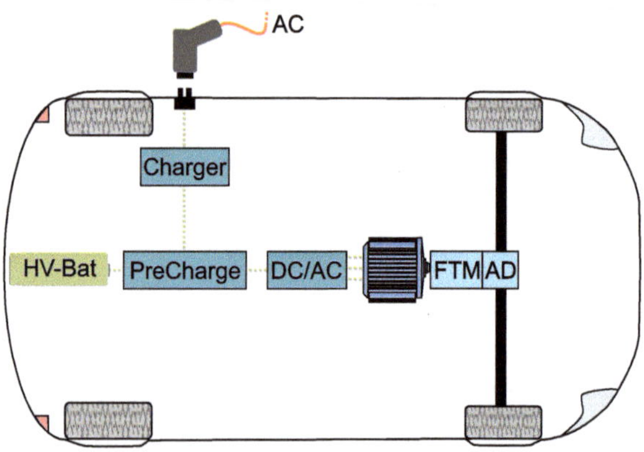


**Fig. 1** Standard drive train topology

presented in roadmaps like [1]. The costs of high-energy Li-Ion cells are estimated to decrease to 300 €/kWh and for high-power Li-Ion cells to 500 €/kWh within the next three years. Power electronics costs will decline more significantly compared to the battery system costs. Thus the main costs are due to battery cells as described in Table 1. For traction inverters costs of 3.7 €/kW, power densities of up to 12 kW/l and an efficiency of 97.5 % are estimated for the year 2015. This paper proposes vehicle concepts based on identical components for different purpose as urban, compact, medium- sized and luxury vehicles. Medium-sized and luxury vehicles are assumed to be composed of urban and compact vehicle components.

With standardized interfaces and scalable components, the amount of identical parts can be increased, which substantially reduces the production price. Additionally, the development and certification costs of electric drive trains can be reduced as the variety of modules is decreased.

## 3 Scalability of Power and Driving Range

Creating standards enables the utilization of economies of scale. The lowest standardization level concerns the auxiliary components, such as electrical connectors and cables as well as cooling systems, which have to be specified and commonly used within the drive trains. Although this aspect is quite obvious, it has not been established so far. However, a significantly higher impact can be achieved with standards concerning voltage levels within the drive train. This approach is only realistic with a dc-to-dc converter as it lifts the restrictions on the battery layout. When the dc-link voltage level is specified, all drive train components can be design accordingly. Ultimately, all car concepts can consist of the same modular building blocks. The amount of battery packs is chosen according to the desired range and drive train power, while the traction power can be adjusted with the number of machines and inverters. Thereby, a common component market can be created, granting access for new market players. The advantages are numerous: new battery concepts or technologies can be immediately tested within cars as only the dc-to-dc converter has to be changed or adapted. Defective devices can be replaced more easily. In particular, instead of one monolithic battery pack there are several medium sized battery modules, which make a replacement more manageable and cheaper. Furthermore, there is no issue concerning different states of

**Table 1** Cost goals for power electronics and Li-Ion cells in electric- and hybrid vehicles in the year 2015

| | Power (kW) | Energy (kWh) | Inverter (€) | Li-Ion cells (€) |
|---|---|---|---|---|
| Urban vehicle | 30 | 20 | 112 | 6,000 |
| Compact vehicle | 50 | 30 | 186 | 9,000 |
| Medium-sized vehicle | 80 | 40 | 297 | 12,000 |

health or even manufacturers of the battery modules. Small modular components can be divided throughout the limited space within the car more arbitrarily.

Scalability creates the important link to introduce identical modules for different vehicle platforms. Figure 2 shows the components of a modular drive train, including low-voltage (LV) batteries, battery-integrated dc-to-dc converter as well as a high-voltage (HV) dc-link bus that feeds the traction inverter and is used for charging.

Modular drive trains allow to build platforms with single machine urban and compact vehicles or multi-machine driven medium-sized and luxury vehicles (Fig. 1). These concepts offer new possibilities in terms of driving comfort and dynamics, which can be realized by controlling the torque of the two to four motors individually. Electronic differential gear and driving dynamic features like torque vectoring are established.

The installed power is scalable by replacing or adding machines in the electric drive train. Driving range is adjusted by adding or removing battery pack modules with integrated dc-to-dc converters.

Urban and compact cars can be powered by single motor modules, identical to those of the multi-machine vehicles. To optimize the drive train for each vehicle type concerning its characteristic driving cycle, different combinations of machine types are reasonable. For urban and compact vehicles a relatively small power is sufficient. Thus, the Interior Permanent Magnet Motor (IPMSM) is beneficial to achieve high power densities and high efficiencies for the low speed driving cycle. The most important alternative cost effective machine types are the induction machine (IM) and the switched reluctance machine (SRM). Especially the SRM achieves high power density and high efficiency comparable to an IPMSM. As the switched reluctance machine has nearly no rotor losses compared to the IPMSM and the IM it is especially suitable for high-speed integrated drive systems with 15,000–25,000 rpm. Thereby, high power-densities about 12–18 kW/l can be reached. The main issue of the switched reluctance machine is the acoustic noise radiated by the machine housing due to its electro-mechanical excitation. Active and passive damping of this noise is a main research topic for this machine type.

Machine powers of up to 50 kW allow an efficient and agile vehicle. For urban use one battery pack is sufficient to meet the driving range and power requirements. In [1] it is shown that the average driving distance per day is below 70 km for about 77.9 % of the private vehicles in Germany. For commercial used vehicles, average driving distances of 130 km and below are reached by about 77.2 %. Thus, the assumption of one medium-sized battery pack is reasonable to keep extra costs of the periphery, like a pre-charging unit and a dc-to-dc converter, low in comparison to a two battery pack solution. It is possible to charge the vehicle at night without any drawback for the general use.

Compact vehicles can be equipped with IMs to enable moderate efficiencies at low and medium speed. The amount of battery packs can be adopted on the demanded range for each vehicle concept. By combining the advantages of both machine types, medium-sized and luxury vehicles can be equipped with IPMSM and IM or SRM to meet demands for long distances and higher agility.

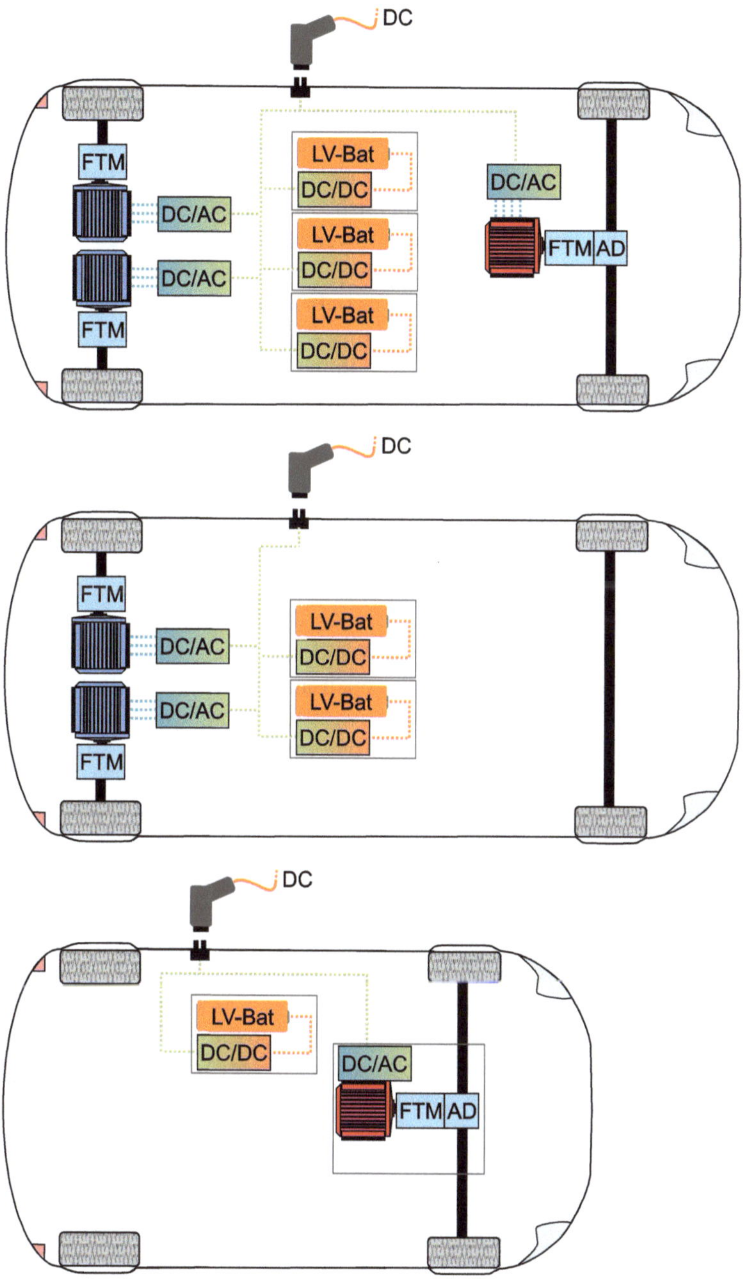


**Fig. 2** Modular drive trains based on identical components

For example the "e performance" Project, funded by the BMBF, realized a multi-machine concept to allow efficient and powerful driving for a sports car [4]. The vehicle compromises two induction machines with 60 kW each at the rear axle and one permanent-magnet machine with 50 kW at the front as a transaxle. This concept affords and includes studies for each presented vehicle concept. It is scalable between 50, 60/120 up to 170 kW in the four wheel driving mode. Detailed analysis and results will be published in future.

By increasing the amount of medium power machines, in a range of 30–60 kW, a redundancy and fault tolerance is ensured. Because of the independent components in the modular drive train most faults only limit the maximum system power. Thus battery packs, dc-to-dc converters or even traction inverters and machines can fail without causing a break down. Range and maximum speed might be limited, but at least a limp home functionality can be offered.

## 4 Low-Voltage Battery

Low-voltage traction battery concepts within electric vehicles are motivated by concerns of safety and reliability. Firstly, the vehicle maintenance is not necessarily carried out by skilled electricians. While this might be avoidable for utility vehicles, it can be expected that people repair and modify their electric passenger cars on their own the same way they do with their combustion engine car. Secondly, accidents potentially lead to major deformations of the car structure, which can result in risks for driver and rescuers due to exposed voltages. These risks have to be minimized with a certain safety effort, such as insulation monitoring and a redundantly insulated battery packs to prevent access from laymen.

Battery concepts with voltages $V_{\mathrm{batt}} < 60$ V fall in a less strict safety category as these voltages are not considered to be dangerous for the human body. Combining these voltages with a high traction power, however, implies the usage of a dc-to-dc converter as otherwise the currents achieved within the drive train are too high.

A structure that depicts a low-voltage battery pack is shown in Fig. 3. The two batteries each have a voltage of 48 V with respect to the vehicle ground potential. By connecting both in series the dc-to-dc converters input terminal voltage is 96 V. The converter creates the specified dc-link voltage at the output terminals. When the system shuts down the dc-link voltage, e.g. the car is parked, all capacitors are discharged and the batteries are disconnected. Thus, only the 48 V battery voltages are left within the system, making it inherently safe in terms of exposed voltages even in failure case.

The integrated dc-to-dc converter decouples the low-voltage battery from the dc-link. Hence, all the mentioned advantages of a modular system with dc-to-dc converter apply. Besides, by defining a modular battery package by its output terminal parameters, the system design can be carried out independent of the battery pack's inner structure.

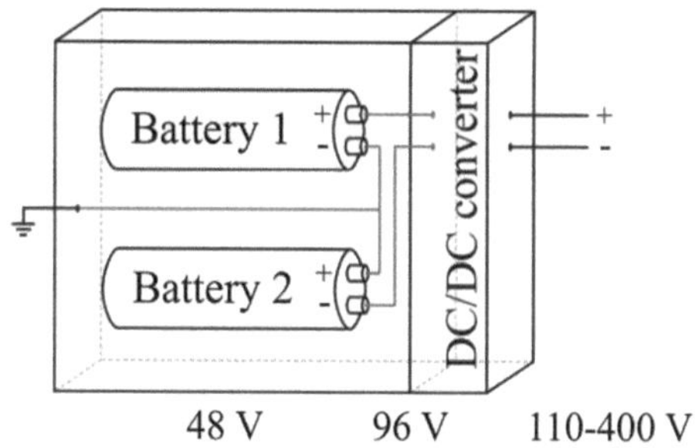


**Fig. 3** Low-voltage battery pack

Table 2 lists the estimated costs for two modular LV and one HV battery pack of 40 kWh. On basis of these values the higher effort for a low voltage battery concept can be assessed. The costs increase by 2.2 %, which is mainly caused by the extra housing and electronics. With respect to the advantages the additional costs of modular LV battery packs are reasonable.

## 5 Increasing Efficiency

The total system efficiency is an important aspect for the electric vehicle design as it directly influences the driving range and the cooling effort. There are several possibilities to increase the system efficiency depending on the vehicle purpose. Driving cycle, range and traction power have a deep impact on the system configuration. In this paper the proposed modular topologies and their advantages are investigated further. Vehicles without dc-to-dc converter have to be designed to provide the rated maximum traction power at minimum voltage. This results in an oversized drive train, because it has to be designed according to the high currents, which are particularly restricted by the thermal limits of the inverter and the machine. By adding a dc-to-dc converter the inverter input terminal voltage can be adjusted independent of the batteries state of charge (SOC). Under partial load conditions multi-phase dc-to-dc converters are able to increase the converter's efficiency by disabling phases. The maximum efficiency of each phase can be established at different multiple operating point. The use of several battery packs equipped with dc-to-dc converters further increase the adaptable efficiency range from low to full load condition.

**Table 2** Estimated costs of HV with 40 kWh and two LV Battery packs with 20 kWh each (without dc-to-dc converter)

| | Costs | |
|---|---|---|
| | HV (€) | LV (€) |
| Cells | 15,350 | 15,350 |
| Periphery | 1,572 | 1,947 |
| Over all | 16,922 | 17,297 |

Dc-to-dc converters can also be used to create a variable dc-link voltage [2, 3]. Thereby, the voltage level can be adjusted according to the drive's operating point to reduce the inverter's switching losses or enable a six step operation mode. The simulation results in Fig. 4 show two efficiency maps of the inverter losses with fixed and variable dc-link voltage assuming space vector modulation. In base speed range of below 5,000 rpm, especially at high torque, the inverter losses are decreased according to the reduced dc-link voltage.

Assessments show that the advantage of a system with dc-to-dc converter depends on the driving cycle. For city cycles like the "Hyzem urban" and the New European Driving Cycle "NEDC" higher efficiencies can be reached. However, driving cycles with higher performance like the "Hyzem rural" or the "Hyzem highway" show higher efficiencies without dc-to-dc converters. An example is given in Table 3, taken from [4].

## 6 Fast High-Power DC Charging

Charging concepts are still an essential discussion aspect concerning vehicle mobility. Ultimately, the electric vehicle is assessed with respect to the combustion engine cars, which has a range of several hundred kilometers and needs about ten minutes to be refueled. However, the battery energy storage capability limits the range of an electric vehicle, while the charging power is limited by the processes within the battery.

Common concepts expect either a single-phase charging process with around 3.7 kW and 230 V at home (Europe) or fast charging with up to 43.5 kW and 400 V at public charging stations. The inconsistent requirements are difficult to combine with a compact charger. A concept with dc-to-dc converter and a

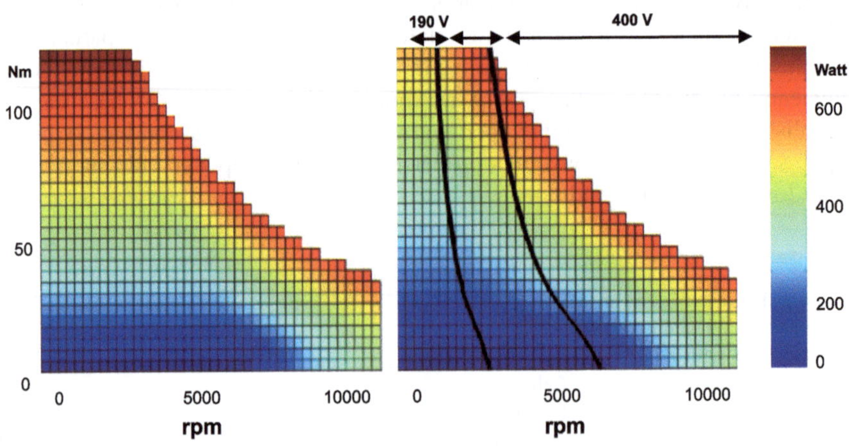


**Fig. 4** Efficiency maps

**Table 3** Total converter losses with and without dc-to-dc converter

| | Without dc-to-dc converter (%) | With dc-to-dc converter (%) |
|---|---|---|
| NEDC | 4 | 3.9 |
| Hyzem urban | 5.4 | 3.2 |
| Hyzem rural | 3.2 | 3.6 |
| Hyzem highway | 2.1 | 4.7 |

standardized dc-link voltage overcomes this issue: There is no need for an additional charger within the car. The charging station is designed according to the power grid at the connection point and does not have to fulfill automotive requirements. It feeds the dc-link with a standardized voltage while the dc-to-dc converter takes over the charging process and realizes the battery management system. As the dc-to-dc converter is designed according to the traction power, the possible charging power is significantly increased, making either the battery itself or the available terminal power the bottleneck.

For future grid structures, dc-grids are currently in discussion for transmission lines as well as smart homes. These concepts can lift the need for a charging station as the vehicle's dc-link can be connected to the dc-supply grid of the house. Furthermore, this concept is inherently bidirectional. Thus, the vehicle's battery can be used as energy storage to support the grid.

## 7 Conclusion

An overview on modular drive train topologies has been presented. It has been described how modularity and standardization can increase the chances for a market breakthrough of electric vehicles by providing a better opportunity for high production volumes. This is achieved by using identical building blocks for all vehicle classes. It potentially leads to a better utilization of economies of scale and can significantly reduce the vehicle costs.

A concept with low-voltage batteries with an integrated dc-to-dc converter has been proposed, which substantially reduces the safety effort. It has also been shown that the additional costs of this battery concept are not significant.

By using an integrated dc-to-dc converter higher system efficiency can be achieved. Furthermore this concept enables the possibility of high-power dc-charging to eliminate additional onboard charger units.

## References

1. ETG (2010) Energietechnischen Gesellschaft im VDE, VDE-Studie Elektrofahrzeuge – Gesamttext, VDE Verband der Elektrotechnik
2. März M, Eckardt B, Schletz A (2007) Mechatronische Integration von Hochleistungs-elektronik in Komponenten des Antriebsstrangs von Hybridfahrzeugen. *Neue elektrische Antriebskonzepte für Hybridfahrzeuge*, p 80

3. Schoenen T, Kunter M, Hennen M, De Doncker R (2010) Advantages of a variable dc-link voltage by using a dc–dc converter in hybrid-electric vehicles. In: Vehicle power and propulsion conference (VPPC)
4. van Hoek H, Boesing M, van Treek D, Schoenen T, De Doncker R (2010) Power electronic architectures for electric vehicles. VDE-Kongress

# Electric Vehicle Preparation for Vehicle Performances Analysis and Battery Evaluation in Normal Operation Conditions

**Alberto Fraile Del Pozo, Sara Sánchez Monclus and Emilio Larrodé Pellicer**

**Abstract** The primary objective of this project is the transformation of a vehicle powered by an internal combustion engine into an electric competition vehicle with zero-emission able to perform long distances and having high performance, as well as establishing a design process model for future electrifications. It determined the technical characteristics and the parameters needed for further transformation. Currently, the developed vehicle is monitored and adequately equipped with measurement equipment and allows evaluating the different types of batteries depending on the technical characteristics of the vehicle and the journey to perform, therefore evaluation of vehicle performances can be done in order to test battery behavior in different driving conditions.

**Keywords** Electric vehicles · Battery · Controller · Electrification · Power sizing

## 1 Introduction

This project was carried out in the company Automotive ZYTEL collaboration with the University of Zaragoza. ZYTEL Automotive is a Spanish company dedicated to the electrification of vehicles and electric vehicle manufacturing. As part of the development process of their electric vehicles and promoting your

A. F. D. Pozo (✉) · S. S. Monclus · E. L. Pellicer
Department of Mechanical Engineering, University of Zaragoza, C/Maria de Luna s/n, Zaragoza 50018, Spain
e-mail: afrailep@unizar.es

S. S. Monclus
e-mail: ssanchez@unizar.es

E. L. Pellicer
e-mail: elarrode@unizar.es

J. Fischer-Wolfarth and G. Meyer (eds.), *Advanced Microsystems for Automotive Applications 2013*, Lecture Notes in Mobility, DOI: 10.1007/978-3-319-00476-1_23,

brand, has decided to participate in the ZERO RACE, race where the electric vehicle will go around the world with zero $CO_2$ emissions. The way in was decided to approach the project was by adapting a vehicle to the requirements of career and regulations imposed by the European Union. The achievement of this objective is performed by meeting a number of objectives are described below. First, there was a work vehicle documentation to know its main systems as well as the design and modeling of its main components, within the field of electric vehicles. Also undertook a market study for the selection of the vehicle, component selection, both main and auxiliary systems, and components prior test. Once the vehicle structure was determinated and the final parameters are known, it was made a process of dynamic performance calculation, modeling components, mechanical analysis, electric circuit design, design protections and vehicle energy efficiency study, for the detection of shortcomings and possible improvements in the vehicle. Upon completion of the above tasks, we proceeded to study the integration of the vehicle components, the assembly process planning and subsequent implementation.

## 2 Technical Characteristics of the Project. Regulations

The starting point was an analysis of the regulations issued by the organization of the race, from which we extracted the points related to project specifications concerning the vehicle, such as distance traveled (30,000 km) or the limitation that the power source should only be the electricity [1].

Established the regulations of the race, there is a need that the vehicle must comply with all legal requirements driving in countries that will be crossed by the ZERO RACE. After studying the rules of movement of each country, was extracted the conclusion that complying with the regulations established by the European Economic Commission (EEC) would be sufficient to ensure compliance with the regulations established in Asian and American countries. Directive 2007/46/EC [2] establishing a framework for the approval of motor vehicles and for trailers, systems, components and separate technical units intended for such vehicles. After the study of that directive and its annexes, concludes that: the vehicle that intends to participate in the race (motor vehicle intended for the carriage of persons, two seats, with 4 wheels and maximum weight exceeding one ton) is framed in category M1 (Directive 70/156/EEC [3], annex I), which according to their category and changes initially raised therein, is affected by the following regulations:

- Regulation 10 (R10) [4] “Radio interference EMC (electromagnetic compatibility)”;
- Regulation 100 (R100) [4] “Battery Electric Vehicle”;
- Regulation 101 (R101) [5] “Electric Motor Consumption”; which will be considered throughout the project.

# 3 Power Sizing

Today it is difficult to store large amounts of energy, especially in the form of electricity. Electric energy storage systems are expensive and heavy, so a previous calculation of electric power that will be used to achieve the vehicle to reach established targets is essential. Through the determination of the equations governing the forces opposing the advance of the vehicle and the initial conditions will be set an algorithm that will result planned energy that would be needed to store. This process is called power sizing. The main function of power sizing process or Calculation Model of the Power Sizing (CMPS) is to determine the energy required to make a particular vehicle perform a certain number of drive cycles. Parallel it is used for various functions such as calculation of the engine power required, the maximum vehicle speed, the acceleration, the selection of the gear ratio and the obtaining of the vehicle's curves among others. Using the CMPS and yield maps of the electric motor together allows a previous analysis of possible modifications to be made to the vehicle in order to obtain maximum performance. The energy necessary for the traction of the vehicle is set by the simultaneous sum of the resistance to advance, the speed and the time elapsed during the movement. Resistance to advance is caused by the friction of the tyres with ground, the friction with the air, inertial resistance and strength, positive or negative, produced in the ascent or descent of terrain slopes. Fixed the methodology for calculating the resistance to the advance of the vehicle, it'll study consumption cycles in the regulation-101 [5].

## *3.1 Driving Cycles*

A drive cycle graphically represents the vehicle's speed versus time during a supposed route. Throughout a cycle can distinguish four driving modes: acceleration, deceleration, constant speed and stop. The standard driving cycles (cycle ECE-15 and cycle EUDC) have been developed to simulate a conduction that is representative of urban and interurban driving. These cycles are used to evaluate the use of vehicles on test benches simulating real operating conditions. The cycle New European Driving Cycle (NEDC) is used in the European Union Regulation 101 [5] to certify passenger vehicles and light trucks. It is composed of four urban cycles ECE-15 and a cycle EUDC thus simulating real conditions daily. It has duration of 1,200 s at an average speed of 33.60 km/h and a total distance of 11,022 m. The result shows figures which are difficult to interpret at first sight therefore are transformed to more conventional units as it is consumption in kWh/100 km, so for this case get consumption per cycle of 17.23 kWh/100 km.

### 3.2 Estimated Battery Weigh

Following the selection criteria through the regulatory requirements of the project, it must set the weight of batteries as a determining factor when selecting the vehicle and batteries. To do so, according to article 9.6 of the regulations of the race's organization [1], it is must daily tackle a first stage of 225 km with a maximum recharge of 4 h and then another stage of 225 km.

As conclusions in this first phase of power sizing, without knowing what the structural components of the project will be, it can say looking at Table 1 that the first election is either lithium or sodium batteries. It will select a light vehicle, large load capacity, a drive system as light as possible and high efficiency. It can be seen as the weight will be a key factor in the development of the project, mainly due to two reasons, one of them is the high percentage of mass which will provide the batteries and the second to increased consumption by the weight which would require an greater number of batteries, converting the calculation in an iterative process.

## 4 Market Research. Components Selection

Established the technical characteristics of the project and conducted a first approximation of the energy needs of the vehicle, it must continue with the selection of the components that will make possible the achievement of the objectives established. For this has been developed a diagram that brand guidelines to be followed for the selection (Fig. 1):

### 4.1 Vehicle Selection

Set of selection criteria which were as follows:

- Vehicle of independent frame, for simplification afforded by this type of vehicle to the structural changes of the same and the high load capacity that allows;
- Vehicle with at least two seats and cargo space for a minimum of 600 kg of batteries;
- Vehicle of low weight to obtain higher final performance;

**Table 1** Batteries possible to incorporate

| Batteries | Lead-acid | Lithium | Sodium |
|---|---|---|---|
| Energy density (Wh/kg) | 45 | 115 | 120 |
| Weight _necessary (kg) | 1,506.96 | 589.68 | 565.11 |

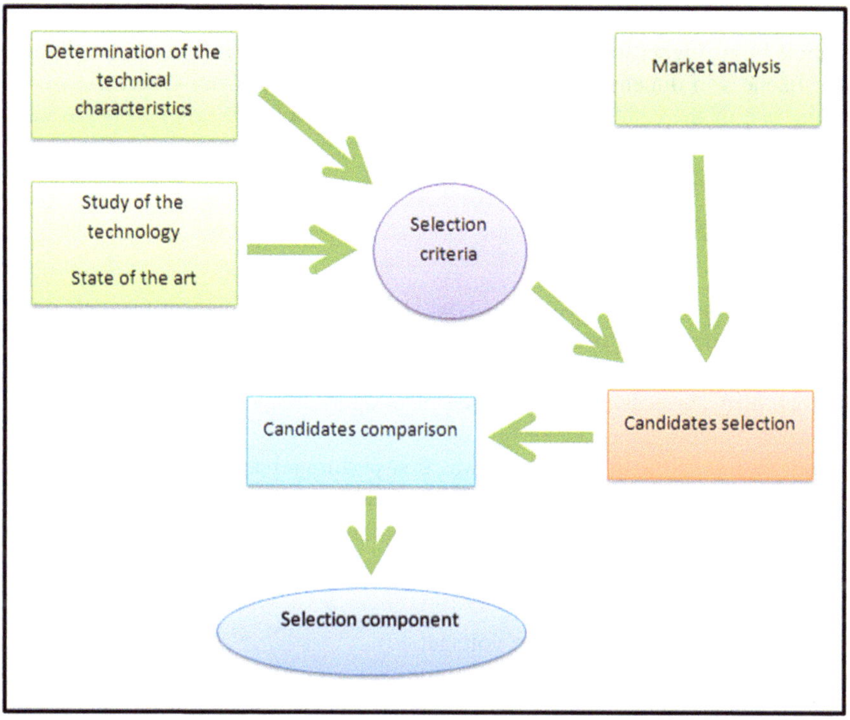


**Fig. 1** Diagram of components selection procedure

- A high resistance and reliability of structural components that guarantee the possible changes thereto, as well as an increase of the load to bear;
- Maximum safety of occupants with a minimum equipment of ABS and Airbag;
- Low acquisition cost.

After establishing selection criteria was elaborated a study of the local and national markets. For economic reasons it chose the second hand market. It was determined that the most appropriate range of vehicles for the electrification would be of those models that have an independent frame design, so that the number of candidates would be presumably minimized, since the trend of the past 15 years in the manufacture of cars has been developed around the self-supporting body (Table 2).

**Table 2** Technical characteristics of selected vehicle

| Model | Suzuki (Jimny-05) |
|---|---|
| Dimensions [mm] | 3,625 × 1,600 × 1,655 |
| Weight [kg] | 1,140 |
| Power [CV] | 81 |
| Load Capacity [kg] | 360 |

In comparison with other similar models, it selected the Suzuki Jimny-05 vehicle by its higher load capacity, greater amount of safety devices, its reliability and likeness in dimensions one of the ZYTEL vehicles (Zytel GORILA), so fulfilling one of the conditions laid down by the company.

## *4.2 Motor Selection*

The chosen motor type was one of type DC Brushless permanent magnets due to its high torque, high speed of rotation, high acceleration, high reliability and a characteristic curve linear torque-speed. Due to the high price of DC Brushless motors, also the possibility of choice of a squirrel-cage asynchronous motor is valued as another option.

For the purchase of the engine, was first performed an analysis of the market. The selection criteria were based on the technical characteristics of the project and existing technology. Set out below are the criteria used:

- High torque and power up to 45 kW;
- Low weight;
- High efficiency and operating voltage;
- Dimensions contained for easy location.

After the search for distributors and manufacturers of motors, a first selection of engine models and a comparison was made between them. All selected motors had a high efficiency over 90 %, with voltages higher than 300 V and maximum speed over 5,000 rpm; being these factors valid for design. The factors that determined the selection of the motor were the specific power (kW/kg) and its dimensions, standing out above the rest and being the final choice, the motor PH145 of UQM with maximum power 145 kW and 94 % of efficiency. In Fig. 2 below is detailed the efficiency's map and torque-power curves running as a motor:

### 4.2.1 Controller

The UQM-PH145 motor has to operate with a controller from the manufacturer, the controller/inverter DD45-500L. This controller is based on half-bridge architecture with 3 IGBT as elements of power and a digital signals processor (DSP). The addition of the DSP allows that the controller, also the motor control, incorporated communication series, CAN Bus support, diagnostic capability, speed reading and reading of the motor's temperature and controller's temperature. The controller allows control the motor from three different modes: torque control, speed control and voltage control. The controller-motor unit is a machine capable of working in all four quadrants of operation, so it will be possible to its operation as a generator, or what is the same, will have regenerative braking function.

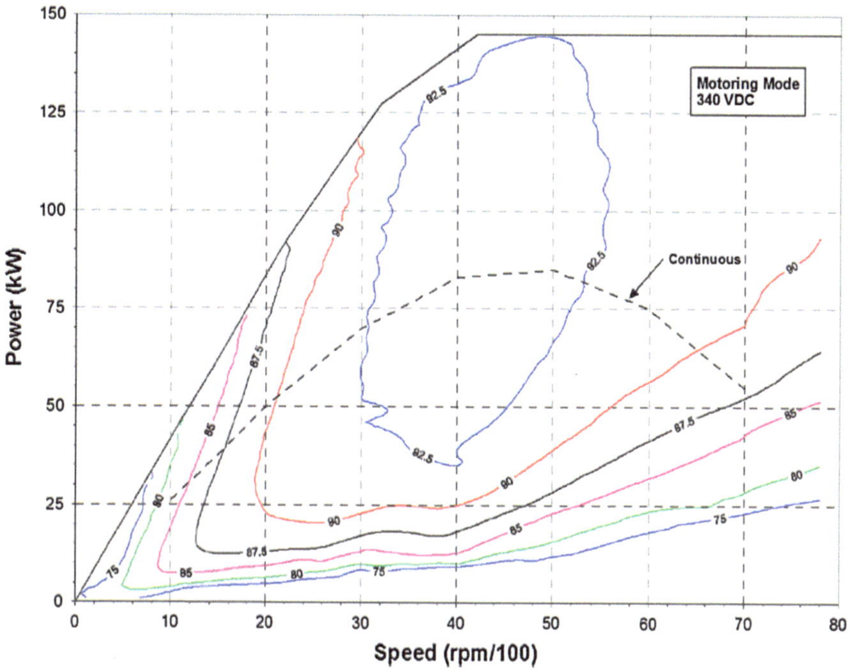


**Fig. 2** Efficiency and torque-power graph of PH145 motor

### 4.2.2 Electrical Connections

The electrical circuit of the motor-controller set describes the connection between the motor, controller and batteries as a power circuit and the controller as a control circuit. The electrical connection between the motor and the controller has three different connection types (Fig. 3). The first one is the connection of power with the wires connecting the motor to the controller installed on the motor. The second is the connection of data transmission using a connector amphenol-21 of the controller. The third is the union of the shells between motor and controller to balance the electrical potential between them and thus reduce pollution emission electromagnetic behaving both as a Faraday cage.

## *4.3 Battery Selection*

The choice of one type of battery or other determines the vehicle's weight, the maximum discharge, the power's motor and cycle life of the batteries. Based on the results of Sect. 3, the batteries should store 70 kWh of energy with a limitation of 700 kg. These values of energy and weight result in the batteries must have a specific energy greater than 100 Wh/kg, so according to the studied batteries is

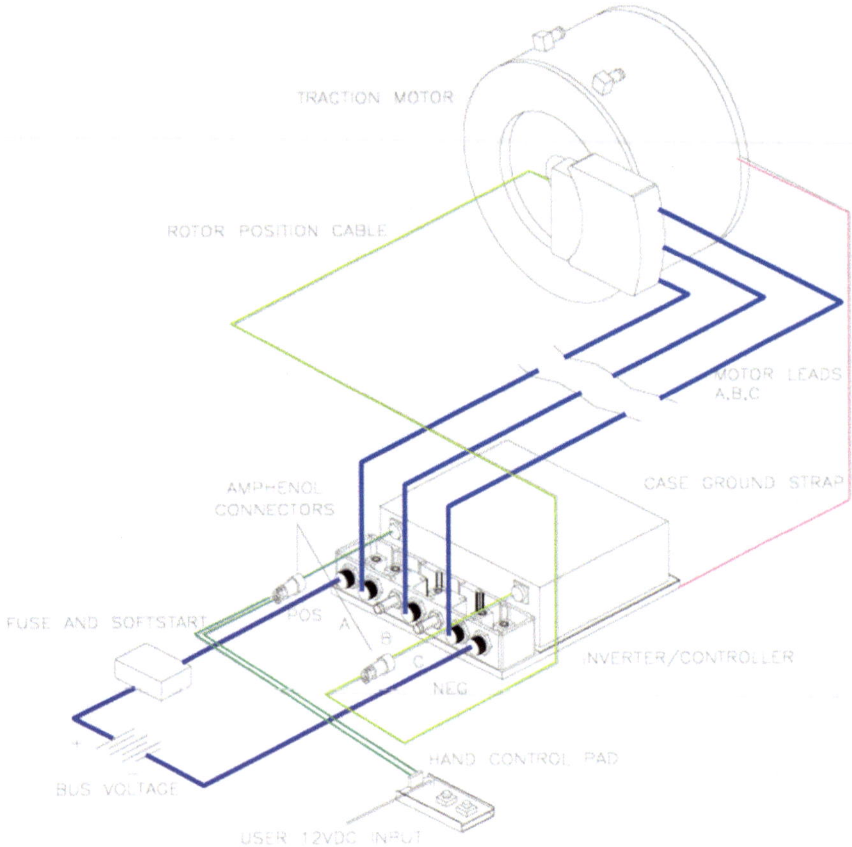


**Fig. 3** Electrical system connections

used two types of batteries based on lithium or sodium. Sodium batteries operate at a very high temperature (>250 °C) which compromises the safety of occupants and the reliability of the whole. This makes this typology of battery selected is based at first in lithium batteries [6, 7].

To purchase the batteries, firstly it performed an analysis of the market where the selection criteria were based on the technical characteristics of the project and existing technology, below are the criteria used:

- High specific energy greater than 100 Wh/kg;
- High specific power greater than 200 W/kg;
- High discharge capacity, higher than 3C (Battery Capacity [Ah]);
- High load capacity than 1C and low internal resistance less than 5 mΩ;
- Low sensitivity to temperature;
- Low self-discharge, less than 1 % a week;
- Cycles of life, over 500 cycles.

### 4.3.1 Properties of the Battery Selected

After analyzing the various manufacturers European, American and Asian, battery select was SE180AHA of the Sky Energy manufacturer since it is the most similar to the requirements. The vehicle traction battery is composed of 120 units connected in series and divided in 4 different locations on the vehicle. The battery pack provides a total energy of 71.28 Wh, 396 V nominal voltage, with a minimum voltage of 240 V and a weight of 672 kg. The internal resistance afforded by the battery assembly is 360 mΩ. The battery system is a crucial component in the development of the project. This motivates mainly perform three tests to define correctly the battery and know its behavior. Initially there is information provided by the manufacturer even so it is important to carry out two tests on the batteries. The first test is aimed at knowing the internal resistance of the battery which is calculated and resulted 3.02 mΩ. Download curves with different load are obtained in the second test. Finally to get the maximum energy stored in the traction battery set each battery should be balanced individually during charging [8]. This action is not possible due to the type of battery so it adopts solution for balancing previously to installing of batteries in the vehicle.

## *4.4 Selection of Auxiliary Components*

Systems that integrate a vehicle equipped with an internal combustion engine vary to incorporate electric motor due to amending or some of its components are removed. In addition, for the operation of the electric traction vehicle is necessary to introduce other new systems. Cooling system, brake system, steering system, auxiliary battery system, electric power system and the system of traction's batteries—Battery Management System (BMS) are new systems that will be incorporated or modified but which here weren‘t detailing descriptively.

# 5 Calculation Process

The dynamics of a vehicle is composed of existing forces in the three axes of space, two axes included in a horizontal plane, and a third axis in the vertical plane. The vertical axis represents the vertical dynamic behavior of the vehicle and is necessary to optimize the reactions to ensure the comfort and safety of passengers. The suspension system is responsible for this. Modification of the vehicle's weight, involves a change in its vertical dynamics. Therefore incorporating batteries to perform the transformation in electric vehicle should redesign the suspension system, taking into account tyres, springs and shock absorbers. These modifications and its satisfactory results shall grant the vehicle one greater safety index, after being amended the overall weight, the width of the axes, unsprung weight and changed the springs and dampers of the suspension to support a greater weight. As seen throughout the project there are many elements that directly affect vehicle safety. The gravity center is one of the essential elements for vehicle dynamics, being closely related to height, so it is particularly important in the design of the vehicle. Logically, a lower gravity center height positively influences the overall balance of the vehicle, improving stability, weight distribution and braking. Other important aspect is linked to the clearance height and the angle of attack, mainly when they are moving by ground badly paved or unpaved. These parameters are changed by the weight and the placement of the batteries. So it should study how have been modified and where necessary raises any appropriate modifications for the new vehicle complies with all legal and functional requirements. To increase the clearance height it could be increased the diameter of the wheels, raise the height of the frame by increasing the length of the suspension and/or raise the height of the bodywork respect to the frame.

## 6 Planning and Implement Electrification Process

The electrification vehicle process consists of 7 different phases in chronological order with the possibility of overlap between them. These phases are distributed as follows:

1. Removing MACI and body elements
2. Adaptation of existing systems
3. Implementation and installation of new components
4. Modifications to the bodywork
5. Installation and adjustment of electric traction
6. Installation of electrical circuit and batteries
7. Test and the revision of the components.

## 7 Conclusions and Possibilities For Continuation

The possibilities for continuation of the project are many although all go in the same direction: increase the degree of optimization of the electric vehicle. On the one hand could verify the accuracy of the calculations of consumption and autonomy. Calculation models, proposing new specifications for the optimization of vehicle and its subsequent verification in real conditions could be extended in this way. It also is studying the possibility of assembling multiple sensors on the vehicle, to perform a very valuable data acquisition to deepen the knowledge of the behavior of the vehicle and its components, especially the traction systems and batteries. Currently, the developed vehicle is monitored and adequately equipped with measurement equipment and allows evaluating the different types of batteries depending on the technical characteristics of the vehicle and the journey to perform, therefore a battery testing vehicle has been performed and evaluation of vehicle performances can be done in order to test battery behavior in different driving conditions.

## References


1. http://www.zero-race.com
2. Directive 2007/46/EC 2007. Parliament and European Council
3. Directive 70/156/EEC 1970. European Council
4. Regulation 10 and Regulation 100. Directive 2007/46/EC
5. Regulation 101. Directive 2007/46/EC
6. Fink A, Beaty H (2007) Standard handbook for electrical engineering. McGrawHill, New York
7. Linden D (2010) Handbook of batteies. McGrawHill, New York
8. Affanni A et al (2005) Battery choice and management for new-generation of electric vehicles. IEEE Trans Ind Electron 52(5): 1343–1349

# Strategy Car Performances Selection for an Efficient in Urban Freight Distribution

**Sara Sánchez, Alberto Fraile and Emilio Larrodé**

**Abstract** In this paper, different strategies of electrical vehicles for urban freight distribution have been defined in function of the capacity of load. The transport operation can be performed through an only electrical vehicle with large load capacity able to perform by itself all the freight distribution process established one journey, or distribution the load among several small electric vehicles with maximum load of transport of 500 kg, or through an optimal combination among an only large electrical vehicle and a determinate number of small electrical vehicles with maximum load of transport of 500 kg. It has been performed several simulations of different strategies of distribution, modifying the load capacity of a large electrical vehicle and modifying the number of small electrical vehicles with a maximum load of transport of 500 kg, until to find the optimal combination for a efficient system of freight urban distribution.

**Keywords** Electric vehicle · ADVISOR · Drive cycles · Energy consumption cost · Logistics cost

S. Sánchez (✉) · A. Fraile · E. Larrodé
Campus Río Ebro, Edificio Betancourt, Universidad de Zaragoza, C/María de Luna s/n, Zaragoza 50018, Spain
e-mail: ssanchez@unizar.es

A. Fraile
e-mail: afrailep@unizar.es

E. Larrodé
e-mail: elarrode@unizar.es

J. Fischer-Wolfarth and G. Meyer (eds.), *Advanced Microsystems for Automotive Applications 2013*, Lecture Notes in Mobility, DOI: 10.1007/978-3-319-00476-1_24, 

# 1 Introduction

Nowadays, the transport systems are considered a vital function but the people know his several negative effects like the emissions of exhaust, particulate matter, the contribution to greenhouse and their consequent deterioration of habitats and climate change, noise and accidents among others [1].

For the economic development of the cities, regions and countries, the freight urban distribution systems are fundamental, but his negative effects should be consider. A feasible solution for decrease of these factors in the environment urban is the design of electric vehicle fleets that perform the freight urban distribution [2]. With them, not only the pollution emissions decreases, also waste materials that can eventually contaminate the environment. These kinds of vehicles, help to the decrease of noise pollution generating a more comfortable and quiet city. Moreover, the quality of the transport favors the environment and intervenes on the life quality of the cities [3].

# 2 Experimental Procedure

The analysis has been performed through the software of simulation Advanced Vehicle Simulator (ADVISOR). It's a tool of flexible modeling with which it can evaluate the operation and economy of fuel conventional vehicles with internal combustion engine, electrical vehicles, fuel cell vehicles and hybrid vehicles among others [4].

ADVISOR has a large database of components of vehicles and several configurations. It has also a details information about internal combustion engines, electric engines, fuel cell and batteries as well as different strategies of control of powertrain, several exhaust systems, transmissions, couplers pair, wheels, axis and several accessories [5, 6].

The software includes also a wide range of drive cycles with the possibility of generate tests in custom cycles. The drive cycles consists in a velocity–time sequence, obtained with statistic information represented in a specific area. This cycles are considerate very helpful for can obtain experimental factors of pollution emissions, performance in vehicle, fuel consumption, etc., of several vehicles. For perform a perfect measurement of the performances in vehicle design and emissions generates by the vehicles, the drives cycles show the dynamic of conduction observed in real operations and represents the actual driving mode. For this, different countries and organizations, perform, define and standardize the cycles, in order to compare different vehicles that they have been tested in the same drive cycles.

The drive cycles used in the research, are urban cycles with a maximum velocity lower to 50 km/h, with several stops and starts that simulate the normal characteristics of conduction in urban environmental like a traffic lights, roundabouts, etc. the cycles present large travel distance, around 20 km, because they are cycles associated to big cities like Madrid, Barcelona, Paris, etc., where there are greater difficulty in the freight urban distribution.

## 3 Results and Discussion

For the analysis of operation costs, it has been considered the costs of the fuel, named energy consumption costs, and the costs of the salary of the driver, named logistics costs.

The analysis of the transport operation costs has been performed to a real case of 6,000 kg of load demanded with two demand points of 3,000 kg, point 1 and point 2 of load demanded, Fig. 1. All the vehicles should begin the system distribution on the point O, which it is the point where there is the company. It has been looked for the optimum combined with which we obtain the best result. It has been established five different cases of distribution, through the combined of large and small electrical vehicles with different loads.

In the first case, it has been established a large electric vehicle able to distribute all the freight by itself and the next cases of study, it has been incorporated small electric vehicles for the transport of the freight. In the different cases, the large vehicle going to transport less load and the small electric vehicles going to increase his number of units required for able to distribute all the load demanded. Of this way, the last case of study of distribution of freight is through twelve small electric vehicles with a maximum load of 500 kilograms each vehicle, Fig. 1.

The first case, consists in an only large vehicle able to perform the system distribution by itself, which transports the 6,000 kg of the load demanded. On the second case, there is one large vehicle, which transports 4,000 kg, and four small electric vehicles with 500 kg how maximum load capacity. The third case consists in a medium electric vehicle, which transports 3,000 kg, and six small electric vehicles with maximum load capacity of 500 kg. On the fourth case, there is a medium electric vehicle that transports 2,000 kg and eight small electric vehicles with 500 kg how maximum capacity of load. Last, the fifth case consists in twelve small electric vehicles with maximum capacity of load of 500 kg. The cases of the small electric vehicle with maximum capacity of load of 500 kg, are a subcontract so as to perform the path imposed therefore their came back to the company has not been considered.

The powertrain configurations of all the vehicles involved on the five cases, are in the Table 1.

All the cases studied, have been simulated with the drive cycle of Fig. 2. It is a cycle with urban character, with several stops and starts simulating in this way the effect of roundabouts, traffic lights, among others, and his maximum speed id of 45 km/h. This cycle, has been simulated in each of the stage of each distribution system of the different cases studied.

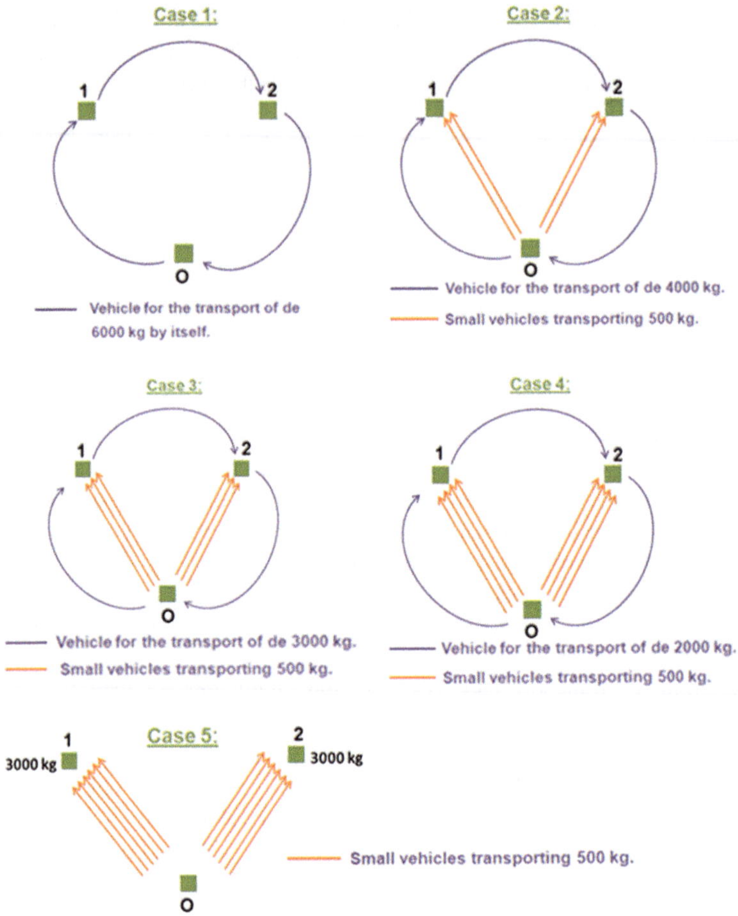


**Fig. 1** Cases of study of the different distribution systems

**Table 1** Powertrain configuration of the vehicles

| Configuration of the vehicles | | Case 1 | Case 2 | Case 3 | Case 4 | Case 5 |
|---|---|---|---|---|---|---|
| Large electric vehicle | Load transported (kg) | 6,000 | 4,000 | 3,000 | 2,000 | – |
| | Batteries (Ah) | Pb (91) | Pb (91) | Pb (91) | Pb (91) | – |
| | N° of modules | 93 | 70 | 43 | 28 | – |
| | Electric Engine (kW) | 219 | 219 | 123 | 87 | – |
| | Weight on empty (kg) | 7,833 | 4,612 | 3,450 | 2,033 | – |
| Small electric vehicles | N° units required | – | 4 | 6 | 8 | 12 |
| | Batteries (Ah) | – | Pb (91) | Pb (91) | Pb (91) | Pb (91) |
| | N° of modules | – | 11 | 11 | 11 | 11 |
| | Electric Engine (kW) | | 75 | 75 | 75 | 75 |
| | Weight on empty (kg) | – | 1,595 | 1,595 | 1,595 | 1,595 |

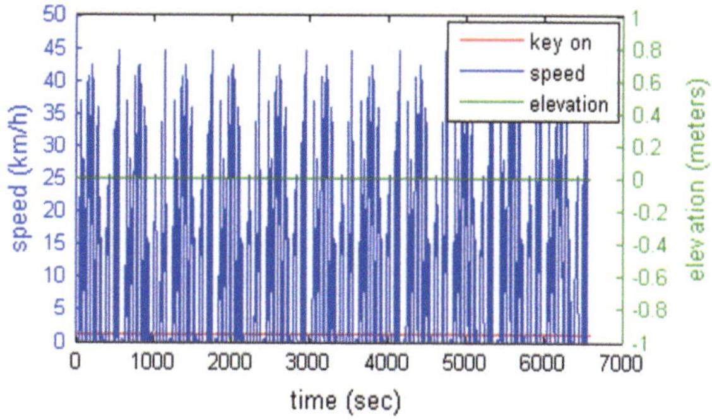


**Fig. 2** Drive cycle of all the cases studied

## *3.1 Energy Consumption Costs*

The energy consumption costs of the vehicles, equation [1], it is based on function of the energy consumed by the vehicles in each one of the cases studied. With the help of ADVISOR, we obtain the energy consumption in kJ and with the applying the current rate of electricity in force in Spain, it has been able to obtain the economic data. It has been established that the vehicles should be recharged by the night therefore the electricity tariff is cheaper.

$$C_i = \sum\nolimits_{j=1}^{n} \left[ (Energy\,consumption)_{\mathrm{j}} * (Electricity\,tariff) \right] \quad (1)$$

Being n the total number of vehicles involved on the distribution system of each case, therefore, the values of i and of n for each one of the case are on Table 2.

On the Fig. 3, it is possible to observe the economic data of the energy consumption costs in function of the number of small electric vehicles, with maximum capacity of load of 500 kg, involved in each one of the cases.

On Fig. 3, it is possible to observe that it exist a minimum on the energy consumption on the second case. Of the five cases studied, the second case, with four small electric vehicles with maximum load of transport of 500 kg each of them, is who presents lowest energy consumption and therefore is who presents lowest energy consumption costs.

**Table 2** Values of i and n for each one of the cases studied

| Case | i | n |
|---|---|---|
| 1 | 1 | 1 |
| 2 | 2 | 5 |
| 3 | 3 | 7 |
| 4 | 4 | 9 |
| 5 | 5 | 12 |

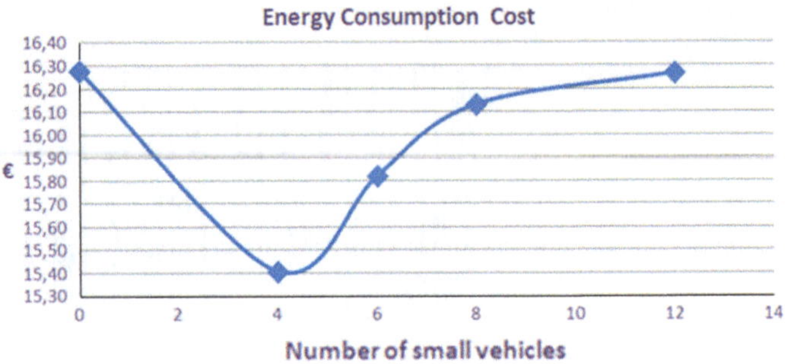


**Fig. 3** Energy consumption costs of the cases studied in function of the number of small electric vehicles involved in each case

## *3.2 Logistics Costs*

The calculating of logistics costs, equation [2], has been based on the total number of vehicles involved in each case and with them, the number of the drivers necessaries, the time of each driver for perform to the route and the time of each operator for loading and unloading of the goods.

$$C_i = \sum_{j=1}^{n} \left[ (Driver\,salary) * (t_{route})_{\rm j} \right] + \sum_{\rm j=1}^{\rm n} \left[ (Operator\,Salary) * \left(t_{l/u}\right)_{\rm j} \right] \quad (2)$$

Being $i$ the cases studied, $n$ the total number of the vehicles involved in each case, Table 1, $t_{route}$ the time of the route and $t_{l/u}$ the time of the loading and unloading of the goods.

In this case, with increasing of the number of small electric vehicles, increases also the logistics costs, Fig. 4.

**Fig. 4** Logistics costs of the cases studied in function of the number of small vehicles involved in each case

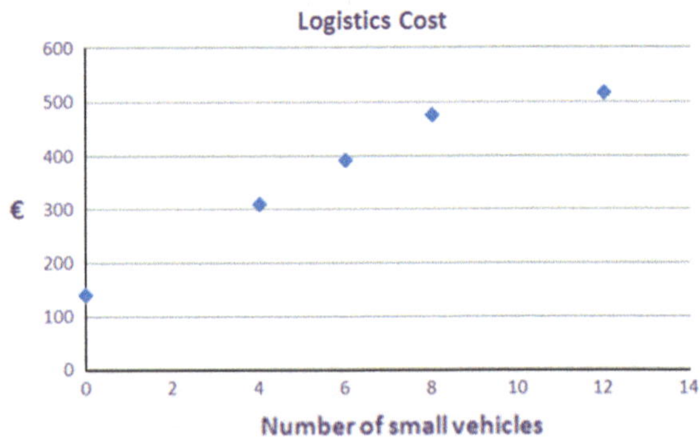

### 3.3 Analysis of the Operation Costs

Observing the energy consumption costs, there is a minimum energy consumption on the second case, that is to say, the optimal energy is obtained with two small electrical vehicles transporting each one 500 kg to the point 1, others with two small electrical vehicles transporting each one 500 kg to the point 2 and a medium electric vehicle transporting 4,000 kg so that unloads 2,000 kg on the point 1, then unloads the others 2,000 kg and last comes back empty to the company, point O.

In return, with the logistics costs there isn't a minimum, because the logistics costs depends on the total number of vehicles involved in the distribution system, so that as the number of vehicles involved increases, the logistics costs also increases (Table 3).

When the total operation costs has been calculated, although the energy consumption costs has a minimum value in the second case, the logistics costs is higher than the energy consumption costs and therefore the total operation costs increases with the increase of the number of vehicles involved in each case, Fig. 5.

Also the different between the energy consumption costs and the logistics costs has been analyzed, Fig. 6 and it is quite high.

The costs function of the transport operation depends on the energy consumption and logistics costs. In this case, the difference between energy consumption costs and the logistics costs, Fig. 6, is quite high because the transport fleet is very small and must be taken in account that when the size of the fleet increases, the logistics costs decreases.

**Table 3** Total operation costs

| Case | Energy consumption costs (€) | Logistics costs (€) | Total operational costs (€) |
|---|---|---|---|
| 1 | 16.28 | 140.07 | 156.35 |
| 2 | 15.41 | 308.43 | 323.84 |
| 3 | 15.82 | 392.61 | 408.43 |
| 4 | 16.13 | 476.79 | 492.92 |
| 5 | 16.27 | 518.88 | 535.15 |

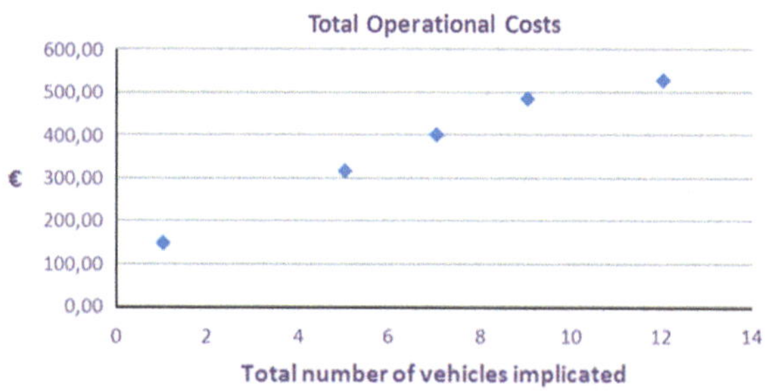


**Fig. 5** Total operation costs in function of the number of vehicles involved in each case

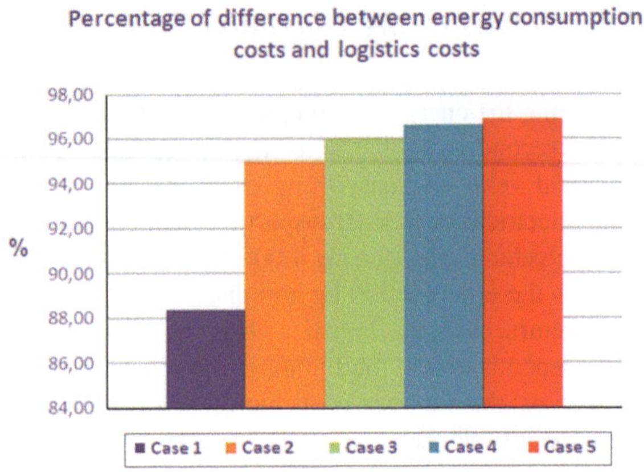


**Fig. 6** Percentage of difference between the logistics costs and the energy consumption costs

When the fleets have a big size, the difference between logistics costs and energy consumption costs will decrease and in this case, it will be very important identify the distribution system with the minimum energy consumption costs.

## 4 Conclusions

To perform an analysis of operation costs, it has been defined five different cases of distribution system of freight. The analysis of the transport operation costs has been performed to a real case of 6,000 kg of load demanded with two demand points of 3,000 kg, point 1 and point 2 of load demanded, and the company located in the center of the city, Fig. 1.

In the first case, the distribution system is performed with an only large electric vehicle able to perform by itself all the distribution system. Along different cases has been increased the presence of small electric vehicles with a maximum capacity of load of 500 kg, so that, in the last case the distribution system is performed with twelve small electric vehicles with a maximum capacity of load of 500 kg.

The operation costs have been calculated with the energy consumption costs and with the logistics costs. In the energy consumption costs, there is a minimum energy consumption in the second case, with four small electric vehicles with a maximum capacity of load of 500 kg and a medium size electric vehicle with the other 4,000 kg of load, Fig. 3.

In return, in the logistics costs, as this costs depends on the number of vehicles involved on the distribution system, the logistics costs increases when the total number of vehicles involved increases also, Fig. 4.

In this way, as the logistics costs is bigger than the energy consumption costs and despite of the minimum energy consumption costs of the second case, the total operation costs increases when the total number of vehicles implicates on the distribution systems increases, Fig. 5.

In this case, the size of the fleet is small and exits a great difference between logistics costs and the energy consumption costs, Fig. 6. It should be taken into account the size of the fleet. When the fleet will be high, with an elevated number of vehicles, the difference between logistics costs and energy consumption costs, will be lower than the difference of this case.

So, in order to select an appropriate number of vehicles for the transport operation in urban freight distribution, it is necessary to evaluate both, the energy consumption costs and the logistics costs.

## References

1. Doll C, Wietschel M (2008) Externalities of transport sector and the role of hydrogen in a sustainable transport vision. Energy Policy 36: 4069–4078
2. Oltra V, Saint Jean M (2009) Variety of technological trajectories in low emission vehicles (LEVs): a patent data analysis. J Cleaner Prod 17:201–213
3. Gaffney JS, Marley NA (2009) The impacts of combustion emissions on air quality and climate-from coal to bio fuels and beyond. Atmos Environ 43:23–36
4. Jhonson VH (2002) Battery performance models in ADVISOR. J Power Sources 110:321–329
5. Markel T, Brooker A, Hendricks T, Jhonson V, Kelly K, Kramer B, O'keefe M, Sprik S, Wipke K (2002) ADVISOR: a system analysis tool for advanced vehicle modeling. J Power Sources 100: 255–266
6. Pesaran A (2002) Battery thermal models for hybrid vehicle simulations. J Power Sources 110:377–382

# Part IV
# Energy Efficiency

# A Bandwidth Enhanced Regenerative Suspension System for Electric Vehicles

**Chen-Yu Hsieh, Bo Huang, Farid Golnaraghi and Mehrdad Moallem**

**Abstract** This paper proposes a novel energy harvesting methodology utilizing a regenerative vehicle suspension system. Unlike conventional linear generator, this new generator is able to bend the frequency response curve; thus extending the bandwidth of the system. A method of realizing nonlinear stiffness force in a vehicle suspension through a switch-mode power converter with hysteresis current control (HCC) has been presented [1]. Adopting the mechanical–electrical analogy the mechanical stiffness constant maps to an inverse electrical inductor, a new method of inductance emulation using Variable Active–Passive Reactance (VAPAR) has been implemented [2]. The proposed technique is analyzed using perturbation method and its effectiveness is verified through numerical simulations.

**Keywords** Electric vehicles · Regenerative suspension · Energy harvesting · Bandwidth enhancement · Variable inductance

## 1 Introduction

Recent attention to vibration-based energy harvesting has resulted in the creation of technique for regenerative vehicle suspension. The technique converts kinetic motion of the suspension into electrical power and then reserves it in the energy

C.-Y. Hsieh · B. Huang · F. Golnaraghi (✉) · M. Moallem
Mechatronic Systems Engineering, Simon Fraser University, Room 4240, Surrey, BC, V3T 0A3, Canada
e-mail: mfgolnar@sfu.ca

C.-Y. Hsieh
e-mail: chenyuh@sfu.ca

B. Huang
e-mail: bha23@sfu.ca

J. Fischer-Wolfarth and G. Meyer (eds.), *Advanced Microsystems for Automotive Applications 2013*, Lecture Notes in Mobility, DOI: 10.1007/978-3-319-00476-1_25,

storing device for further uses, such as supplying power for wireless sensor networking and active control, etc. To date the linear resonant generator is the most common device used in energy harvesting application. Kawamoto et al. proposed an electromagnetic damper consisting of DC motor and the ball screw mechanism [1]. It can be used as an active suspension and energy generator. Li et al. [2] introduced an electromagnetic shock absorber based on a rack-pinion mechanism for the purpose of energy harvesting and vibration damping. In these designs, the generator will capture the maximum power at its resonant frequency. However, a drawback occurs in that the performance will be sacrificed significantly if the difference between two frequencies exists because of the narrow bandwidth of this system.

To overcome the disadvantage of the linear resonant generator, the nonlinear strategies for the purpose of extending operating frequency region is studied. Ramlan et al. [3] investigated a nonlinear mechanism with hardening stiffness, which provided a greater bandwidth. Burrow et al. [4] added magnetic reluctance forces to the linear spring compliance to provide a wider bandwidth of operation. All of above utilized the similar nonlinear characteristics of Duffing's equation, which contains cubic nonlinear spring force. Therefore, adding cubic nonlinearities into suspension system are supposed to bend the frequency response curve; thus extending the bandwidth of the system.

A new method of inductance emulation using Variable Active–Passive Reactance (VAPAR) has been implemented to control the existing reactance in a switch-mode power converter [5]. The produced active reactance works as a passive reactive component, even in transient state. The method of emulating linear time varying inductance is expected to be a fast and high efficiency/power rating alternative to conventional op-amp based active inductor, providing a possible application to simulate the dynamics of a mechanical mass-spring-damper system with variable stiffness.

The available power analyses involving both methods of linear and nonlinear have been demonstrated. The linear method involves merely connecting a discrete power resistor across the back EMF terminal of dc motor. The nonlinear method includes not only terminal resistance but also a nonlinear variable inductor (NVI) to introduce cubic nonlinear dynamics in the electrical domain similar to those of Duffing's equations [7].

## 2 System Description and Modeling

The regenerative suspension system can be simplified by a quarter-car model, which is shown in Fig. 1. The model consists of sprung mass, physical spring, linear to rotational conversion mechanism, dc motor/generator, and back-end switch-mode power converters.

First, we define relative displacement to be: $z = x - y$ with $x$ indicating the movement of the sprung mass and $y$ indicates the movement of the base excitation.

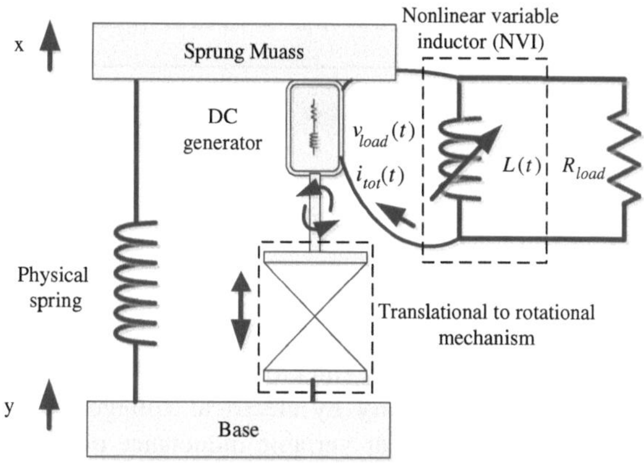


**Fig. 1** Base excitation model of the proposed model incorporating nonlinear force with power electronics circuitry

The governing equation of the suspension model while implementing either linear or cubic nonlinear method are shown in Eqs. (1a) and (1b), respectively,

$$m\ddot{z} + kz + \left(\frac{k_b k_t}{R_{load} d^2}\right)\dot{z} = F_{road} \tag{1a}$$

$$m\ddot{z} + \frac{k_b k_t}{R_{load} d^2}\dot{z} + \left(k + \lambda_L \frac{k_b k_t}{d^2}\right)z + \lambda_N \frac{k_b^3 k_t}{d^4} z^3 = F_{road} \tag{1b}$$

where

$$F_{road} = my\omega^2 \cos(\omega t + \tau) \tag{2}$$

The coefficients of cubic nonlinearities are indicated by $\lambda_N$ and $\lambda_L$. Moreover, $R_{load}$ is the load resistance connected to the motor, which provides the equivalent electrical damping to the vehicle's suspension system. $d$ is the constant converting translational motion to rotational, which is determined by the reference point and properties of the translational to rotational mechanism. $F_{road}$, $k_t$ and $k_b$ are the road force introduced to the proposed vehicle suspension model, torque and voltage constant of the motor, respectively.

In the case of adopting the linear method, the total current drawn from the motor only flows through the load resistance, thus, resulting the equivalent electrical damping coefficient $\frac{k_b k_t}{R_{load} d^2}$ as shown in Eq. (1a). In the case of adopting the cubic nonlinear method one has to consider the total current as the sum of load current and the current flowing through the nonlinear variable inductor (NVI). Thus, as shown in Table 1 would have the equivalent damping and stiffness terms $\frac{k_b k_t}{R_{load} d^2}\dot{z} + \lambda_L \frac{k_b k_t}{d^2} z + \lambda_N \frac{k_b^3 k_t}{d^4} z^3$ as shown in Eq. (1b). The resonance frequency is modified to be $\sqrt{\frac{k}{m} + \lambda_L \frac{k_b k_t}{m d^2}}$, the equivalent damping force introduced to the system would be $\frac{k_b k_t}{R_{load} d^2}\dot{z}$, and the equivalent stiffness force introduced to the system, by the switch-mode power converter, would now be $\lambda_L \frac{k_b k_t}{d^2} z + \lambda_N \frac{k_b^3 k_t}{d^4} z^3$.

**Table 1** Stiffness and damping in linear and nonlinear systems

| System | Linear | Nonlinear |
|---|---|---|
| Equivalent linear stiffness | $k$ | $k + \lambda_L \frac{k_b k_t}{d^2}$ |
| Equivalent nonlinear stiffness | n/a | $\lambda_N \frac{k_b^3 k_t}{d^4}$ |
| Equivalent damping | $\frac{k_b k_t}{R_{load} d^2}$ | $\frac{k_b k_t}{R_{load} d^2}$ |

As previously mentioned, we have adopted the mechanical–electrical analogy to represent mechanical force by electrical current, mass displacement by voltage flux, and mass velocity by electrical voltage. Moreover, recalling the previous derivation on nonlinear variable inductance one will realize in order to correctly supply cubic nonlinearity to the system the HCC controller should always maintain both the instantaneous current $i_L(t)$ and voltage flux $\varphi(t)$ to the following relationship:

$$i_L(t) = \lambda_N \varphi(t)^3 + \lambda_L \varphi(t) \tag{3}$$

Also, if the controller is able to satisfy the relationship as indicated in Eq. (3) the instantaneous nonlinear variable inductor will be presented by the function of the following:

$$L_{ref}(t) = \left(3\lambda_N \varphi(t)^2 + \lambda_L\right)^{-1} \tag{4}$$

While knowing the NVI current $i_L(t)$, the resistance current and total current would be

$$i_{tot}(t) = \lambda_N \varphi(t)^3 + \lambda_L \varphi(t) + \frac{v_{load}(t)}{R_{load}} \tag{5}$$

where $v_{load}(t)$ is the instantaneous load voltage.

## 3 Theoretical Analysis

In this section, the perturbation method will be utilized to determine the effect of the cubic nonlinearities. Especially, the method of multiple scales will be used. The dynamic equation of the suspension system as shown in Eq. (1b) can be rewritten as

$$\ddot{z} + 2\omega_1 \zeta_1 \dot{z} + \omega_1^2 z + \alpha \omega_1^2 z^3 = Y_1 \cos(\omega t + \tau) \tag{6}$$

where

$$\omega_1 = \sqrt{\frac{k}{m} + \lambda_L \frac{k_b k_t}{d^2 m}}, \zeta_1 = \frac{k_b k_t}{2\omega_1 R_{load} d^2 m}, \alpha = \lambda_N \frac{k_b^3 k_t}{\omega_1^2 d^4 m}, Y_1 = y\omega^2$$

### 3.1 Perturbation Analysis

Since most energy introduced from the road surface to the generator will be located in the region around resonance, we need to analyze the nonlinear effect on the primary resonance where $\Omega^* \approx 1$. Firstly, to make the solution independent we need to make $z$ dimensionless by using the initial displacement $z_0$. By using the method of averaging, the steady-state solution can be obtained as

$$\sigma = \frac{3}{8}\alpha a^2 \pm \left(\frac{f^2}{4a^2} - \mu^2\right)^{1/2} \tag{7}$$

where $\sigma$ is introduced as a detuning parameter since we consider the case $\Omega^* \approx 1$, $a$ is the amplitude of the dimensionless solution. It follows that the approximation of the solution is

$$z = z_0 a \cos(\omega t + \tau - \gamma) + \frac{1}{32} \in z_0 \alpha a^3 \cos(3\omega t + 3\tau - 3\gamma) + \cdots \tag{8}$$

where $\varepsilon$ the amplitude of motion and $\varepsilon \in (0, 1)$, $\gamma$ is the phase of the dimensionless solution.

### 3.2 Frequency–Response Curve

A representative curve of a as a function of $\sigma$ can be obtained for given $\mu$, $\alpha$ and f. By solving (7), the amplitude curve corresponding to the singular points in different state plane can be plotted. The solution is valid only if $a \leq \frac{f}{2\mu}$, so the peak amplitude is given by $a_p = \frac{f}{2\mu}$. Figure 2 shows the a – $\sigma$ curve with the different selected values of nonlinear factor $\alpha^*$.

From the figure we can see that the effect of the nonlinearity is to bend the amplitude curve to the right for positive nonlinearities or left for negative nonlinearities. As a result, the bandwidth of the generator is widened. The extent of bending is dependent on the value of $\alpha$.

The portions indicated by red circles are unstable and the corresponding steady-state motion cannot be physically realized by the system.

## 4 Principle and Realization of Variable Inductance

The NVI is essentially a switch-mode full-bridge switching topology power converter as shown in Fig. 3. It is consisted of four discrete power MOSFETs, discrete fixed power inductor, and a dc load, such as a dc motor. The converter is an active inductance emulator where the duty ratio of the MOSFETs are controlled by HCC

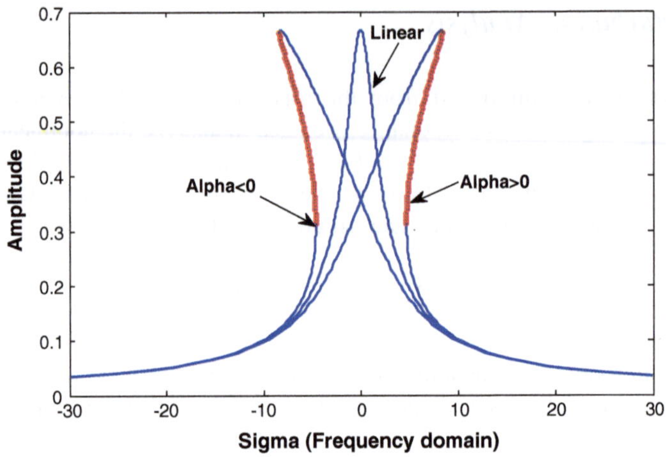


**Fig. 2** Frequency response curves of linear and nonlinear systems

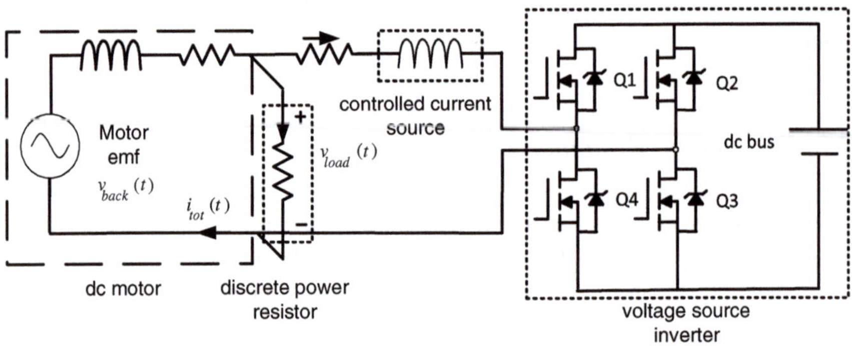


**Fig. 3** Switch-Mode NVI emulator

in order to maintain the correct relationship between $i_L(t)$ and $\varphi(t)$ as mentioned in the previous sections. Thus, according to the mechanical–electrical analogy adopted it is possible to vary the stiffness of a vehicle's suspension with the help of dc motor.

A method called hysteresis current control (HCC), as shown in Fig. 4, is well known for controlling the output voltage of a switching converter or inverter. It is implemented for regulating the actual measured instantaneous current, by means of comparators with hysteresis. Unlike the PWM current controller, the HCC controls the current on an instantaneous basis rather than an average basis. The switching instants of MOSFETs Q1-4 are determined by comparing the integral of the voltage error signal with the limits of a tolerance band placed about the desired reference current [3]. For the switching operation of the NVI, as the instantaneous current reaches the top boundary of the tolerance band MOSFETs Q1 and 3 are

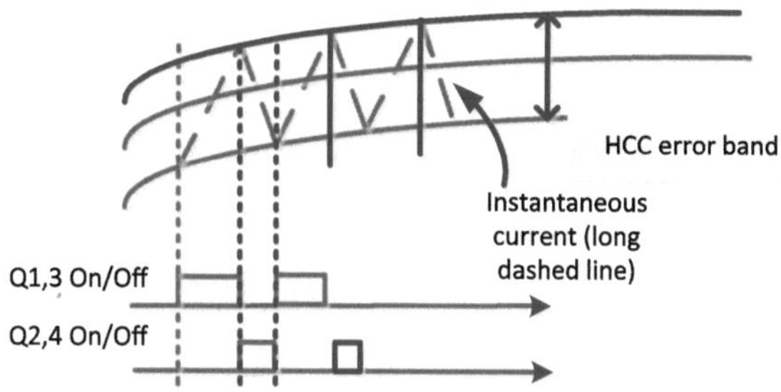


**Fig. 4** Power MOSFETs switching according to HCC

turned on, thus, causing the current to ramp down. When the decreasing current reached the pre-determined tolerance band MOSFETs Q2 and 4 are turn on, causing the instantaneous current to ramp up.

The main values of a quarter-car model adopted for simulating in MATLAB/ SIMULINK SymPowerSystem are: mass = 5.4 (kg), stiffness constant = 7,358.2 (N/m), d = 0.4, $R_{load} = 1\Omega$, and dc motor torque and speed constant $k_b = k_t = 2$ (N-m/A and V-sec/rad).

The frequency response curves shown in Fig. 5 were simulated assuming constant amplitude base excitation while forward sweeping across entire frequency spectra (normalized). The figures showed available power and relative displacement of both linear and nonlinear system with different degrees of nonlinearity. When the system is purely linear (i.e. $\lambda_N = 0$) there is no effect of "bending". Indeed, similar to the Duffing's equation the "bending" of the frequency curve appears to be more significant with the increasing value of nonlinear coefficient (i.e. $\lambda_N$), resulting higher available power and relative displacement in higher frequency spectra, thus, effectively enhancing the available power bandwidth.

## 5 Standard Road Profile Simulation

Taking the real-life application into account, a standard road excitation is generated from ISO 8608:1995 for verification. The road profile is shown in Fig. 6. Since the tire can be considered as a low-pass filter, the original road data is filtered and the smoothed data is used as the excitation. A Grade-B road profile is generated, and the speed of the vehicle is 70 km/h. Soft damping/stiffness is used where spring constant is 322,752 N/m and damping coefficient is 3,145 Ns/m. The natural frequency is 4.9 Hz and damping ratio is 0.15.

The responses of the linear and nonlinear systems are indicated in Fig. 7. The chassis displacement, chassis acceleration, relative displacement and instantaneous power are all displayed in the figure. The blue curves indicate the linear responses and the red curves indicate the nonlinear responses. All the simulation values are shown in Table 2.

It is known that the relative displacement is a proper measuring of road handling and the absolute acceleration is a proper measuring of ride comfort. The

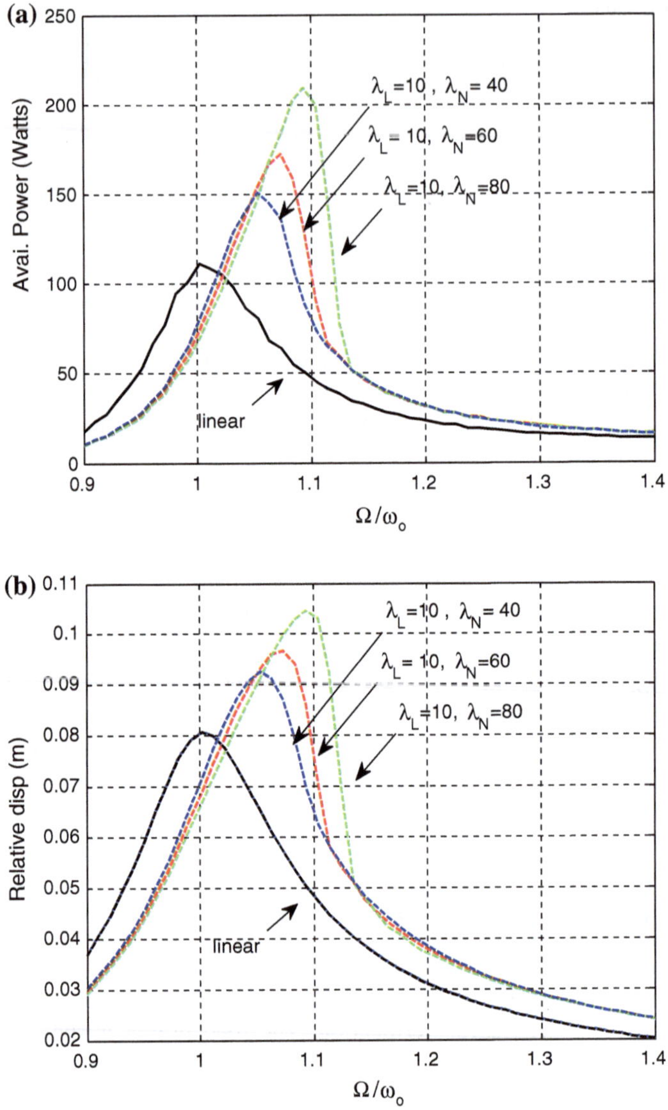


**Fig. 5** Bandwidth enhancement of (**a**) available power and (**b**) relative displacement with different levels of cubic nonlinearities generated by switch-mode NVI converter

comparison of average absorbed power, relative displacement and absolute acceleration of the linear system and nonlinear systems with low and high nonlinearities is listed in Fig. 8, where the values have been normalized with respect to the results of linear system so that their relative changes can be seen clearly, neglecting their absolute values. The results suggest that road handling is almost

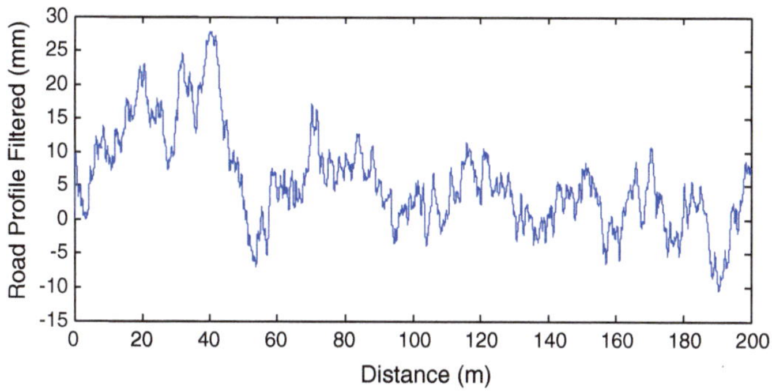


**Fig. 6** Standard Grade-B road surface profile

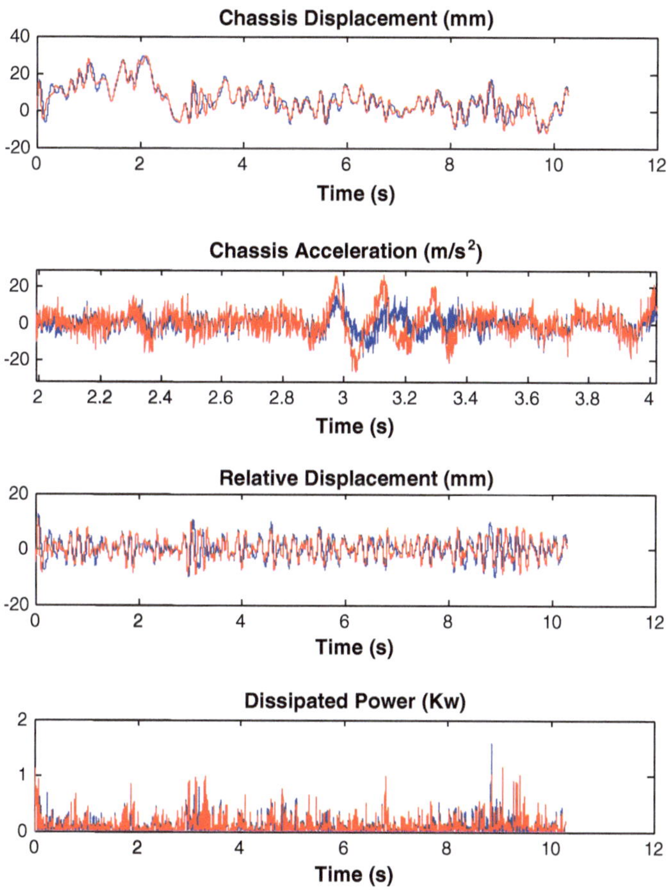


**Fig. 7** Comparison between responses of linear (*blue*) and nonlinear systems (*red*)

**Table 2** Simulation results of different systems

| | Linear | Alpha = 0.05 | Alpha = 0.1 |
|---|---|---|---|
| Average absorbed power (w) | 85.2 | 101.2 | 126.1 |
| RMS of Relative displacement (mm) | 3.6 | 3.7 | 3.5 |
| RMS of Absolute acceleration (m/s$^2$) | 5.8576 | 7.4693 | 9.4831 |

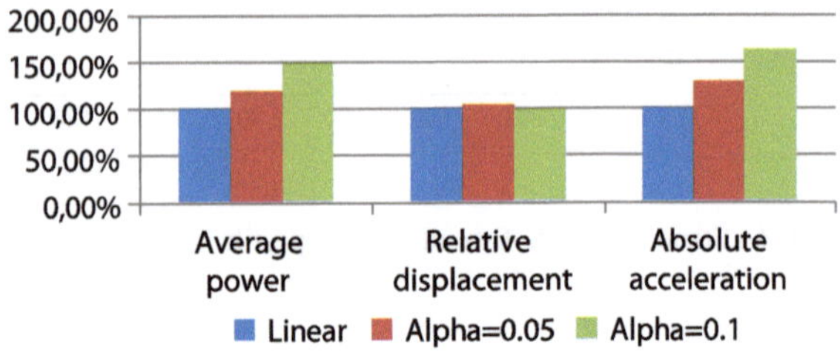


**Fig. 8** Comparison of the systems performance with different nonlinear factors

not affected by the nonlinearities, and the energy harvesting capability of the regenerative suspension system can be improved by dozens of percentage while the ride comfort can be sacrificed at an accepted level.

## 6 Conclusion

Incorporating cubic nonlinearities into the regenerative suspension system has the potential to enhance the bandwidth. The system will benefit from different frequency regions by tuning the nonlinear factor based on the different road conditions. The work of the VAPAR has not only simulated the linear and non-linear variable inductive operations but the capacitive and negative inductive operations were also successfully simulated [5]. The method of varying inductance is expected to be a high efficiency/power rating alternative to conventional op-amp based active inductor, providing a possible analogy to simulate both linear and nonlinear dynamics of a suspension system with time varying mechanical stiffness force.

## References

1. Kawamoto Y, Suda Y, Inoue H, Kondo T (2007) Modeling of electromagnetic damper for automobile suspension. J Sys Des Dyn 1(3):524–535
2. Zuo L, Li Z, Luhrs G, Lin L, Qin YX (2012) Electromagnetic energy-harvesting shock absorbers: design, modelling and road tests. Accepted by IEEE Transactions on Vehicular Technology
3. Ramlan R, Brennan MJ, Mace BR, Kovacic I (2010) Potential benefits of a non-linear stiffness in an energy harvesting device. Nonlinear Dyn 59(4):545–558

4. Burrow SG, Clare LR, Carrella A, Barton D (2008) Vibration energy harvesters with non-linear compliance. Proc SPIE 6928:692807
5. Funato H, Kawamura A, Kamiyama K (1997) Realization of negative inductance using variable active-passive reactance (VAPAR). IEEE Trans Power Electron 12(4):589–596
6. Nayfeh AH, Mook DT (2004) Nonlinear oscillations. Wiley, New York

# Optimal Energy Efficiency, Vehicle Stability and Safety on the OpEneR EV with Electrified Front and Rear Axles

**Stephen Jones, Emre Kural, Kosmas Knödler and Jochen Steinmann**

**Abstract** This paper relates to the European publicly funded project OpEneR. A central field of research and development in this co-operative project, is energetically optimal operation of the twin electric axle drivetrain of the fully electrified OpEneR prototype cars. Advanced operation strategies for the two 50 kW e-machines are being developed, which consider both energetically optimal propulsion and energetically optimal braking of the vehicle. Both e-machines are used in such a way, that the maximum possible amount of energy can be electrically recovered. However, during electrical recuperation the impact of the brake force distribution on the vehicle's stability, e.g. during braking on split–$\mu$ surfaces, and thus safety, has to be carefully considered. In this work, an advanced co-simulation approach which allows the virtual evaluation of different electrical recuperative braking, frictional braking and related control strategies is described.

**Keywords** Electric vehicle · Predictive energy management · Twin axle · Electric powertrain · Regenerative braking · Electrical recuperation · Friction

---

OpEneR: Optimal Energy consumption and Recovery based on a system network, grant agreement number 285526, funded by the EC Seventh Framework Programme FP7-2011-ICT-GC [1]

---

S. Jones (✉) · E. Kural
AVL LIST GMBH, Hans List Platz 1 A-8020 Graz, Austria
e-mail: Stephen.Jones@avl.com

E. Kural
e-mail: Emre.Kural@avl.com

K. Knödler · J. Steinmann
ROBERT BOSCH GMBH, Robert-Bosch-Allee 1 D-74232 Abstatt, Germany
e-mail: Kosmas.Knoedler@de.bosch.com

J. Steinmann
e-mail: Jochen.Steinmann@de.bosch.com

J. Fischer-Wolfarth and G. Meyer (eds.), *Advanced Microsystems for Automotive Applications 2013*, Lecture Notes in Mobility, DOI: 10.1007/978-3-319-00476-1_26,

braking · Vehicle dynamics · Traction · Stability · Safety · ESP · VDC · ABS · TCS · Split–$\mu$ · Climbing · Co-simulation

## 1 Introduction

The range anxiety of potential buyers of fully electric vehicles (EV) needs to be addressed, in order to give e-mobility the chance of a broader market entry in the future. Many different approaches are being pursued by industry, research institutes and universities to address this. The European publicly funded project OpEneR represents a cooperation between industry and research institutes. Unlike other attempts focussed on new technology, e.g. novel lithium based high voltage batteries, or new more efficient e-machines, the OpEneR project envisages a networked system of on-board and off-board systems, fusing data together in highly sophisticated energy saving driving strategies or algorithms [1]. The networked systems in OpEneR are so numerous, that from the beginning the project was split into three major levels, the so called OpEneR Technology Levels L1 to L3. Figure 1 shows this and level L0–the baseline vehicle definition and build-up.

The baseline prototypes L0 were derived from the production Peugeot 3008 Diesel Hybrid, which is driven by a diesel combustion engine on the front axle with 120 kW, and by an e-machine on the rear axle with 27 kW power. Both motors are not directly coupled to each other. The fully electric OpEneR prototypes were built up with twin e-axles, each with its own 50 kW e-machine and dog clutch, two inverters, four high voltage battery packs, a high voltage charging system and a completely adapted cooling and heating system (see Fig. 2).

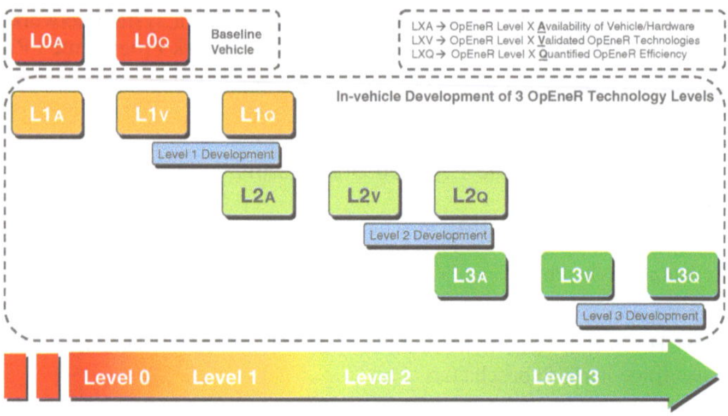


**Fig. 1** OpEneR technology levels

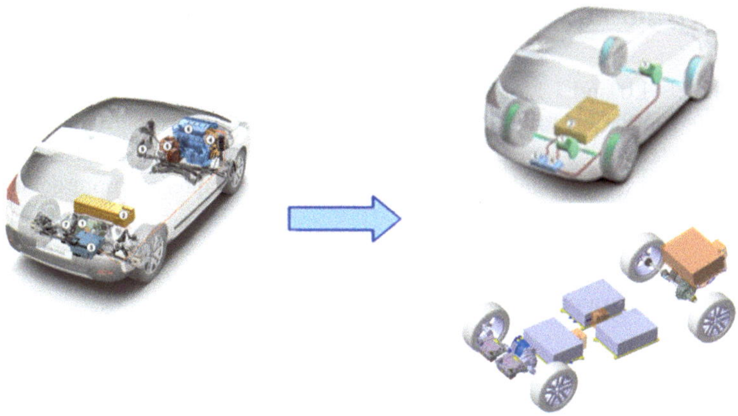

**Fig. 2** Fully electric OpEneR prototype derived from Peugeot 3008 Diesel Hybrid, by replacing the diesel engine with a second e-machine and adding four high voltage battery packs, etc

Within OpEneR Technology Level L1, information from inside the vehicle and information from the GPS SatNav system are used for predictive energy management functions, e.g. an enhanced vehicle coasting assistant [2–4]. In Level L2 further systems for surround sensing will be included into the network, e.g. a long range radar sensor and a multi purpose camera in the front of the vehicle and two mid range radar sensors in the rear. Units for car-to-infrastructure communication will be included in L2, and in the final OpEneR Technology Level L3 car-to-car communication will be added. This process of stepwise integration is shown in Fig. 3. Alongside this a virtual development methodology relying on the usage of advanced vehicle and powertrain simulation tools has been developed and extensively used to front load project development.

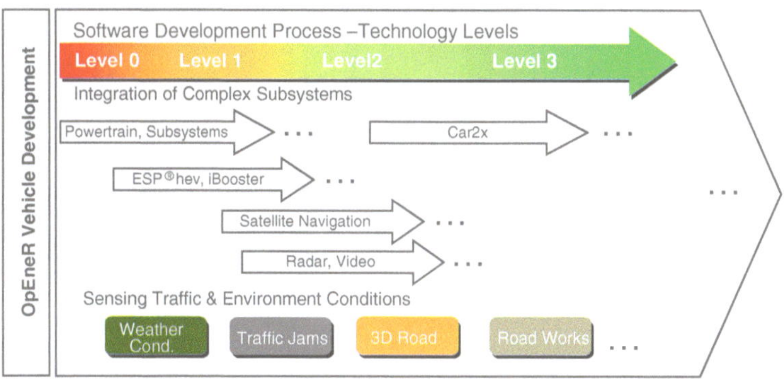


**Fig. 3** OpEneR vehicle development over technology levels L0 to L3

# 2 Powertrain and Regenerative Braking System

Within levels L0 and L1 of the OpEneR project, the main integration activities were related to the electric powertrain, including the high voltage battery system and power electronics and the regenerative braking system. The electric powertrain and the braking system being described in this chapter in more detail.

## *2.1 Electric Powertrain*

The prototypes employed in the project are pure electric vehicles and were developed from the production Peugeot 3008 Diesel Hybrid vehicle. They offer e-4WD via two e-machines. Each e-machine motors acts on one axle, front or rear, and features 50 kW of nominal power and 200 Nm of nominal torque (see Fig. 4). Aside from the cost benefits from using two small production e-machines, the main advantage of the e-4WD topology is that it offers the possibility of optimally distributing the traction and the regenerative braking torque between the front and the rear e-machines, which results in a higher overall energy efficiency, by load point switching, for example to one e-machine at lower loads and speeds [2].

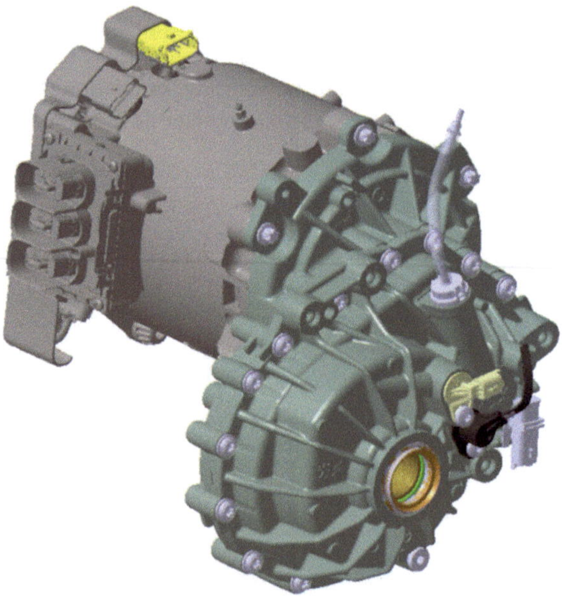

**Fig. 4** One of the two 50 kW e-machines employed in the OpEneR prototypes

## 2.2 *Regenerative Braking System ESP®*hev

The braking system components depicted on the left side of Fig. 5 are commonly used in conventional combustion engine driven vehicles. A pneumatic brake booster handles the vacuum based braking force amplification and a standard ESP® system—including Anti Blocking System (ABS), Traction Control System (TCS) and Vehicle Dynamics Controller (VDC)—modulates the brake pressure to each wheel individually, to enhance vehicle traction and stability. On the right side of Fig. 5, the alternative and so-called *iBooster* system, for hybrid and electric vehicles, which uses electromechanical brake force amplification and the regenerative braking system known as ESP®*hev* is shown.

The advanced regenerative braking system ESP®*hev* used in the OpEneR prototypes is characterised by a rear axle brake circuit that is fully decoupled from the brake pedal during normal operation. In other words, driver brake pedal application does not directly result in a brake pressure increase on the rear axle brakes. Within a dead stroke in the master cylinder, and based on a signal from a pedal stroke sensor, the braking demand given by a driver is covered by the e-machines' regenerative braking only. Beyond the dead stroke, the brake pressure in the front axle brake circuit will be directly built up via the master cylinder. The amount of brake torque at the rear axle being covered by the e-machine as long as the recuperation potential of the system is not exceeded, and no stability critical driving situation is reached. Since the recuperation potential of e-machines is not constant over vehicle velocity, the appropriate supplementary frictional braking pressure to achieve the desired vehicle deceleration is generated by the ESP®*hev* system within the rear axle brake circuit. This compensation is called brake torque blending. Shown over the different domains of *actuation*, *power supply*, *modulation* and *foundation*, the transition between conventional brake system components and the ones used in the OpEneR project is visualised in Fig. 6.

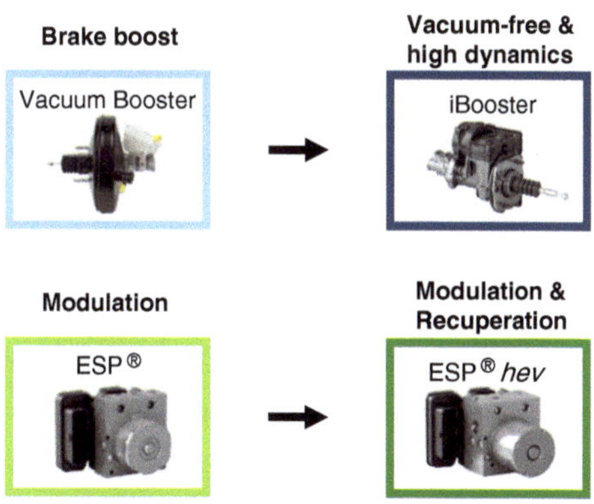


**Fig. 5** Transition from conventional brake system components vacuum booster + ESP®, to *iBooster* + ESP®*hev* systems used in OpEneR EV

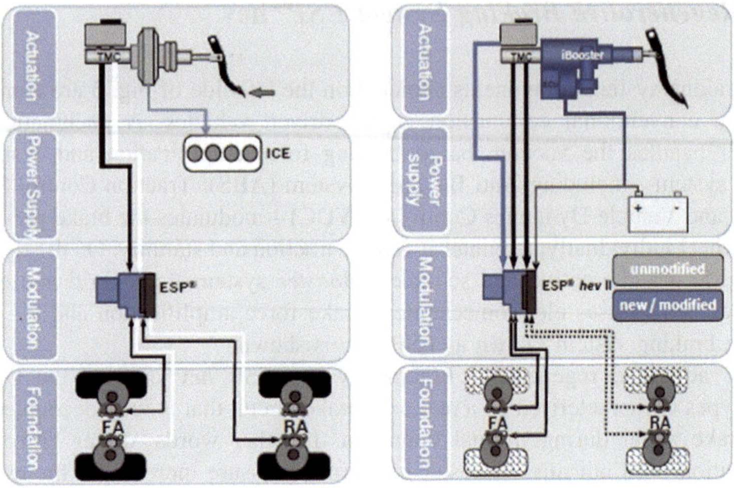


**Fig. 6** Conventional and OpEneR EV brake systems over different domains

In case of system degradation (i.e. failure) or a critical vehicle stability situation, the ESP®*hev* system is able to decrease, or even completely switch-off the brake torque derived from the e-machines, in order to guarantee safe vehicle operation. Figure 7 describes the difference between simple braking systems without blending, and the cooperative regenerative braking system ESP®*hev*. In the latter case the driver will experience the expected constant deceleration when constantly applying the brake pedal. In simpler systems, the driver will need to compensate for variations in brake torque from the e-machines, by himself via the brake pedal.

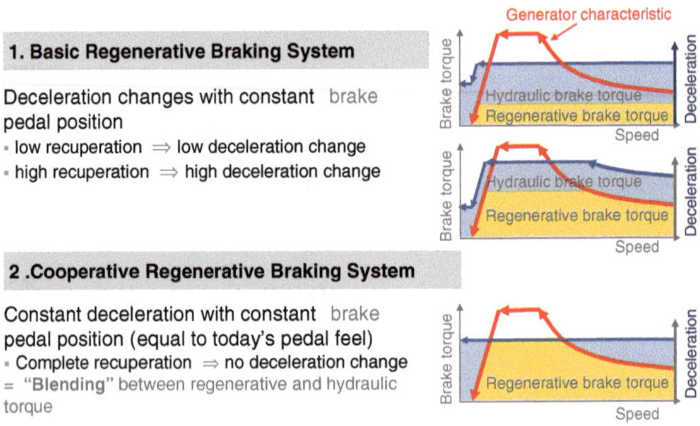


**Fig. 7** Principle of the blending between conventional hydraulically actuated friction brake torque and electric brake torque provided by e-machines

# 3 Simulation Environment

Various diverse OpEneR vehicle and powertrain subsystems, need to be considered in complex new algorithms, related to energy saving driving strategies and vehicle stability and thus safety. And whilst the stepwise project approach due to the OpEneR Technology Levels and the modularity of the whole system, help to manage this complexity, the development challenge is nevertheless considerable and thus precludes the use of wholly conventional development techniques.

## *3.1 Simulation Tools*

One important approach that directly addresses the complexity of the system during the development of new control functions or algorithms, thereby helping to ensure vehicle efficiency and safety, is the application of advanced model based simulation techniques. However, their successful employment in a demanding project like OpEneR, requires the development of an accurate simulation platform, consisting of realistic vehicle and powertrain models, 3D real world driving routes and sensor, controller, actuator and human driver models. In the OpEneR project, co-simulation techniques are deployed to combine the best features of the following simulation tools:

**AVL Cruise**, to include a detailed description of the electrified powertrain topology, subsystem and controller interfaces.

**IPG CarMaker®**, to simulate the vehicle dynamics, 3D road and complex human driver functionality. This enables the virtual vehicle to interact dynamically with traffic objects and its surroundings.

**Matlab/Simulink®**, to include existing complex sub-system models from the project partners e.g. Bosch ESP®*hev*, and various operational control strategies e.g. energy management functions.

## *3.2 Integrative Simulation Environment*

Figure 8 depicts this AVL developed co-simulation environment as it is implemented at the moment in the OpEneR project.

This detailed and realistic system co-simulation model allows the development of advanced energy saving functions, and the direct evaluation of the benefit in terms of efficiency. Furthermore, it enables the mandatory investigation of the influence of the energy saving functions, e.g. energy optimised torque distribution, on vehicle dynamics and safety. As mentioned above, the model contains also the Bosch ESP®*hev* as a subsystem. Thereby simulation approaches similar to the ones described by Lutz et al. [5] for conventional vehicle ESP® functions, can now also

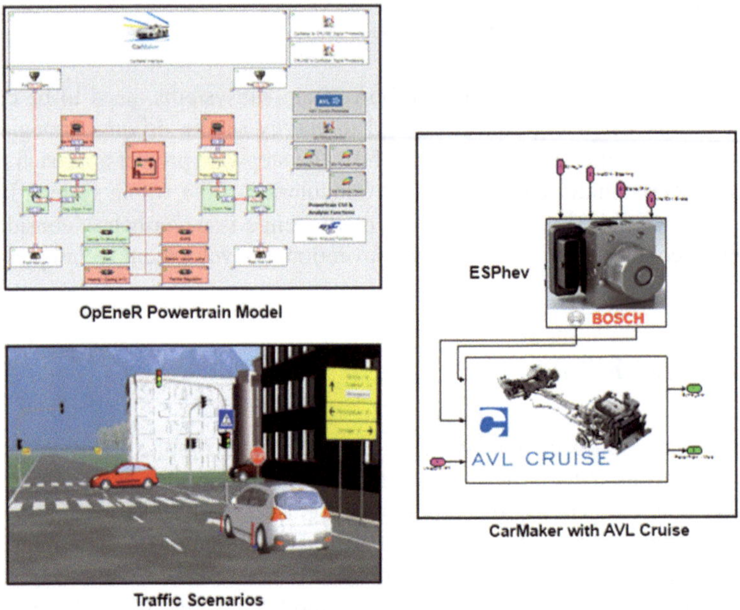


**Fig. 8** OpEneR co-simulation environment including the AVL CRUISE powertrain and Bosch ESP®*hev* model

be undertaken for advanced fully electric or hybrid vehicles with novel powertrains. The ESP®*hev* system used in OpEneR, is an advanced cooperative regenerative braking system on one hand, but it also contains vehicle stabilising functions on the other hand. To address this, the interaction of the advanced energy saving functions, vehicle stability and the vehicle stability controller ESP®*hev*, are virtually investigated using the co-simulation methods described above, and advanced vehicle stabilizing functions on both axles (each independently driven by its own e-machine) thereby be developed. This simulative approach allows early virtual testing of the new algorithms, thereby reducing the risk to persons and material during later real testing. Thus, vital safety aspects of the project are economically taken into account during the early development process.

## 4 Simulation Use Cases

In 2012 the OpEneR project progressed to level L1. As mentioned above, this level focuses on the powertrain, the regenerative braking system and the optimal interaction between them. This section shows a set of use cases that are addressed using the co-simulation environment in level L1. One example, being a standard braking manoeuvre with the cooperative braking system ESP®*hev*. Further use cases (i.e. combinations of vehicle manoeuvres, traffic scenarios, environmental conditions,...), that address relevant energy management functionality and specific safety aspects of the OpEneR vehicle and its electric powertrain, are built up and virtually tested in the developed co-simulation environment. Such scenarios can be used to virtually investigate new energy manager software safety features and margins in terms of vehicle stability.

### *4.1 Standard Braking Manoeuvre with the Cooperative Braking System*

The first simulation use case shown is the test and development of advanced energy saving and recovery functions. In this section the simulation of a normal or standard braking manoeuvre with the regenerative braking system ESP®*hev* is presented.

A sample plot of co-simulation results obtained is depicted in Fig. 9. The *Brake Pedal [%]* signal shows that the brake pedal is activated to 40 % at a vehicle speed of 80 km/h. As explained in the description of the ESP®*hev* system in Sect. 2.2, the brake pedal activation results in hydraulic pressure at the wheel brakes at the front axle, when the dead stroke of the master brake cylinder has been exceeded. The signal *Brake Press FL [bar]* shows that there is a pressure of 12 bar from the beginning of the braking manoeuvre in the wheel brakes at the front axle, while the

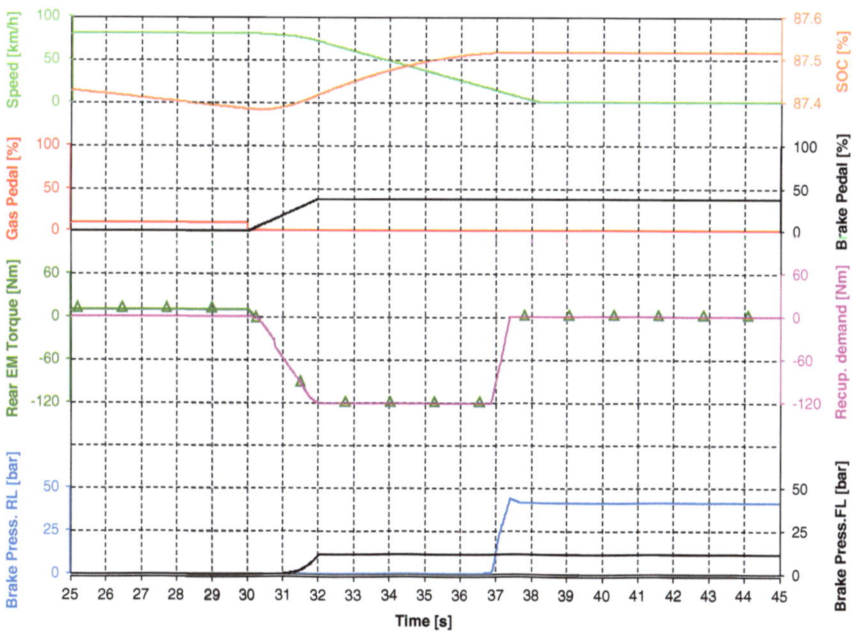


**Fig. 9** Normal braking manoeuvre with the cooperative regenerative braking system ESP®*hev*

signal *Brake Press RL [bar]* shows that there is no pressure in the wheel brakes at the rear axle. The driver's brake request is derived from the brake pedal stroke by the ESP®*hev* system, and the regenerative brake torque and the friction brake torque is thus coordinated. The regenerative brake torque request is routed to the e-machines by the signal *Recup. Demand [Nm]* which goes down to −120 Nm. Thus the major part of the vehicle deceleration is realised by the electrical regenerative brake torque and a high amount of energy is thus recovered by the system. The signal *SOC [%]* shows that the battery state of charge is increasing during the braking manoeuvre. Towards the end of the braking manoeuvre at low speeds the recuperation potential of the e-machines is decreasing towards zero. The ESP®*hev* system receives the decreasing recuperation potential and the decreasing regenerative torque is progressively replaced by the mechanical friction brake torque. This so called torque blending is visible at 37 s in the signals *Rear EM Torque [Nm]* and *Brake Press RL [bar]*.

## 4.2 Split-$\mu$ Braking

One manoeuvre that shows a limitation of regenerative electrical braking due to vehicle stability, is a braking manoeuvre on a road surface with the so called split–$\mu$ friction coefficient condition. i.e. the road-tire friction coefficient is different for the wheels on the left side of the vehicle and those on the right side of the vehicle. This situation is visualised by the screenshot from the co-simulation environment shown in Fig. 10.

In the powertrain configuration employed on the OpEneR vehicle, each e-machine is acting on both wheels of one axle via a differential gear. Thus, the

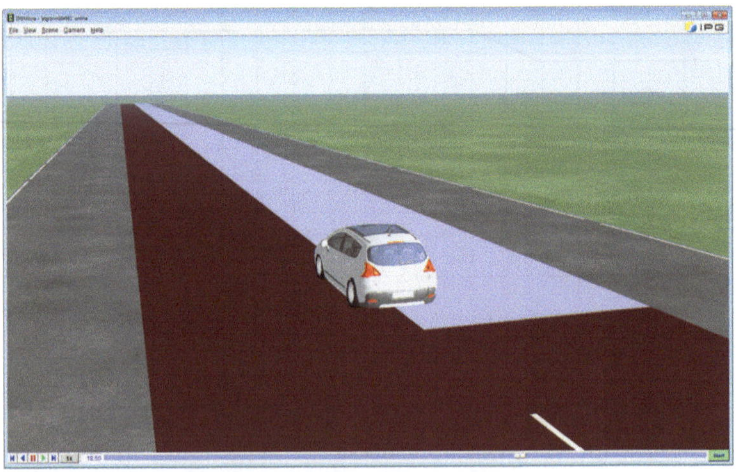

**Fig. 10** Visualisation of split-$\mu$ co-simulation OpEneR test case

torque distribution of the e-machines between the left and the right wheel *cannot* be controlled by the system. During regenerative braking on split-$\mu$ this leads to a torque around the z-axis of the vehicle that turns the vehicle to the high-$\mu$ side, because a higher braking torque can be generated on the high-$\mu$ side.

This situation is recognised by the vehicle stability control system ESP®*hev*. As mentioned above, this system is able to perform torque blending between regenerative braking and friction braking. However, during the split–$\mu$ braking manoeuvre the regenerative brake torque is disabled and replaced by frictionally derived brake torque. The friction torque *can* be controlled at each wheel individually by the ESP®*hev* system, and conventional vehicle dynamic controller functions are applied to ensure vehicle stability and safe operation of the vehicle.

The simulation results are depicted in Figs. 11 and 12. The signal *Mu Split Zone* indicates the split-$\mu$ area. As can be seen in the *Brake Pedal* signal, the braking manoeuvre already begins before the vehicle has entered the split-$\mu$ area. In this part of the manoeuvre the conditions are identical to the standard braking manoeuvre. Thus, the signal *Recup. Demand [Nm]* is routed as a recuperation request to the electrical machines which follow the request directly. However, once the vehicle has entered the split-$\mu$ area, the slip of the wheels on the right low–$\mu$ side increases, whereas the wheels on the left high-$\mu$ side remain stable, visible on the signals *WheelSpd_FL [km/h]* and *WheelSpd_FR [km/h]*. The situation is recognised by the ABS controller, which is indicated by the flag *ABS active*. Because the distribution of the regenerative brake torque between the left and the right wheel on one axle *cannot* be controlled by the system, the regenerative brake torque is disabled as soon as ABS is activated. The regenerative

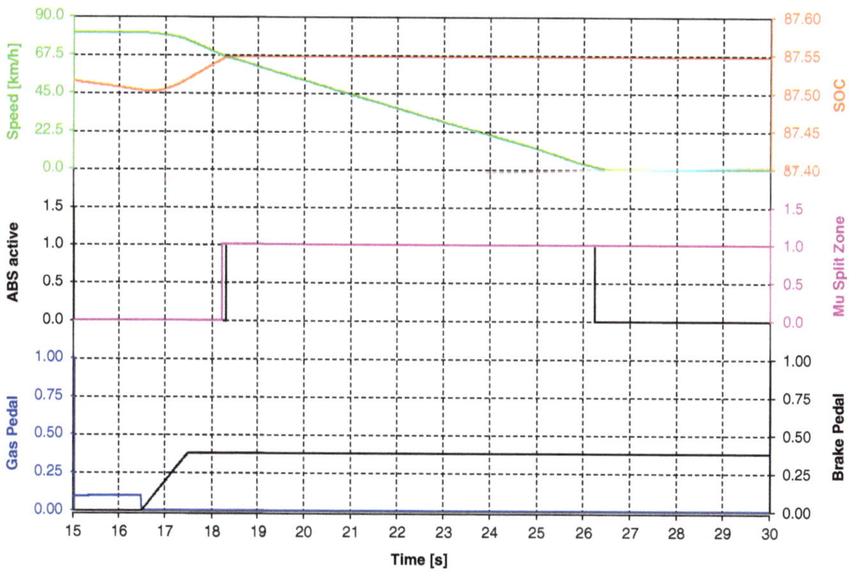


**Fig. 11** Plot A of the split-$\mu$ manoeuvre

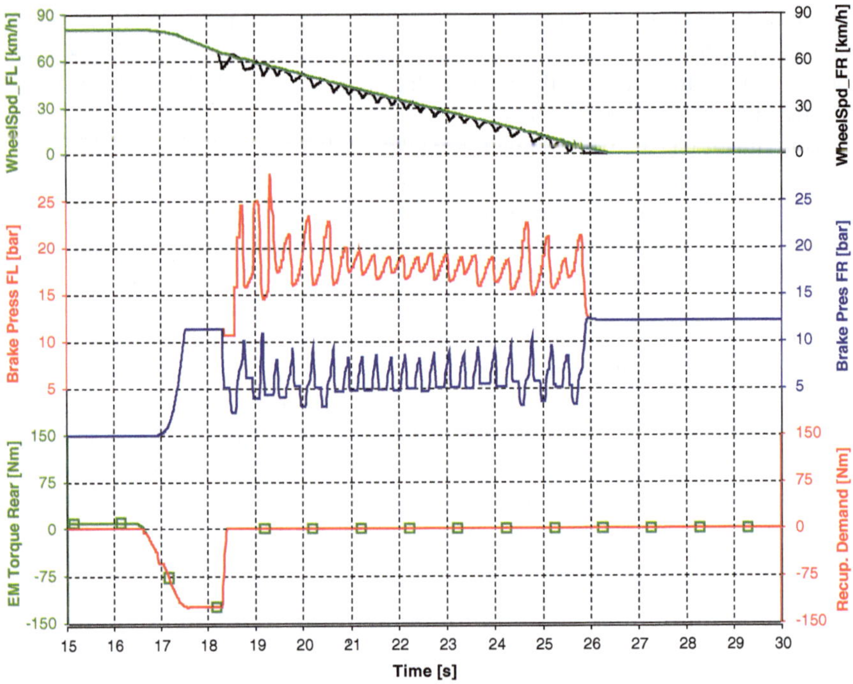


**Fig. 12** Plot B of the split-$\mu$ manoeuvre

brake torque is replaced by friction brake torque via the ESP® *hev* system, since the friction brake torque *can* be controlled individually at each wheel brake. The system adjusts the hydraulic system brake pressures in such a way that the vehicle remains steerable for the driver, and does not turn to the high-$\mu$ side in an uncontrollable way. The different pressure levels on the left high- and right low-$\mu$ sides are visible in the signals *Brake Press FL [bar]* and *Brake Press FR [bar]* respectively.

## 4.3 Hill Climbing on a 10 % Hill on Split-μ

During a hill climbing manoeuvre, on either low-$\mu$ or split-$\mu$, it is necessary to control the traction of the wheels, to ensure that the slip of the wheels stays in an area where the vehicle remains steerable and the best possible grip level is maintained. To support this the TCS function is part of the ESP®*hev* system. On a electrified axle there are two actuators, the hydraulic wheel brake and the e-machine, that can be used for control of wheel traction. In the powertrain of the OpEneR vehicle there is one e-machine on each axle, which means that the ESP®*hev* system has to control the torque of each e-machine individually.

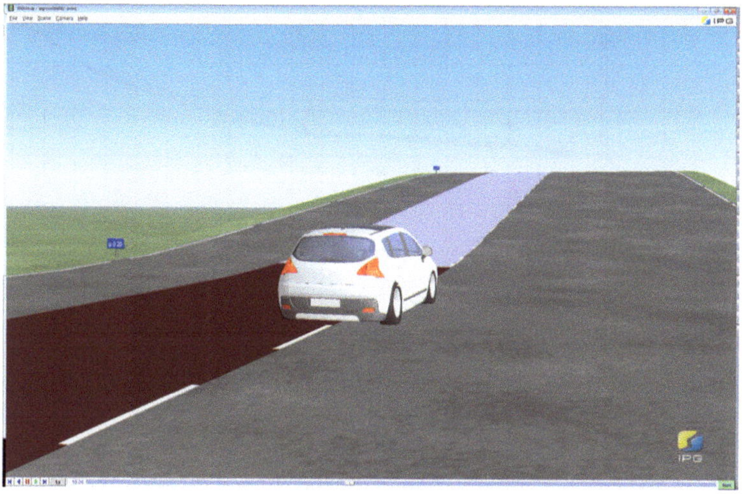

Fig. 13 OpEneR car hill climbing on split-$\mu$ surface

Of course, in order to increase overall vehicle efficiency the frictional wheel brakes should be used as little as possible.

Figure 13 shows a simulation screenshot of the hill climbing manoeuvre. The manoeuvre begins on a flat high-$\mu$ surface and the vehicle rolls slowly up the 10 % hill with the split-$\mu$ surface. The simulation results of the manoeuvre are depicted in the plots shown in Figs. 14 and 15. As soon as all four wheels of the OpEneR

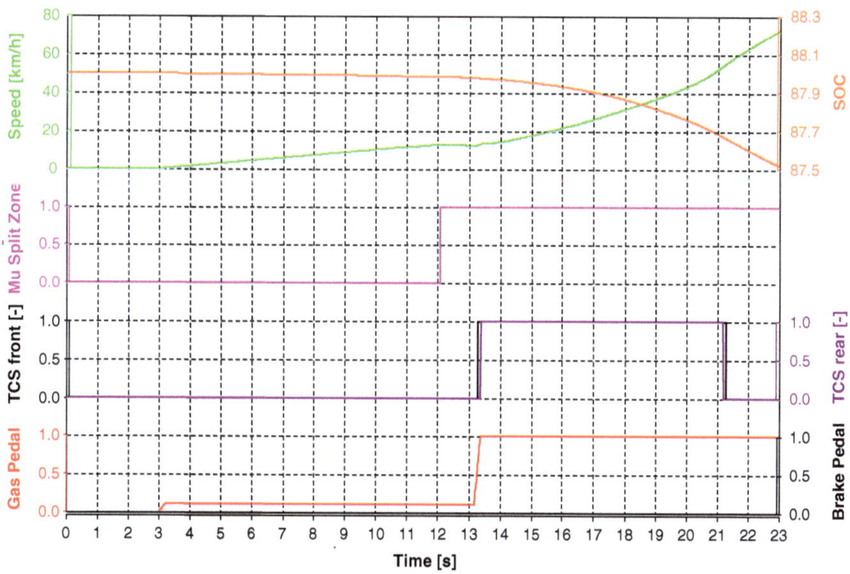


Fig. 14 Plot A of the hill climbing manoeuvre on split-$\mu$ surface

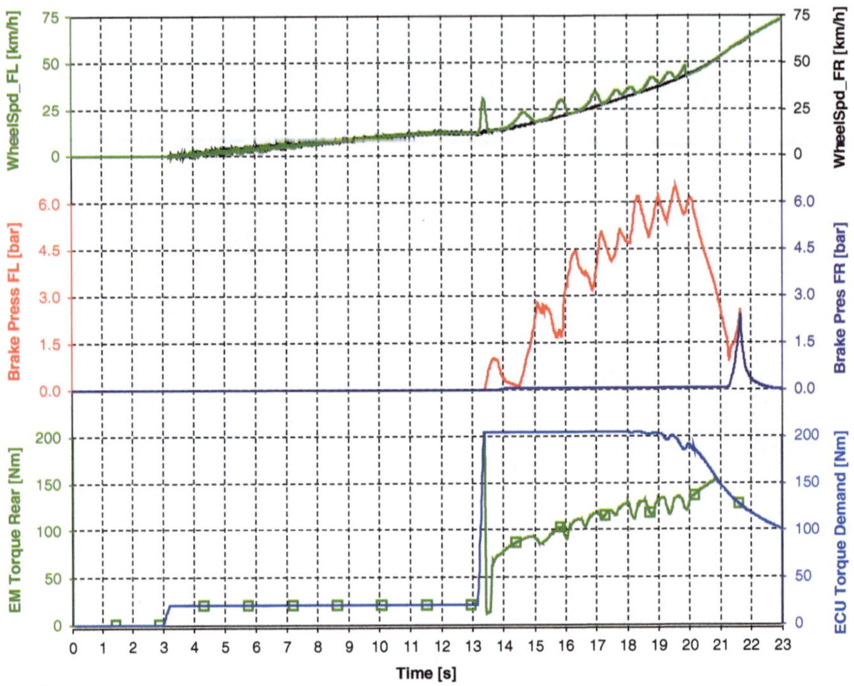


**Fig. 15** Plot B of the hill climbing manoeuvre on a split-$\mu$ surface

electric car are on the split-$\mu$ uphill slope, the accelerator pedal is applied fully, as indicated in the signals *Mu Split Zone* and *Gas Pedal [%]* in Fig. 14.

When the accelerator pedal is applied fully, the wheels on the left low-$\mu$ side go to slip and TCS becomes active, Fig. 15. That the TCS control function is axle specifically is visible in the signals *TCS front* and *TCS rear*, which do not switch on at the same time. For reasons of clarity only the torque signals of the rear e-machine are shown in Fig. 15. The signal *ECU Torque Demand [Nm]* is derived from the accelerator pedal position and shows the high torque requested by the driver. As soon as TCS function is activated, the e-machine torque is reduced by the ESP®*hev* system, to an appropriately lower level that can be applied in the current driving situation. This is visible on the signal *EM Torque Rear [Nm]* that follows the TCS torque request.

In the split-$\mu$ situation it is necessary to apply a modest amount of friction braking at the wheels on the left low–$\mu$ side, because the differential gear transfers torque from the high–$\mu$ side, where higher torque can be applied, to the low-$\mu$ side. Thus the wheels on left low-$\mu$ side have to be held by the friction brake slightly to control their slip and their traction. This is automatically performed by the ESP®*hev* system, without intervention by the driver i.e. the *Brake Pedal* signal is zero throughout the whole manoeuvre.

For comparison, note that in a vehicle pullaway or launch manoeuvre on a flat low-$\mu$ surface, the traction control function is performed through modulation the e-machine torques only, without any use of the friction brakes and thus energy is not unnecessarily and wastefully dissipated in the friction brakes.

## 5 Conclusion and Outlook

In this paper, the European OpEneR project is used to demonstrate the development of technology for a twin axle pure EV, fitted with a predictive energy manager for use in future Intelligent Transportation Systems. The developed system is implemented on a vehicle that will utilise diverse sources of information, including on-board sensors, radar, video, GPS satellite navigation, car-to-infrastructure and car-to-car communication systems, in order to extend the electrical range, compared to an equivalent electric vehicle with a conventional energy manager. The development of such a complex vehicle and powertrain requires realistic co-simulation models to support system design, and to provide accurate simulation plant models for virtual energy management control function development, and also to ensure vehicle stability and safety. To describe the interaction of powertrain energy flows and vehicle dynamics effects, an integrated co-simulation toolchain comprising IPG CarMaker® and AVL CRUISE has been developed, whilst Matlab/Simulink® has been deployed to integrate the Bosch ESP®*hev* system vehicle stability controller, which interacts with the powertrain energy management, because strong recuperation via e-braking is possible on both axles. This co-simulation platform, allows the user to virtually test various use cases and scenarios, to virtually find the best compromise between the most efficient brake force distribution between the e-machines and friction brakes, whilst ensuring vehicle traction, stability, and thus safety. This is demonstrated with two simulated EV use cases, split-$\mu$ braking on level road and climbing on a split-$\mu$ slope.

As the OpEneR project proceeds reproducible testing of predictive energy management and related safety functionality, will be performed on AVL testbeds fitted with InMotion vehicle simulation software, where the real OpEneR car will be virtually driven over the same test cycles and manoeuvres as used in the pure office simulation environment described above. In parallel track testing and finally public road testing, will be used to further validate the simulation models developed as well as to test and refine the OpEneR system with respect to efficiency and safety.

**Acknowledgments** The work has been supported by the OpEneR project, grant agreement number 285526, funded by the EC Seventh Framework Programme FP7-2011-ICT-GC.

## References


1. Homepage of the OpEneR project: http://www.fp7-opener.eu
2. Jones S, Huss A, Kural E, Albrecht R, Massoner A, Knödler K (2012) Optimal electric vehicle energy efficiency and recovery in an intelligent transportation system. In: Proceedings of 19th ITS World Congress, Vienna (Austria) 2012
3. Köhler S, Viehl A, Bringmann O, Rosenstiel W (2012) Optimized recuperation strategy for (hybrid) electric vehicles based on intelligent sensors. In: Proceedings of 12th international conference on control, automation and systems, Jeju Island (Korea)
4. Knoop M, Kern A (2013) Deceleration profiles for optimal recuperation and comfort. Stuttgarter Symposium 2013, Stuttgart
5. Lutz A, Macaire F, My W (2012) Virtual test drive in the application process of ESP®-systems to ensure performance and robustness, FISITA 2012, Beijing

# Position Sensor for Brake System Designed for Energy Recuperation

**Bernhard Schmid, Frank Grunwald, Sören Lehmann and Heinrich Acker**

**Abstract** This paper describes a concept and product design of a sensor to detect the position of an actuation piston in a highly integrated and compact electronic brake system. This sensor is part of the highly integrated brake system MK C1 which can easily be integrated into a kinetic energy recuperation system for re-charging the batteries of (hybrid-) electric vehicles. The concept "Linear Inductive Position Sensor" (LIPS), which measures the position of the brake actuation piston contactless, copes with the challenges of the integration into such a compact and safety relevant brake system.

**Keywords** Electric hydraulic brake system · Energy recuperation · Linear position sensor · Differential transducer · SPICE simulation · Goertzel filter

## 1 Introduction and Motivation

In an increased globalized and individualized market the numbers of different vehicle architectures are continuously increasing. Ranging from the smallest A-category like Smart, the well known Golf-class up to the luxury class of

B. Schmid (✉) · F. Grunwald · S. Lehmann · H. Acker
Continental, Division Chassis and Safety, Business Unit Passive Safety and Sensorics, Guerickestr. 7 60488 Frankfurt, Germany
e-mail: Bernhard.Schmid@continental-corporation.com

F. Grunwald
e-mail: Frank.Grunwald@continental-corporation.com

S. Lehmann
e-mail: Soeren.Lehmann@continental-corporation.com

H. Acker
e-mail: Heinrich.Acker@continental-corporation.com

J. Fischer-Wolfarth and G. Meyer (eds.), *Advanced Microsystems for Automotive Applications 2013*, Lecture Notes in Mobility, DOI: 10.1007/978-3-319-00476-1_27,

Mercedes Benz, BMW, AUDI etc., we count already 27 different categories. This plurality will increase further due to the introduction of electrified vehicles ranging from small hybrids up to full electric propulsion available in many vehicle classes. To accomplish a broad band of all the different technical requirements deducted from these categories but also to present a cost attractive product, Continental developed the highly integrated brake system MK C1 which can be easily integrated into a kinetic energy recuperation system for re-charging the battery of (hybrid-) electric vehicles but also fits into classical combustion vehicles. However, in such a highly compact and encapsulated environment, it is a challenge to integrate a sensor which measures the position of the actuation piston contactless. This measurement is crucial to apply the desired brake force and therefore, has to comply with the highest safety requirements of an ASIL D system.

The sensor concept bases on a well known transformer principle where a permanent magnet placed on the actuation piston will saturate the transformer and therefore, a contactless variable differential measurement principle is created, which depends on the travel distance of the actuation piston.

## 2 Regenerative Brake Systems

Since the beginning of the auto-mobility epoch electrified propulsion technologies are in usage. In the early days the electric propulsion was in advantage due to the availability of batteries and electric motors, but the same challenges as today put all the effort in the development of propulsion systems which are large in travelling range and low in cost of ownership. Therefore, engineering pioneers like Benz, Otto, Diesel, Opel etc. and hundreds of engineers after them pushed the fossil driven motor technology to the comfortable and reliable level we can experience today. However, fossil energy resources are limited and their byproducts causes tremendous environmental problems, especially $CO_2$ is known as one of the root causes of the global warming.

With the increasing availability of regenerative energy, electric driven vehicles seem to be the solution for individual mobility in a post oil age. Nevertheless, the amount of stored electric energy in the vehicle is still very limited and thus the travelling range.

### *2.1 General Performance*

Since electric motors can also be employed in the generator mode it opens the possibility to transform the kinetic energy back into electric energy instead of the termination into heat by conventional friction braking. In such a system the

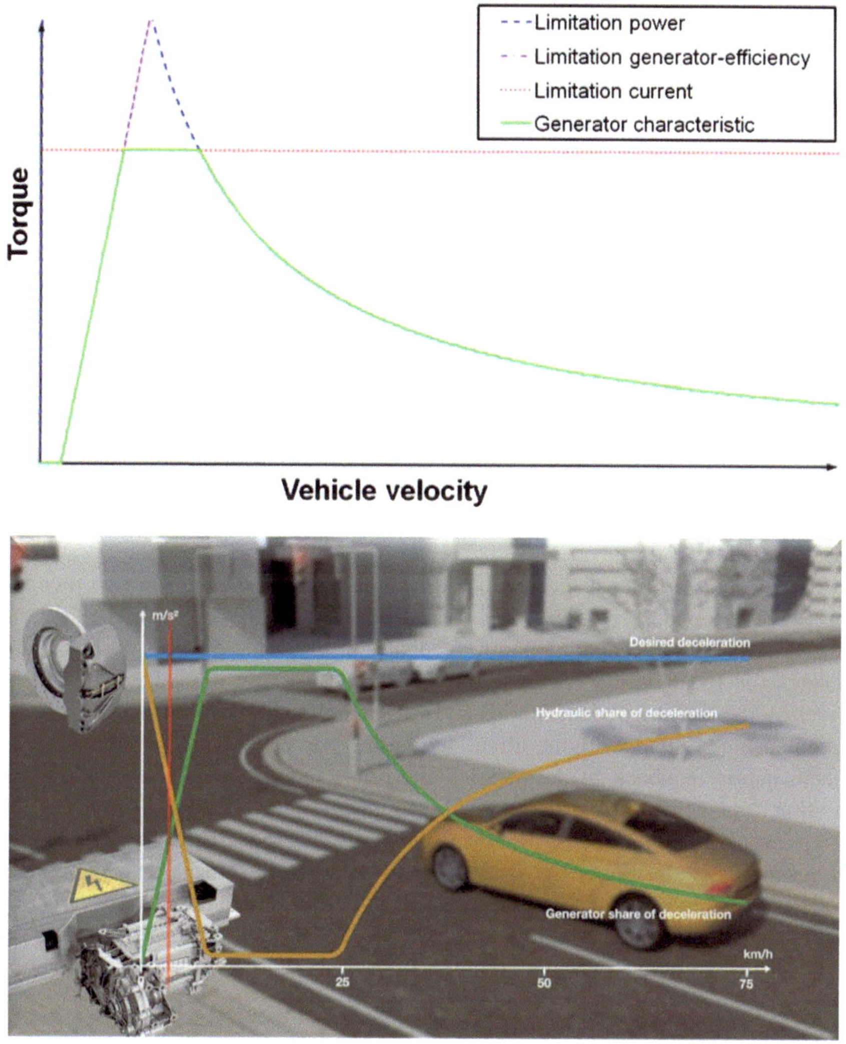


**Fig. 1** *Top* A typical torque characteristic of a battery-generator system is shown. The graphs displays battery charging and generator torque performance limitation in dependence of the vehicle speed [1]. *Bottom* Transition of regenerative braking to friction braking [2]

recuperation is limited by the maximum charging capacity of the battery and the motor-generator torque characteristic which depends on the vehicle speed. As shown in Fig. 1 these effects limit the maximum deceleration torque in dependence of the vehicle speed [1]. In case constant or higher deceleration is requested, the lack of deceleration torque is compensated by conventional friction braking.

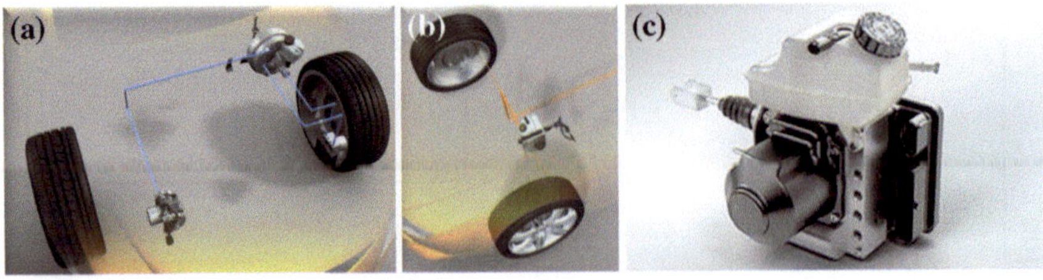


**Fig. 2** System components of a conventional brake system (**a**) and the new brake system MK C1 of continental (**b** and **c**) with fewer components, tubes and cabling harness [2]

## 2.2 MK C1 Brake System Driver Interface

During the development of the high-end brake system MK C1 of Continental beside the targets reduced installation effort, high packaging density, less complexity and reduced weight (Fig. 2), also the capability to ensure maximum recuperation braking with gear shift compensation and zero drag have been accomplished.

The mission profile of the recuperation braking is defined by low to medium speed where low deceleration levels are reached. In this driving mode the driver is usually very sensitive against haptic feedback of the vehicle. With normal brake systems the recuperation deceleration wouldn't cause any force feedback to the driver by the braking pedal. The MK C1 raises the feedback force by an actuation piston-spring system without direct hydraulic pressure control which gives a decent haptic feedback to the paddle. The brake request is detected by a linear position sensor, measuring the position of a magnet placed on the piston (Fig. 3).

The sensor signal feeds the control setting of the electric motor, which in- respectively decreases the hydraulic brake pressure. Since the recuperation deceleration is to weak to ensure an emergency braking when the electric actuation fails, the actuation piston employs at the very end of his traveling range directly the increase of hydraulic pressure. On the other hand the false positive situation

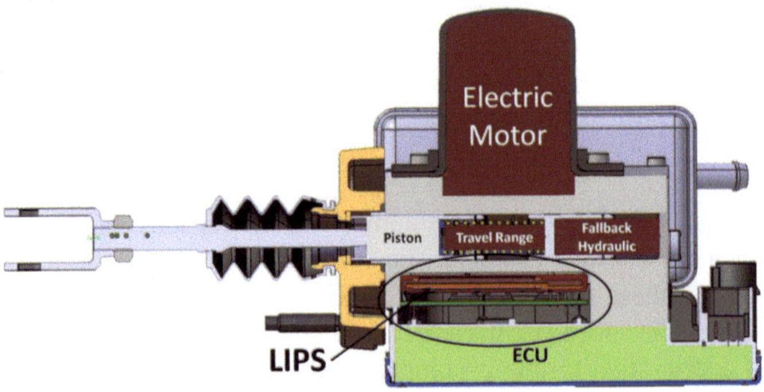


**Fig. 3** Integration of the linear position sensor (LIPS) in the MK C1

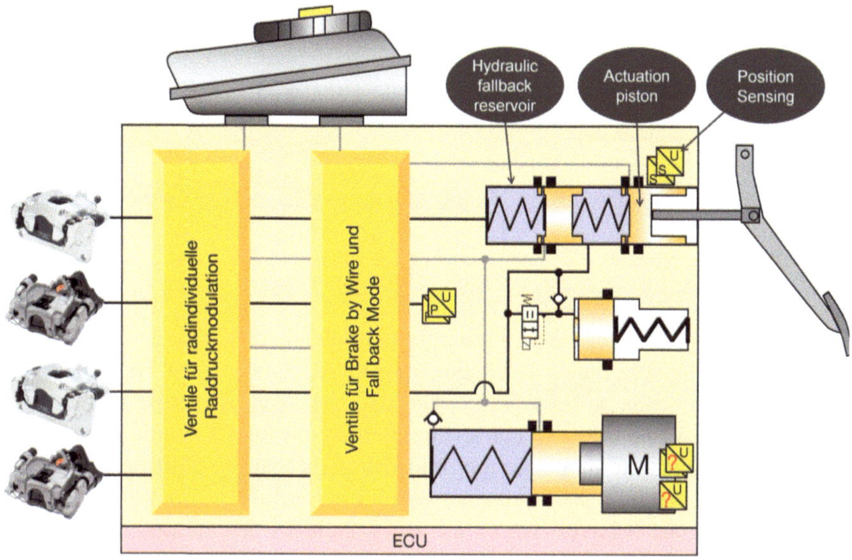


**Fig. 4** System design of the MK C1 driver brake request part [2]

(braking without driver brake request) can only be avoided when the operation of the linear position sensing is detected fault free by the system. Therefore, the sensor is rated as an ASIL D component, respectively decomposed into two times ASIL B(D) sensing paths. The MK C1 system configuration is shown in Figs. 3 and 4 [2].

# 3 Linear Position Sensor Principe

## *3.1 Sensor Concept*

The Linear Inductive Position Sensor (LIPS), belongs to a large family of differential transformer based sensors. Other members of this family are the LVDT (Linear Variable Differential Transformer) and the PLCD (Permanent Magnetic Linear Contactless Displacement Sensor).

The transformer consists of a soft-magnetic core. Three coils are placed on this core, the primary coil, covering nearly the complete core, and two secondary coils, one at each end. To close the magnetic circle a second core is used. In Fig. 5 the principle components of a LIPS is shown.

The transformer can be run either with an alternating current and with constant amplitude (current mode) or with an alternating voltage (voltage mode) and with constant current amplitude. Both methods are possible and have their advantages.

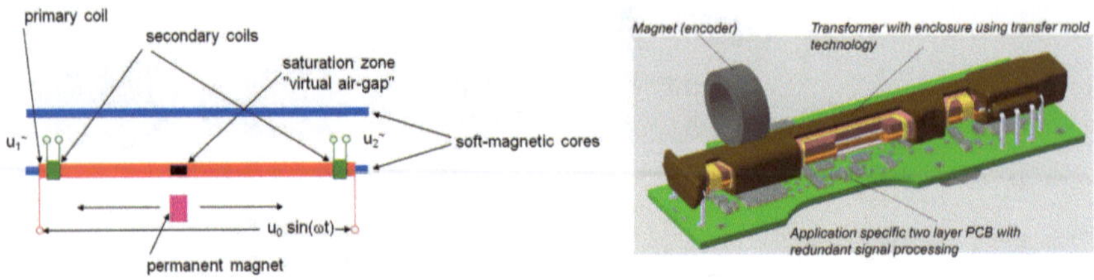


**Fig. 5** The sensor principle (*left*) and product design (*right*) of LIPS

The magnetic field magnetizes the core periodically. The alternating magnetic flux inside of the core induces voltages into the secondary coils. The voltages in both secondary coils are the same because the magnetic flux is the same throughout the whole core. To convert the transformer into a position sensor it is necessary to divide the magnetic flux into two parts that are position dependant.

This is achieved by a strong magnet which saturates the magnetic core locally. The saturated zone behaves like an air-gap, which divides the differential transformer into two transformers with a variable number of primary windings. Thus, in its secondary coils each of these transformers induces voltage, depending on the position of the magnet. The differences of the amplitudes are already a measure of the magnet position. In case of running the sensor in current mode, the correlation between magnet position and amplitude difference is already quite linear. For the case of running the transformer in voltage mode it is necessary to divide the voltage difference by the sum of the voltages to get a linear correlation.

$$P = \frac{\bar{u}_2 - \bar{u}_1}{\bar{u}_1 + \bar{u}_2} \tag{1}$$

This ratio cancels all disturbances that influence the amplitude of both secondary voltages in the same way like temperature effects, fluctuations of the primary amplitude etc. Therefore, we decided to employ the sensor in the voltage mode.

## 3.2 Functional Core

The magnetic cores are made of high permeability, low coercive force material. The magnetization hysteresis curve of such material is generally highly non-linear. This causes a non-linear transformer which induces new frequencies in the secondary coils. However, due to the open magnetic circuit the hysteresis curve of the core material is highly sheared. Therefore, as long as the magnetization of the core is not saturated, the hysteresis curve becomes more linear, whereas the slope is more flat. As a result also the transformer responses more linear.

### 3.3 Redundancy Concept

For safety relevant applications it is necessary to supply redundant sensor signals to the control unit. Therefore, the secondary coils are doubled at each end. To detect external disturbances the sensor is driven at two frequencies at the same time. Due to the linear correlation between the magnetic field applied to the cores and the resultant flux densities inside of the cores, it is possible to run the sensor at two frequencies simultaneously without affecting the amplitude of the other frequency. Only the difference between these two frequencies must be large enough so that the signals of the secondary coils can be separated within reasonable time.

## 4 Signal Conditioning

General requirements which were considered during the entire digital signal processing concept development:

- Signal processing topology should fulfill predefined safety requirements.
- Usage of digital signal processing to guarantee flexibility and endurance.
- Reuse of an established 16 bit μC family for low cost and high volume.
- Displacement measurement results should be provided with an update rate faster than 2.25 ms.
- Communication interfaces are digital.
- Accuracy improvement by using the A/D‘s oversampling capability which compensates for the low 10 bit resolution.

### 4.1 Basic Structure of Digital Signal Processing

To meet the predefined safety requirements a two path fully redundant signal processing solution was chosen. Each of the paths has its own narrow carrier frequencies and each signal processing path should operate selectively such that no cross talk form the other carrier frequency exists.

The simplified signal processing basic structure is shown in Fig. 6 which determines the amplitude of the voltages induced in the secondary side of the differential transformer (S.1.1 and S.1.2 path 1, S.2.1 and S.2.2 path 2). The differential transformer is excited on the primary side with two carrier frequencies. Thus, the amplitude detector must have a frequency selective behavior to distinguish the path-related excitation frequency. By subtracting the amplitudes the desired displacement information is gained.

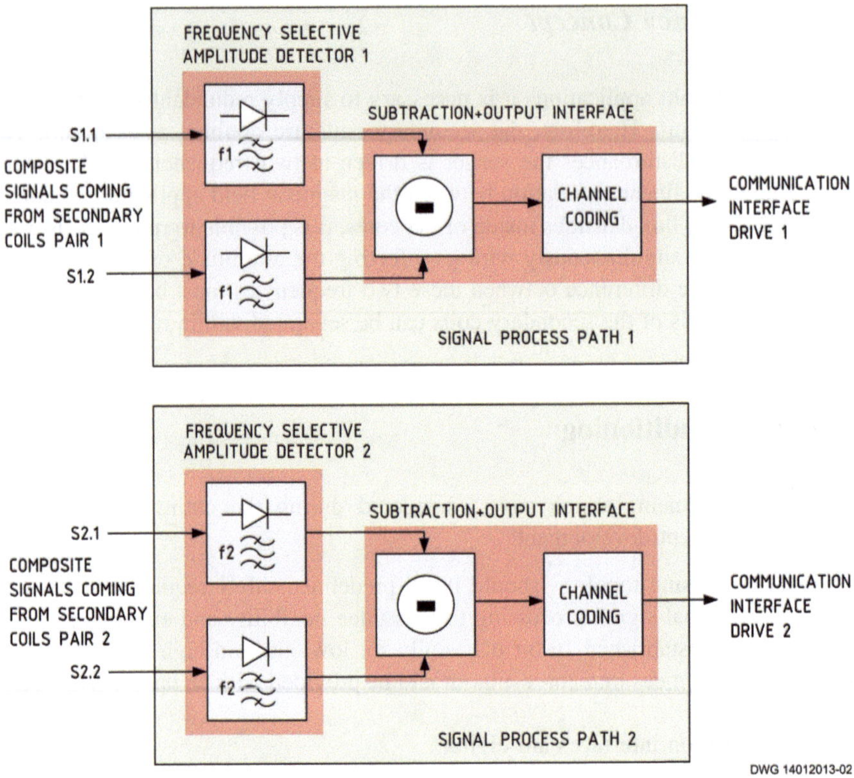


**Fig. 6** Basic structure of digital signal processing

## *4.2 Selective Amplitude Detector: Methods and Characteristics*

The best known selective operating amplitude detector principle, the synchronous demodulator, multiplies the input signal with the oscillator signal of the same frequency. The subsequent low-pass filter characteristic determines the selectivity of this principle. One major disadvantage of the synchronous demodulator is it's phase shift sensitivity between input signal and excitation (oscillator) signal.

A concept was chosen that is not affected by the disadvantage stated above. With the Goertzel algorithm (Discrete Fourier Transformation–DFT) the amplitude of a particular spectral component can be analyzed. In contrast to the well known fast Fourier transform (FFT), the Goertzel algorithm is capable to calculate just one selectable frequency. Thus and because of its simple structure it operates very efficiently and fast in applications where only a few spectral components are required. Since the LIPS signal processing requires only the excitation frequency, it is well suited to be operated on the small μC that should be used [3].

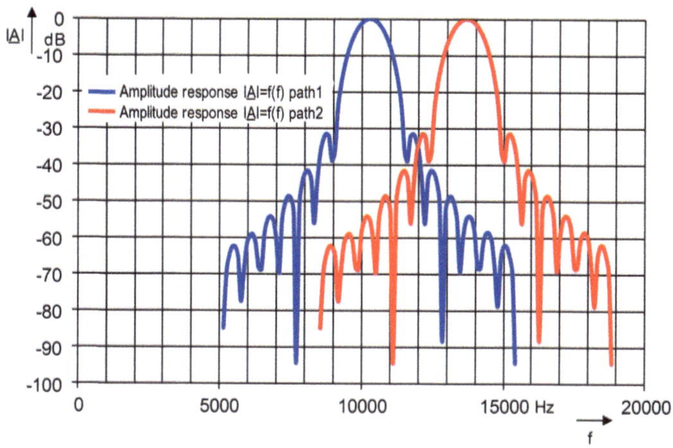


**Fig. 7** Amplitude response of Goertzel filters with parameters: fs = 164.38 kHz, N = 256, fk = 10.273 kHz (*blue*); fs = 164.38 kHz, N = 240, fk = 13.698 kHz (*red*). *fs* sampling frequency, *fk* predefined center frequency, *N* number of samples within a block length

From another point of view the Goertzel algorithm can also be understood as a decimating band pass filter [4]. Through calculating of *N* input samples to *one* output sample (principle of oversampling and decimation) the application of the Goertzel algorithm contributes significantly to the improvement of the signal to quantization noise ratio (S/N) of the whole signal chain. In other words: The effective number of bits (ENOB) of the native 10 bit A/D converter used is raised significantly.

Dimensioning of the Goertzel filter is a compromise between the processing speed (update rate) and the amplitude response (Q-factor). Also the decimation ratio (no. of input samples/no. of output samples) should be as high as possible, since this improves the overall signal quality.

The parameters chosen in the LIPS sensor result on one hand in a filter amplitude response that provides enough selectivity to distinguish reliably between the two excitation frequencies and hence resulting in autonomous operation of the single paths (refer to Fig. 7) and on the other hand provides high processing speed of approx. 1.7 ms which is far below the required 2.25 ms. The remaining processing time is used for computing the measurement results, diagnostic failsafe and encoding of the digital output interface.

## *4.3 Detailed View on LIPS Digital Signal Processing*

A more detailed digital signal processing block diagram is shown in Fig. 8. To simplify the diagram it is reduced to path 1 only. Signal path 2 operates similar but at the excitation frequency f2. The analog composite signals (f1 + f2), coming

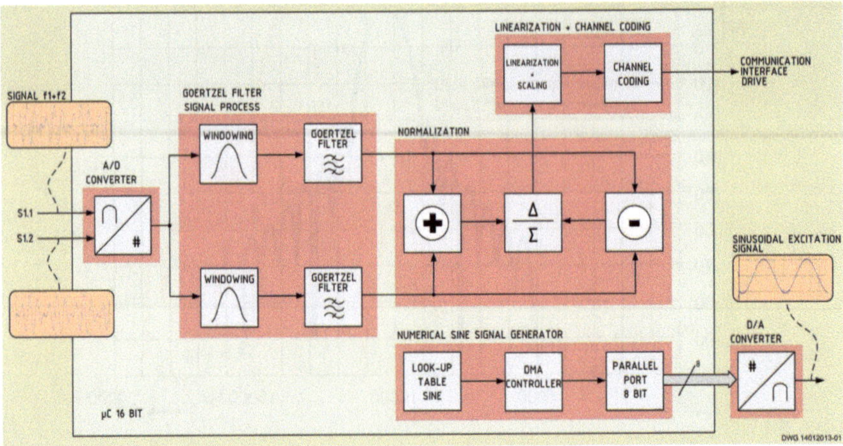


**Fig. 8** Detailed digital signal processing within LIPS displacement sensor. One signal path is shown

from the secondary windings, are converted into digital format by the A/D converter. The two signals S1.1 and S1.2 and all other analog port signals are processed by a window function. In the WINDOWING block all input signals are artificially modulated for the subsequent spectral component calculation. Possible failures as the leakage effect are significantly reduced by this method.

The output signal of the GOERTZEL FILTER blocks represents the amplitude of the voltage induced in the secondary windings of the differential transformer.

In the following NORMALIZATION sub function the amplitude results are further processed: Compared to Fig. 6 not only a subtraction of the amplitude results is carried out but also the quotient of the subtracted and the sum of the amplitude of the secondary signals S1.1 and S1.2 is calculated. With this normalization process the displacement value, which is also the result of the quotient block, becomes independently from e.g. variations in primary excitation amplitude which might vary due to aging or component tolerances.

The signal leaving the NORMALIZATION sub function is the raw displacement signal. It will not meet the requirements in terms of linearity or measurement range scaling and is therefore post processed in the LINEARIZATION + SCALING block. The linearization data applied is unit specific and it is gained within a calibration routine in the factory.

Then the CHANNEL CODING block prepares the agreed digital communication protocol to be transmitted to the ECU subsystem.

Generation of the primary excitation signal is also done by the µC in a digital way: The DMA CONTROLLER is fed with appropriate digital sinusoidal data originating from a LOOK-UP-TABLE. The parallel clocked out DMA CONTROLLER output data is then converted in an external D/A converter to provide the sinusoidal excitation signal.

Since the sine signal generators and the Goertzel filter center frequency parameters share the same master clock oscillator they always match perfectly. Relative frequency drift between sine signal generator and Goertzel filter center frequency, which could result in displacement deviation, is impossible with this procedure.

# 5 Sensor Concept Simulation

Development of the Linear Inductive Position Sensor requires simulation, because many effects and parameters influence its characteristics and simulation is supposed to aid optimization in complex parameter fields. An FEM model was set up, which included hard and soft magnetic materials in various degrees of saturation, coils with excitation currents and induced voltages, and eddy currents flowing in the vicinity of the sensor as well as in its components. It turned out that this model was not sufficient for optimization, because the computational requirements were too high. On a high-end simulation PC, finding the characteristics for one parameter set required about 24 h, which does not allow for high productivity in optimization.

SPICE, the classic circuit simulation program, was studied as an alternative. The model mixes structural and behavioral modeling techniques to reduce the computational effort by reducing the number of degrees of freedom. The price to pay is a moderate loss of accuracy and resolution, while the computing time drops by a factor of about 1,000. The accuracy is deemed insufficient for a prediction of the properties of the sensor (model to product) but sufficient for comparing parameter sets (model to model), which is the key in optimization.

## 5.1 Sensor SPICE Model Concept

The basic idea is to have a *unified* model for the magnetic circuit, the electric signal conditioning circuit and those parts of the software that are on the data path to the position output. SPICE is used for the electric circuits, the magnetic circuit, and the FFT calculation that models the Goertzel algorithm, all in one model. Figure 9 depicts the workflow: The control script starts a number of SPICE runs with their respective parameter sets. SPICE creates text files that include the result of the FFT

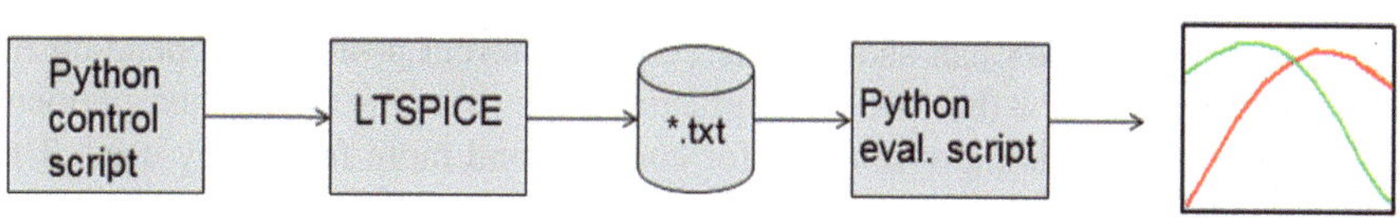


**Fig. 9** Workflow for SPICE simulation

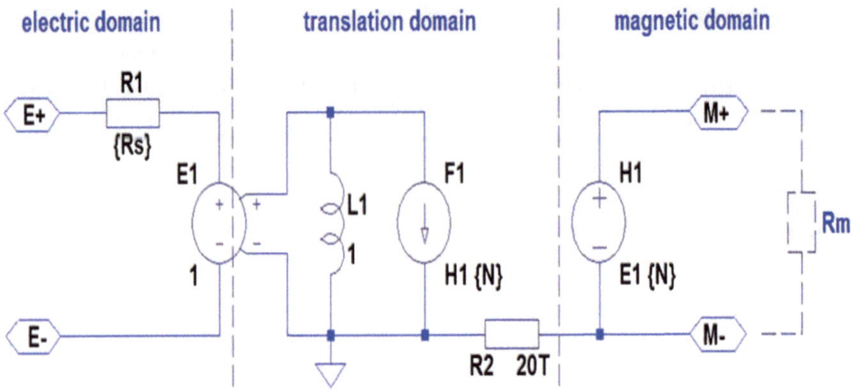


**Fig. 10** Magnetic coupling element for SPICE model

at the operating frequency. An evaluation script performs the ratiometric calculation and creates the position output, eventually plotting the characteristic curve. The scripts were created in the Python programming language; LTSPICE was the SPICE flavor used for simulation.

Since SPICE knows only voltage and current, the magnetic parts of the model need to map field and flux to these quantities. This is done via a magnetic coupling element, shown in Fig. 10. It connects the usual electric circuit domain with the new magnetic domain. The coupling element must model the coupling between an inductor and its magnetic field by two effects:

1. Electric to magnetic: current causes a magnetic field (Ampere's law).
2. Magnetic to electric: flux change induces voltage (law of induction).

This convention results in a magnetic resistance $R_m$ in Ω. Ampere's law is implemented by the source H1 of Fig. 10: It provides a magnetic voltage which is the current in the inductor (E1) times the number of coil turns (N). The law of induction is implemented by F1, L1, and E1: The current in the magnetic domain (H1) times N is copied to the current source F1 and differentiated at the inductor L1. The coupling element itself is linear; all non-linearity is modeled in the magnetic domain, by creating a non-linear magnetic resistance $R_m$.

The model makes heavy use of sub-circuits, which means that showing the entire model is out of scope. Instead, the model of a core slice is displayed (Fig. 12, left), which shows how the coupling element (X1) is actually used. It has an electrical port at the top and a magnetic port at the bottom. The position magnet (not shown) is a current (flux) source connected to two nodes that connect adjacent slices, such that its flux can saturate the core at different positions, depending on the slices selected. For the MK C1 application, the first part of the measurement range (0 … 10 mm) is more relevant for comfort, and more frequently used. The SPICE model was used for a parameter study on how to increase the slope of the characteristic curve of the sensor.

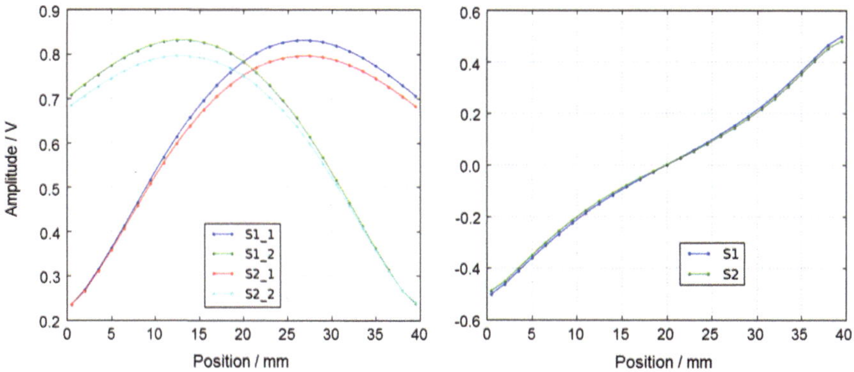


**Fig. 11** Result plots: outputs of Goertzel algorithm (*left*), ratiometric output (*right*)

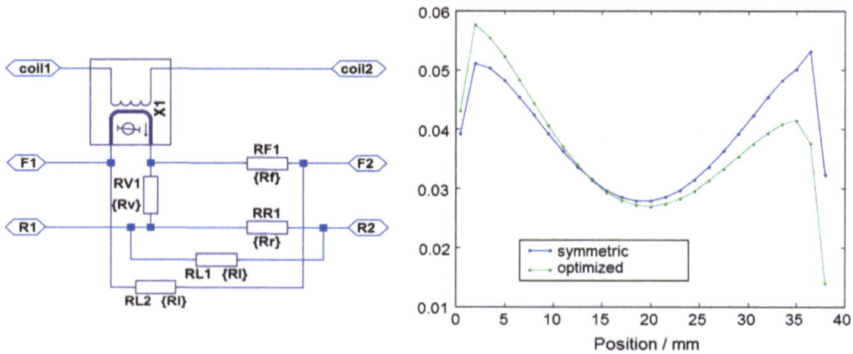


**Fig. 12** Core slice schematic (*left*), ratiometric output slopes (*right*)

## *5.2 Results*

The evaluation script outputs two graphs like those in Fig. 11. The left figure shows the result of the Goertzel algorithm for the two coils of the redundant sensor parts S1 and S2. The right figure shows the characteristic curve as the result of the ratiometric evaluation $(V_1 - V_2)/(V_1 + V_2)$ with the coil voltages $V_1$ and $V_2$.

The result for the slope is plotted in Fig. 12 (right). The green curve (optimized) shows a higher slope in the desired region (compared to the previous, symmetric design) at the expense of a smaller slope at the end of the measurement range, which does not hurt in the MK C1 application. To achieve this result, many design modifications have been tested. Both successful and possible was a reduction of the core length at the end of the position range by 1.5 mm.

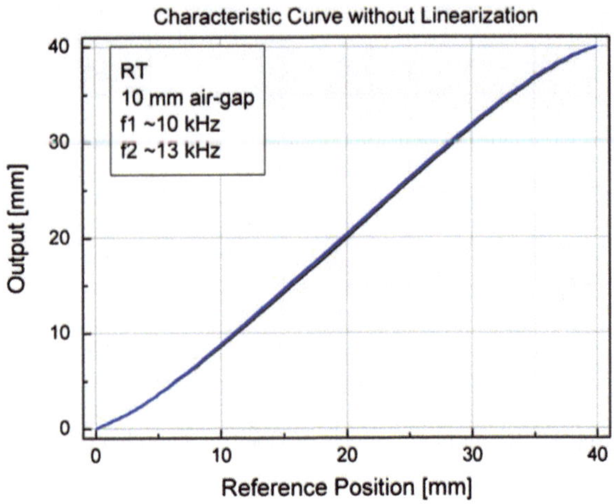


**Fig. 13** Output characteristics of LIPS without any linearity error compensation

# 6 Measurement Results

## *6.1 Raw Output Curve*

Without any additional measures the output characteristic is already fairly linear. For applications with higher requirements a linearization can be easily achieved.

The raw output curve has an S-shape with a linearity error of a few percent. Figure 13 shows two curves measured simultaneously with working frequencies of ~10 and 13 kHz. They are practically identical.

## *6.2 Linearized Output Curve*

Figure 14 shows the output characteristic after linearization.

Both output curves are indistinguishable. Due to the ratiometric evaluation of the output data the curves are practically temperature independent and show nearly no change within a temperature range of −40 °C up to 120 °C.

# 7 Conclusion

An overview of a high performance brake system is given which copes on one hand with the raising cost pressure by a highly integration approach and on the other hand with the enormous variety of different vehicle architectures, especially for electric energy recuperation. Further, the function principle, the signal conditioning and its magnetic simulation of the EHCU integrated driver brake request

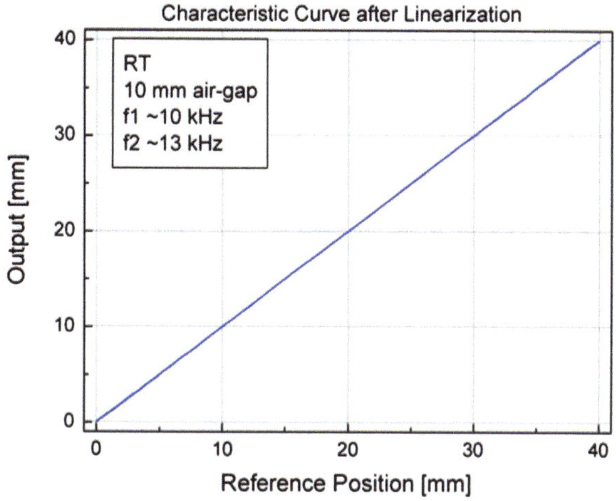


**Fig. 14** Linearized output function of LIPS

sensor LIPS is described in detail. The advantages of the inductive differential transducer function principle and the Goertzel filter, as well as the dual channel and dual frequency safety concept to suppress disturbances and to cope with ASIL D requirements are explained. In sum this sensor concept is well designed for such cost sensitive and harsh application condition and can be applied also for other precise linear position measurement tasks in automotive and industry.

## References

1. Bletz M (2012) Chapter ‘electric regenerative braking’. In: Breuer B, Bill KH (eds) Brake technology handbook 2012, 4th edn. Vieweg + Teubner Verlag
2. Dr Feigel H-J (2012) Integrated brake systems, IQPC Wiesbaden, 5–6 Nov 2012
3. Kurz K, Wagener W (1991) Signalprozessoren-Praxis Algorithmen, Baustein, Systeme, Applikationen. Franzis-Verlag GmbH, München
4. Grüningen D (2008) Digitale Signalverarbeitung. Carl Hanser Verlag, München

# High Temperature Heat Exchanger for Rankine Cycle based Exhaust Waste Heat Recovery

**Jean-Paul Janssens and Robert Cloudt**

**Abstract** A high temperature heat exchanger has been developed for application in a Rankine cycle. (Organic) Rankine Cycle technology is considered for exhaust Waste Heat Recovery from heavy-duty vehicles and other high power applications. The developed exhaust evaporator has a modular and durable design and can operate at high temperature and pressure range. The design has been tested and validated and has been certified according to European pressure equipment and German AD 2000 directives. Units have been accumulating in-field operating hours since 2007 without failure. The modular design allows to tailor an evaporator for specific applications. This process is supported by a developed dimensioning tool and simulation models.

**Keywords** Waste heat recovery · Rankine cycle · Heat exchanger · Evaporator · Simulation

## 1 Introduction

The past decade has seen an increased attention for the fuel consumption of internal combustion engines. This attention is caused by an increased awareness for sustainability, $CO_2$ emissions, fossil fuel energy security and increasing oil prices. This trend is also reflected in new legislative documents affecting internal combustion engines in passenger cars and commercial vehicles (e.g. [1]). Where up till recently attention was geared towards reducing pollutants, the focus is shifting to the reduction of fuel consumption and $CO_2$ emissions.

J.-P. Janssens (✉) · R. Cloudt
Bosal International, Dellestraat 20, Lummen, Belgium
e-mail: dr.Janssens@eur.bosal.com

R. Cloudt
e-mail: robert.cloudt@eur.bosal.com

J. Fischer-Wolfarth and G. Meyer (eds.), *Advanced Microsystems for Automotive Applications 2013*, Lecture Notes in Mobility, DOI: 10.1007/978-3-319-00476-1_28,

The larger internal combustion engines consume a considerable amount of fuel. For that reason their efficiency has received continuous development over the years and their fuel efficiency has been kept as low as economically viable within the set limits for emissions. To further reduce the $CO_2$ emissions for these engines new technologies are considered. One candidate technology is the application of exhaust Waste Heat Recovery (WHR) through the application of a Rankine Cycle for a heat to power application (H2P). While also investigated for passenger car applications [2], this technology is particularly suited for larger internal combustion engines running at considerable average loads, such as in long-haul heavy-duty trucks and Combined Heat and Power (CHP) generator applications.

In a Rankine Cycle (exhaust) heat that is otherwise wasted is used to generate steam in an evaporator (Fig. 1). Superheated steam is fed to an expander device that uses the steam expansion to generate mechanical power. After expansion the stream is fed to a condenser which condenses the stream to liquid water stored in a tank. A pump is used to feed the water to the evaporator.

Numerous variants on this cycle exist. The heat recovery cycle may be better compatible with the waste heat source temperature if a working fluid other than water is used. This is for example the case for exhaust Waste Heat Recovery for heavy-duty trucks. Various working fluids are considered for these applications such as ethanol and refrigerant R245fa [3]. With this choice for working fluid the cycle is called an Organic Rankine Cycle.

Also, different options exist for the expander device [3]. It may for instance be of a piston machine, scroll expander or turbine type. Depending on the application the generated power may be fed to the engine crankshaft or converted to electricity.

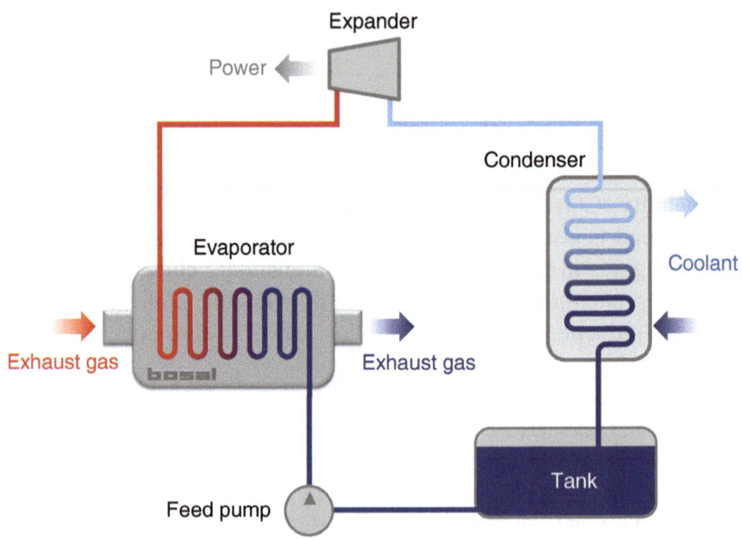


**Fig. 1** Rankine cycle for exhaust waste heat recovery

Specifically for heavy-duty diesel truck application there are generally two waste heat sources of consideration. One is offered by the main exhaust gas stream. The other source is the heat of the exhaust gas that is recirculated to reduce Nitrogen Oxide emissions. This Exhaust Gas Recirculation (EGR) stream is normally cooled by a specific EGR cooler, but may be replaced by an EGR evaporator for Waste Heat Recovery. Experimental studies have already demonstrated fuel savings up to 6 % for systems utilising both sources [4, 5].

With the introduction of the newest Euro VI trucks some sentiments of abandoning EGR to more or less extent have been observed [6, 7]. Engines with no or low EGR tend to have lower fuel consumption, less heat rejection and a cost-effective design. When Waste Heat Recovery would be applied to those (European) configurations, only an exhaust evaporator would be present.

A robust and modular evaporator for Rankine Cycle applications has been developed at Bosal. The evaporator has low back-pressure in combination with a high heat exchanging surface per volume and thus can be used in an exhaust. Its design, validation and application will be discussed in this article. Furthermore some ongoing developments will be highlighted that aid the integration of the evaporator product in future applications.

## 2 Evaporator Design

As exhaust manufacturer, the company has broad experience in high temperature stainless steel applications. For Rankine Cycle applications an exhaust evaporator has been designed that is based on a vertical tube stack concept as depicted in Fig. 2. The stack is constructed from cells of parallel U-tubes. The flow from the U-tubes is distributed and gathered in collector caps before passing to neighbouring cells.

This design has the following desirable properties:

- Low backpressure: The vertical tube stack has considerable open frontal area that allows gas flow without excessive restriction.
- Effective heat transfer: The gas is in intensive contact with the tubes and the vertical tube orientation offers highest heat transfer to the boiling fluid. This allows compact evaporator design.
- Modular design: Variation of the number of cells and the tube length allows the evaporator core to be tailored to a desired performance specification.
- High pressure operation: The construction and material thickness have been optimised to withstand forces occurring at operating fluid pressures up to 60 bar.
- Capable of running dry: Withstands high temperatures in the absence of fluid flow in case of less sophisticated controls and issues with the fluid loop.
- Built-in partial phase separation: Partial phase separation takes place in the collector caps. This ensures a constant output stream quality and prevents propagation of violent phenomena caused by the chaotic nature of boiling.

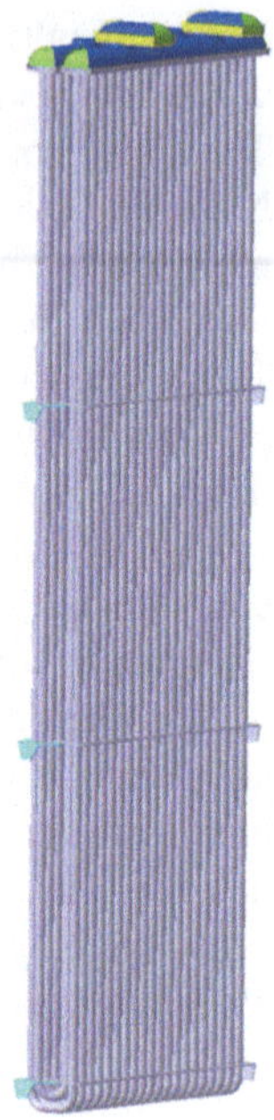

**Fig. 2** Modular tube stack evaporator design concept

- Robust: The design foresees expansion in vertical direction while offering a strong rigid structure in horizontal direction. This construction has to cope with severe temperature gradients and high pressures.

The tube stack is closely fitted with an outer plate such that intensive gas-tube contact remains. The complete core is fitted in a housing that is engineered for the specific application and ensures good flow uniformity (Fig. 3). Well thought material choice results in an evaporator that is able to operate in an aggressive exhaust environment and that is compatible with most working fluids.

## 3 Validation and Certification

The evaporator design has been validated through Finite Element Method (FEM) based Low Cycle Fatigue analysis in combination with in-house thermal cycling and vibration tests. This validation exercise has been input for obtaining European Pressure Equipment Directive and German AD2000 certification. Alongside design certification, also the production process has been proven compliant with certification rules which include traceability and welder certification (Fig. 4).

A test rig has been implemented for evaporator testing on a burner test setup. The setup comprises a water tank, fluid preheater, feed pump, large condenser, valves and all necessary instrumentation. Optionally an expander coupled to a dynamometer can be included for full Rankine Cycle testing as can be seen in

**Fig. 3** CFD Flow uniformity validation for a 1,000 kWth evaporator unit with four parallel cores

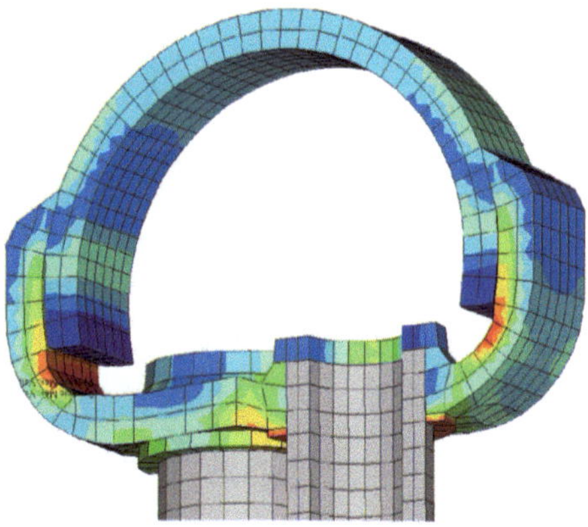

**Fig. 4** FEM design validation

Fig. 5. The test setup allows evaluation of evaporators, expanders and their integration into complete systems.

## 4 Application

Bosal evaporators have seen in-field application since 2007. One early application integrated the evaporator for waste heat recovery in a railway power pack (Fig. 6). Other evaporators have accumulated operating hours in CHP generator sets for over 3 years now. An example of such an application is depicted in Fig. 7. In this application exhaust gas temperatures are on the order of 450 °C and water is used as working fluid. The generated steam is fed to a piston expander coupled to the crankshaft of a 15L V8 engine.

The evaporators have proven very durable, even though some engines were running on biofuel. For the CHP application the evaporator units are intermittently

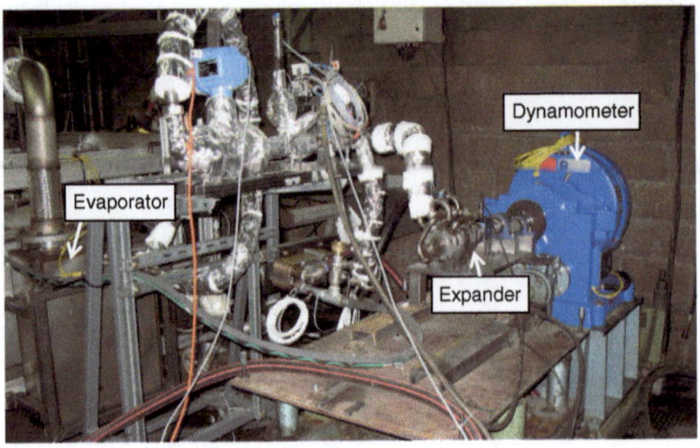


Fig. 5 Rankine cycle test setup with expander and dynamometer

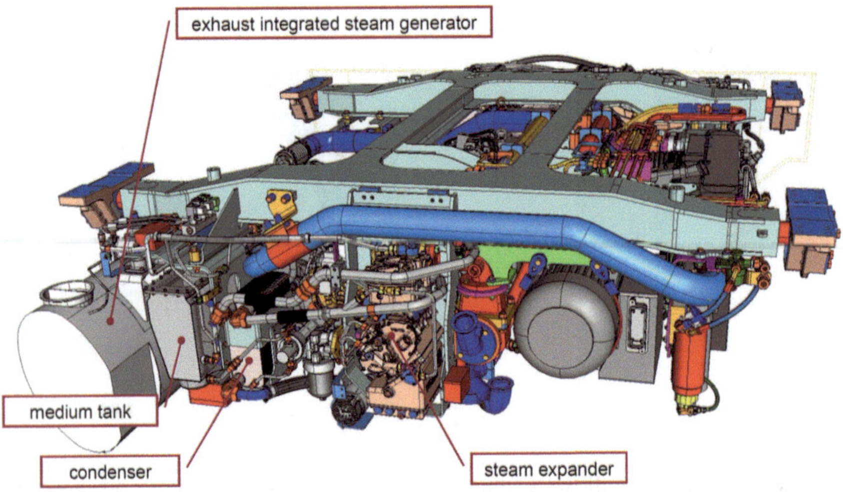


Fig. 6 Railway power pack with Rankine cycle waste heat recovery

Fig. 7 CHP generator set with evaporator mounted above engine

operated under high temperature dry conditions (absence of fluid flow) without sustaining damage.

## 5 Tools and Development

The near future will see more applications of Rankine Cycle Waste Heat Recovery. To offer suitable product proposals, a spreadsheet evaporator dimensioning tool has been developed and extended. The spreadsheet allows to find a suitable evaporator design configuration given an operating condition, desired heat exchange performance and exhaust back-pressure limitation. To meet the demand for alternative working fluids in Organic Rankine Cycle applications a calculation method has been devised based on the so-called effectiveness-NTU method. This approach based on non-dimensional numbers reveals key parameters: heat capacity ratio, dimensionless subcool temperature, Stanton heat transfer numbers and the Jakob number (sensitive to latent heat ratio). The method requires a minimal amount of fluid properties limited to: boiling point, vaporization heat, and specific heat capacity and transport properties at inlet, outlet and saturation conditions. It is also extended to allow predictions with working fluid mixtures.

A second software tool oriented development is the development of a dynamic evaporator simulation model. Current (Organic) Rankine Cycles are typically applied for stationary or mildly transient operation. Future systems, when applied on a heavy-duty truck for instance, will have to operate under transient conditions. The evaporator has its impact on the WHR system response that needs to be accounted for in system integration, optimisation and controls development [8, 9]. A transient simulation model of the evaporator is offered to support this system engineering task. The model is implemented in Simulink[1] and can be integrated into other simulation environments like GT-SUITE [2] or AMESim[3] .

Great care has been taken in the numerical implementation of the model. Simulation of two-phase flow and phase change phenomena can be challenging. Certain discretisation schemes and solution methods can lead to chattering; a phenomenon that causes oscillatory solutions [10]. It is intended to further develop a version of the simulation model that allows to study pulsive flow that can occur for certain setups. Insight into the nature of these highly dynamic phenomena will allow further improvement of the already durable design of the evaporator. For this purpose it is important that numerical simulation artefacts are not mistaken for real system dynamics.

1 www.mathworks.com/products/simulink

2 www.gtisoft.com.

3 www.lmsintl.com.

## 6 Conclusion and Outlook

An exhaust evaporator has been developed for Rankine Cycle Waste Heat recovery applications. It is based on a modular and robust concept and can be operated at high pressure up to 60 bar. Design and production have been certified. Test setups have been developed for component testing and system integration testing. Evaporator units have seen in-field operation since 2007 and proved very durable.

For a broad application of (Organic) Rankine Cycle Waste Heat Recovery it is important that the system efficiency and investment together result in an acceptable earn back period. The efficiency needs to be optimised through a systems engineering approach that can be supported through developed simulation models. The system cost should be optimized through cost-effective production and system integration. The latter could for instance be accomplished through integration of exhaust gas aftertreatment and evaporator. Future work on the evaporator design will focus on further improvement of long-term endurance through study of potential pulsive flow with simulation methods.

## References

1. Environmental Protection Agency 40 CFR Parts 85, 86, 600 et al (2011) Greenhouse gas emissions standards and fuel efficiency standards for medium- and heavy-duty engines and vehicles, 15 Sep 2011
2. Freymann R, Strobl W, Obieglo A (2008) Der Turbosteamer: Ein System zur Kraft-Waerme-Kopplung im Automobil, Motortechnische Zeitschrift (MTZ), issue 5, pp 404–412
3. Seher D, Lengenfelder T, Gerhardt J, Eisenmenger N, Hackner M, Krinn I (2012) Waste heat recovery for commercial vehicles with rankine process, 21st aachen colloquium automobile and engine technology, Aachen, 8–10 Oct 2012
4. Bredel E, Nickl J, Bartosch S (2011) Waste heat recovery in drive systems of today and tomorrow. MTZ Worldwide Edition, Issue 4:52
5. Park T, Teng H, Hunter GL, van der Velde B, Klaver J (2011) A rankine cycle system for recovering waste heat from HD diesel engines—experimental results, SAE paper 2011-01-1337
6. Ellensohn R (2011) Real world emission of a EURO VI long haul truck. In: 6th AVL international commercial powertrain conference, Graz, 25–26 May
7. Volvo Trucks (2012) New Volvo engine for Euro 6, Press Release, 5 July 2012
8. Quoilin S, Aumann R, Grill A, Schuster A, Lemort V, Splietho H (2011) Dynamic modeling and optimal control strategy of waste heat recovery organic Rankine cycles. Appl Energy 88:2183–2190
9. Willems F, Kupper F, Cloudt R (2012) Integrated energy & emission management for heavy-duty diesel engines with waste heat recovery system. In: IFAC workshop on engine and powertrain control, simulation and modeling (ECOSM'12), Rueil-Malmaison, 23–25 October 2012
10. Bonilla J, Yebra L, Dormidob S (2011) A heuristic method to minimise the chattering problem in dynamic mathematical two-phase flow models. Math Comp Model 54:1549–1560

# DHC-Vehicle: Towards Energy Efficiency and Safe Riding in Open Electric Vehicles

González Alonso Ignacio and Palomo Díaz Felipe

**Abstract** We are getting close to a hyperconnected future in which all the devices will share, analyze and manage all the information using the energy more efficiently, improving people's quality of life. This paper presents the DHC-Vehicle Project which intends to provide a real solution to the society needs related to the electric vehicle and device manufacturers. The proposed solution is based on a universal and open communication protocol, the standard Digital Home Compliant, which allows the interoperability among different devices regardless of its brand or model. DHC-Vehicle poses answers to the strategic agendas of the different technological platforms related to electric vehicles.

**Keywords** Interoperability · Electric vehicles · DH compliant · Safety · Energy efficiency

## 1 Introduction

In the recent decades there has been a saturation of the transportation infrastructures. The number of vehicles is steadily increasing and this affects society in different areas of life: emission of pollutants, accidents, traffic congestion, transportation delays and safety. Therefore the automotive sector is investigating in different solutions to address these problems [1].

Environmental and safety issues in driving are being improved with the emergence of electric vehicles (EV) and the development of different technologies.

G. A. Ignacio (✉) · P. D. Felipe
Escuela Politécnica de Mieres, C/Gonzalo Gutiérrez Quirós, Mieres, (Asturias), Spain
e-mail: gonzalezaloignacio@uniovi.es

P. D. Felipe
e-mail: palomofelipe@uniovi.es

J. Fischer-Wolfarth and G. Meyer (eds.), *Advanced Microsystems for Automotive Applications 2013*, Lecture Notes in Mobility, DOI: 10.1007/978-3-319-00476-1_29,

We are getting close to a hyperconnected future in which all the devices will share, analyze and manage all information using the energy more efficiently and improving people's quality of life.

Modern cars are becoming integrated with many electronic devices, sensors, controllers, actuators, improving the comfort and safety to the driver. Therefore, the active safety, connectivity, and intelligence have become a main direction of research and development on the future cars [2]. These are the base of the complex communication system established inside and outside the vehicle, involving a network.

## 2 State of the Art

The first electric vehicle appeared in the 1890s. It is only relatively recently, however, there has been a large increase in the use of the EV, due to increased environmental awareness and technological advances. In recent years, EV have become one of the preferred means of improving the transport system mainly because of being a cleaner transportation. The biggest obstacle to the EV sector is related to the batteries: duration, recharging times and infrastructure currently available for such vehicles.

Now electric vehicle manufacturers are mainly focusing on three fronts:

- Improve battery life: the autonomy of electric vehicles is still far from the autonomy of combustion vehicles.
- Decrease the battery recharging times by means of quick reloads, induction charging, etc.
- Facilitate connectivity between vehicles machine to machine (M2 M), between the VE and the grid (V2G), between the VE and the infrastructure (V2I); and improve driving safety and comfort subjects.

Several organizations are dedicated to the objective of further increasing road traffic safety and efficiency by means of cooperative Intelligent Transport Systems with inter-vehicle communication. For instance, Car to Car (C2C) [3], is developing the 'architecture' to improve mobile communications among cars.

Some relevant projects in this sector are listed below.

- COMeSafety. The main purpose of this initiative is to develop a set of standards to support European Community wide implementation and deployment of Co-operative Intelligent Transport Systems [4].
- Coopers "CO-Operative SystEms for Intelligent Road Safety". The goal of the project is the enhancement of road safety by direct and up to date traffic information communication between infrastructure and motorized vehicles on a motorway section [5].
- CVIS "Cooperative Vehicle-Infrastructure Systems". The CVIS objectives are to create a unified technical solution allowing all vehicles and infrastructure

elements to communicate with each other in a continuous and transparent way, using a variety of media and with enhanced localization [6].

- GeoNet "Geographic addressing and routing for vehicular communications". This is a networking mechanism distributing the information to nodes within a designated destination area. A novel routing protocol is in charge of information dissemination over multiple hops until every vehicle has received this information within the destination area. Each vehicle evaluates whether re-transmission is required and executes it with proper timing if needed [7].
- SAFESPOT "Smart Vehicles on Smart Roads". It creates dynamic cooperative networks where the vehicles and the road infrastructure communicate to share information gathered on board and at the roadside to enhance the drivers' perception of the vehicle surroundings [8].
- SEVECOM "Secure Vehicle Communication". This initiative focuses on providing a full definition and implementation of security requirements for vehicular communications [9].

## 3 DHC-Vehicle

The starting point of the DHC-Vehicle project is the universal communication protocol and open, DH Compliant (DHC) [10]. A brief overview of DHC is shown in Fig. 1. The DHC architecture is based on sub-modules which are the solutions for each of the following requirements:

- Devices must work in groups.
- A system for transmitting location information must be provided.

**Fig. 1** DHC characteristics as general standard

- The user should be able to restrict the actions of the system by creating rules.
- A security framework transmitting sensitive data across a network must be developed.
- The growing interest in efficient energy use raises the need of a specific energy management sub-module.

Interoperability among systems requires more than the simple exchange of information. All the devices involved must share a common interface which enables the cooperation among them [11].

The DHC-Vehicle project is based on the following sub-protocols of DHC, as represented in Fig. 2.

**DHC-Localization**

The localization subprotocol [12] has the purpose of laying the foundations for the concept of location in the DHC protocol. DHC-Localization provides the necessary data and it describes the actions in which the vehicle must follow to be able to reach its destination.

By this subprotocol, the DHC-Vehicle will be capable of indicating to the driver the nearest charging points, calculate alternative routes that pass through the vicinity of said charging points according to the charging needs. Besides, schedule may be synchronized such that the system can notify us of the tasks depending which area you are.

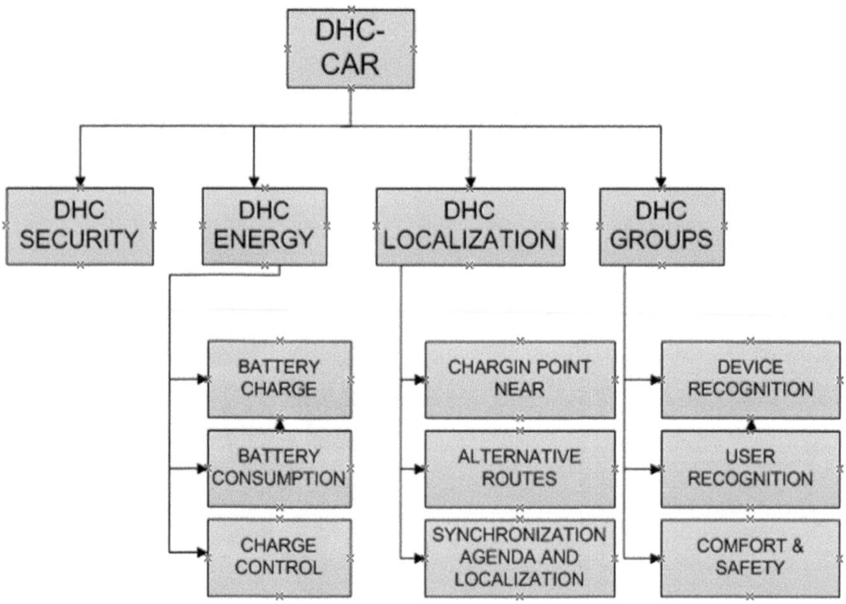


**Fig. 2** Scheme of DHC-Vehicle

**DHC-Groups**

This specification [12] aims to address the collaboration among devices, in order to accomplish a task in the shortest time possible. When a device receives a task, it could use a call to other device/s that implement the same functionality, so the task could be made in the shortest time possible.

By this sub-protocol the DHC-Vehicle devices will recognize both driver and passenger. DHC-Vehicle will be able to identify the driver and tailor the driving position to your preferences. All these capabilities are designed to improve comfort and safety in driving.

**DHC-Energy**

The aim of this submodule [12] is to lay the concept of energy efficiency and Smart Grids in the DHC protocol. The DHC-Energy services were defined as a set of saving and energy management concepts to be used for by different devices. Moreover, through DHC-Energy, the user will be able to know in detail the energy accounting within the electric vehicle environment. By means of the data gathered, the customer will be able to manage, prioritize, reduce or change the global energy behavior and the profiles of the DHC associated devices.

The DHC-Vehicle subprotocol will enable the customer to know the battery level to the driver. These data will be accessible from any user's device, and will be protected thanks to DHC-Security module [12].

**DHC-Security**

The fact that user's data travel within an electric vehicle context could compromise the driver's privacy. The DHC-Security specifies a set of data encryption and authentication procedures in order to avoid unauthorized access to that information or to avoid the system to be controlled by someone different from the EV system owner. Only, the devices that are able to authenticate themselves into the network are capable of inter-operating with other devices. Accordingly, external devices cannot access to the data on the electric vehicle network.

A SysML diagram of the system is represented in Fig. 3.

The diagram explains the functioning of DHC-Vehicle, which is basically an adapter through the DHC protocol will be compatible with all smart devices regardless of model and brand (tablets, smartphones, etc.). This adapter will control all parameters of the vehicle and its environment through the various subprotocol implemented in the system. All this information is stored in the "cloud" so that will be accessible to the user from any location. The driver will be able to know the status of the vehicle and set parameters according to the statistics observed in the application.

The DHC-Vehicle functionalities are designed to improve safety and driving comfort as well as the energy efficiency. The final goal of DHC-Vehicle is to develop a "Vehicles Management System" in order to improve energy efficiency, traffic conditions, interoperability and communication between vehicles, charging infrastructure and people.

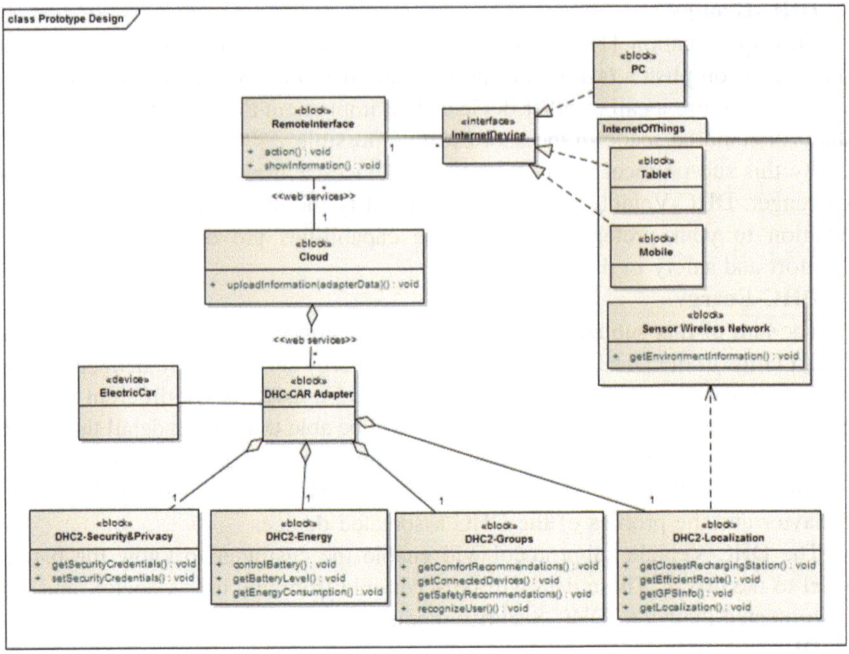


**Fig. 3** SysML diagram

# 4 Objectives

The motivation to pose DHC-Vehicle with different control functions is to create an improved driving experience for the customer.

Control functions will take care of the energy system and the charging state of the battery, so information could be known in advance for the user and facilitates decision taking. Also they will provide security solutions like the user recognition and safe data transmission due to the Security and Privacy module. It will supply intercommunication among environment devices and user's devices so it can favor efficient and safe driving. Besides, it will take advantage of the Cloud Computing and Internet of Things to store information in regard to vehicle localization, to transmit information in real time and to calculate more efficient routes.

This is so because we do not want that the driver has other concern but driving.

## References

1. Figueiredo L, Jesus I, Machado JAT, Ferreira JR, Martins de Carvalho JL (2001) Towards the development of intelligent transportation systems. IEE Intelligent Transportation Systems Conference Proceedings, Oakland (USA) pp 1206–1211

2. Chen Y, Yu P, Liu Z (2011) Research on vehicle intelligent communication based on supernetwork system. In: 2011 international conference on business management and electronic information (BMEI) 5, 94–98
3. "Car to Car (C2C)" [Online]. Available: http://www.car-to-car.org. Accessed: 31 Jan 2013
4. "Comesafety" [Online]. Available: http://comesafety.org/
5. "Coopers: Home" [Online]. Available: http://www.coopers-ip.eu/. Accessed: 03 Mar 2013
6. "CVISproject.org Home" [Online]. Available: http://www.cvisproject.org/. Accessed: 03 Mar 2013
7. "GeoNet Project" [Online]. Available: http://www.geonet-project.eu/. Accessed: 03 Mar 2013
8. "Safespot" [Online]. Available: http://www.safespot-eu.org/. Accessed: 03 Mar 2013
9. "Sevecom" [Online]. Available: http://www.sevecom.org/. Accessed: 03 Mar 2013
10. "Dhcompliant" (2013) [Online]. Available: http://www.dhcompliant.com. Accessed: 31 Jan 2013
11. Alonso IG, Alcalá MRF, Fuente MPAG (2010) Interoperability standard used by service robots: study to develop an open standard bye Infobotica research group. In: Proceedings of the 2010 fifth international conference on systems, Washington, DC, USA, 18–23
12. González Alonso I, Alvarez Fres O, Alonso Fernandez A, Gomez del Torno P, Maestre JM, Garcia Fuente MdPA (2012) Towards a new open communication standard between homes and service robots, the DH Compliant case. Robot Auton Syst 60(6):889–900 Jun 2012

# Optimized Regenerative Friction Braking Distribution in an Electric Vehicle with Four In-Wheel Motors

**Manuel Ignacio González Hernández, Blanca Araujo Pérez, Juan Sabas Martín Sánchez and Esteban Cañibano Álvarez**

**Abstract** One of the most important features of electric vehicles (EV) is their ability to recover significant amounts of braking energy. The electric motors can be controlled to operate as generators in order to convert the kinetic or potential energy of the vehicle into electric energy that can be stored in the battery and then reused. In a hard braking manoeuvre, the braking torque is much larger than the torque that an electric motor can produce. So, mechanical friction braking systems have to coexist with electrical regenerative braking. In order to reach the target of a highly efficiency electric braking, authors have analyzed different strategies to find an optimized braking distribution between axles and between regenerative and friction braking, recovering a huge percentage of the available energy during braking manoeuvres.

**Keywords** Regenerative · Braking · Electrical vehicle · Friction · Optimization · Motor

M. I. González Hernández (✉) · B. Araujo Pérez · J. S. Martín Sánchez · E. Cañibano Álvarez
Cidaut Foundation, Parque tecnologica de Boecillo, P 209, Boecillo, Spain
e-mail: mangon@cidaut.es

B. Araujo Pérez
e-mail: blaraa@cidaut.es

J. S. Martín Sánchez
e-mail: juasan@cidaut.es

E. Cañibano Álvarez
e-mail: estcan@cidaut.es

J. Fischer-Wolfarth and G. Meyer (eds.), *Advanced Microsystems for Automotive Applications 2013*, Lecture Notes in Mobility, DOI: 10.1007/978-3-319-00476-1_30,

# 1 Introduction

As it has already been mentioned, the motivation for the present work comes from the opportunities provided by electric motors to recover energy and then reuse it for different purposes. In order to do this, the motor works as generator and brakes the vehicle.

Generally, the braking torque is much larger than the torque that an electric motor can produce. So, mechanical friction braking systems have to coexist with electrical regenerative braking.

The work, which is shown in this article, was carried out in the frame of FP7 research project E-VECTOORC, a research project, where strategies to control and optimize independent motors are being developed. The focus of this work is to control the electric motors/generators for the vehicle dynamics enhancement.

The final goal of the design and control of the hybrid braking system is to ensure the vehicle's braking performance and its ability to recover as much braking energy as possible.

Previous to the control design, the brake strategy that has to be applied to obtain the largest energy charge was analyzed. A specific Matlab/Simulink tool has allowed determining the regenerated power for different braking conditions (speed and deceleration). It was also able to calculate the total input energy into the batteries when a standard driving cycle was carried out, taking into account the characteristics of the demonstrator vehicle within the project E-VECTOORC.

The present document shows the analysis of the brake calculations carried out for the prediction of the regenerative/friction distribution. The tool allows us to implement different power train configurations: one motor in each wheel or just two motors in just one axle (front or rear). In this article, a four motors configuration has been chosen.

# 2 General Theory

The idea of the developed tool is to help find the best point of distribution of braking force that makes it possible to regenerate the maximum of energy. In the vehicle, the braking forces are determined by the mass distribution in the vehicle and the baseline friction brake.

For a vehicle, the optimum distribution is described by the equiadherence parabola (blue curve in Fig. 1). If the real braking force distribution is below the ideal braking force distribution curve (equiadherence parabola), the front wheels will be locked earlier than the rear wheels. This situation leads to stable behaviour of the vehicle, according to the ECE Regulation [1].

$$\frac{Fbf}{Wf} \geq \frac{Fbr}{Wr} \tag{1}$$

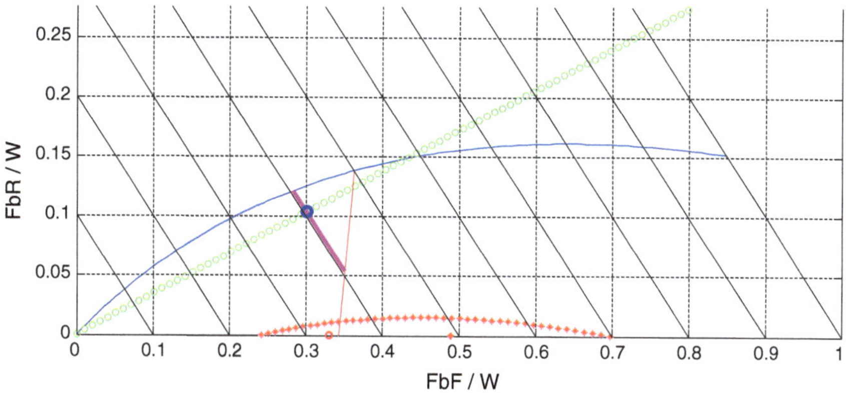


**Fig. 1** Representation of the braking calculations with friction limit

where *Fbi*, is the Braking Force, in each axle (N) (f—front axle; r—rear axle) and *Wi*, is Vehicle Weight on each axle (N).

However, when the working point is much below the ideal braking force distribution curve, most of the braking force will be applied to the front wheel and a very small force will be applied to the rear wheels. This design will cause the problem of reduced utilization of the road adhesive capability.

In order to avoid this situation, additional brake regulations have been developed. So, there exists a maximum possible braking force on the front wheels, limited by the ECE regulation (red curve in Fig. 1).

According to this regulation, for values between 0.2 and 0.8, the braking distribution must satisfy the following expression [1]

$$z \geq 0.1 + 0.85(\mu_{ROAD} - 0.2) \tag{2}$$

where z is Braking rate of vehicle (dimensionless), and

$\mu_{ROAD}$ is Road adhesion (dimensionless)...

Then the allowed zone of the braking forcc distribution is between both curves in the Fig. 1.

According to a conventional vehicle, the baseline friction brake system is defined by the slope of the straight dotted line in the Fig. 1. In this specific case, a possible braking distribution according to the physical characteristics of the friction brakes of the vehicle is calculated as the ratio of the rear and front axles braking forces, according to the geometrical data of the friction braking system.

The tool uses additional lines over the graph shown in the Fig. 1 in order to take into account the road adhesion. The friction limit is implemented as a group of lines in order to represent the relationship of the braking forces on the front and rear wheels when the front wheels are locked and the rear wheels are not locked [1], taking into account the tyre-road adhesion ($\mu_{ROAD}$).

$$FbR = \frac{wb - \mu_{ROAD} \cdot h}{\mu_{ROAD} \cdot h} \cdot FbF - \frac{W \cdot L_b}{h} \quad (3)$$

According to the previous indications, for a specific deceleration z, all the appropriate combinations of rear/front braking forces will be located on the pink line (shown in Fig. 1, case: deceleration = 0.4 g, $\mu = 0.5$). The possible braking distribution points will be located along the pink curve (according to a constant deceleration line, with road adhesion limitation).

The developed tool will allow determining the most interesting rear/front brake distribution and also the regenerative/friction split among all the possible combinations (taking into account different strategies and assumptions).

## 2.1 *Braking Strategy*

In the tool, in order to find the most efficient regenerative/friction brake distribution, from the energetic point of view, three different strategies have been implemented. In all strategies, there is a common rule to find the force distribution. This rule is that the tool starts in the front axle when making calculations to recover energy. So, the front motors try to recover all the available energy in the axle, taking into account the limitation given by the battery capacity. If the front axle motors are not able to cover all the battery capacity, also the rear motors recover energy. According to the dynamics of the vehicle in a braking manoeuvre, the biggest amount of energy that can be recovered is available in the front axle (static + dynamic mass).

The three different strategies that were implemented are:

### 2.1.1 Maximum Regenerated Power Strategy

The selected criterion in this strategy is to regenerate the maximum amount of energy in all the motors. This strategy selects the defined number of points over the constant deceleration line (pink curve in Fig. 1) and calculates the regenerated power in each point. Finally it chooses the point with the maximum regenerated power.

### 2.1.2 Optimal Braking Distribution Strategy

This strategy is the ideal combination between front and rear braking force. A working point over the parabolic curve (blue curve in Fig. 1) means using the maximum braking capability in the vehicle, within a safety performance (both axles will block at the same time).

### 2.1.3 Bilinear Regenerative + Friction Braking Strategy

This strategy is a physical simplification of the optimal braking distribution strategy. In this strategy the parabola curve is approximated by two lines. This strategy is the rule that conventional braking systems apply. This strategy was simulated to know the increase of recovered energy when one of the two first strategies is applied (Fig. 2).

# 3 Tool Algorithms

## *3.1 Braking Force Calculations*

For all the implemented strategies, in function of the braking deceleration equi-adherence parabola and the ECE curve, the working points on the braking diagram (Fig. 1) are located. For the different working point, the braking torque for front and rear axles are calculated.

## *3.2 Regenerative Power Calculation*

The following step in the tool begins identifying if the motor can provide the demanded braking force in the working point or not.

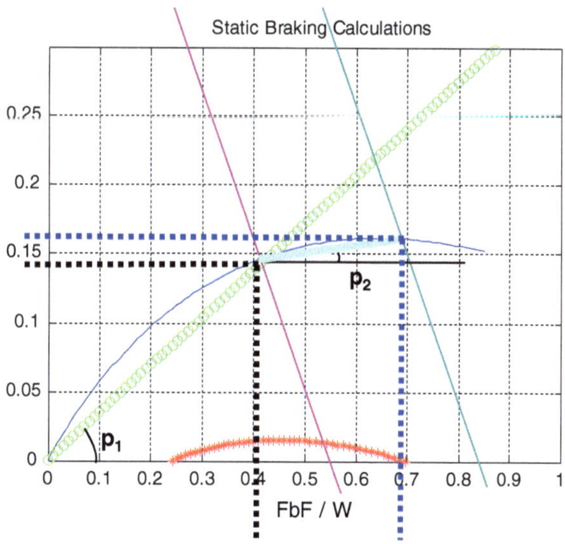


**Fig. 2** Braking force distribution of bilinear strategy

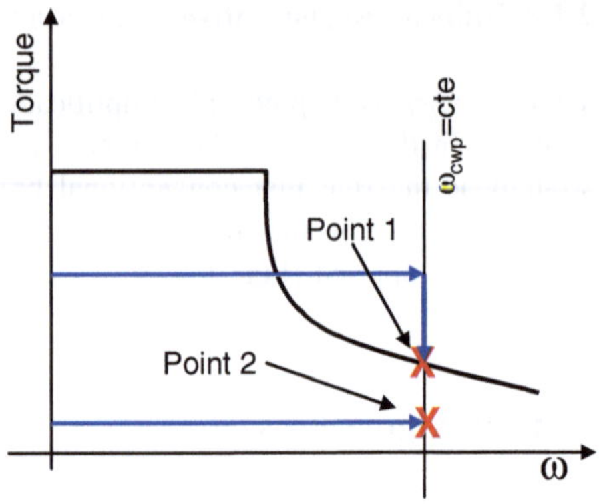


**Fig. 3** Checking the value of the required braking torque in the working point

For each motor, in the working condition (speed), two alternatives are possible:

- The "Required braking torque" to achieve the needed deceleration in the braking manoeuvre is higher or equal than "Maximum braking torque" (according to characteristics of the motor): In this case, the motor gives its maximum torque.
- "Required braking torque" is lower than "Maximum braking torque": In this case, the motor gives the required braking and friction torque is not required (Fig. 3).

## 3.3 *Limitations*

At the end of the mentioned steps, the performance of the motor has been analysed. The tool has defined the braking capacity of the motor. However, according to the working condition, this capacity could not be used because of two introduced limitations:

### 3.3.1 Motor Efficiency

When the motor is working as generator, at low speeds, the recovered energy could be negligible because of its reduced efficiency. In this case, using the motor as generator is not interesting, and the vehicle controller must decrease the speed using the friction braking system instead of the regenerative braking. So, a transition mechanism must be introduced in the tool to implement this split between regenerative and friction braking.

In order to define the transition starting point between regenerative and friction brake at low speeds, during the initial stages in the definition of the brake distribution script, two options were managed:

- Choosing a constant motor speed as borderline.
- Choosing a constant motor efficiency as borderline.

For this work, the second option was implemented in the developed tool, thinking that it is a more interesting way to manage the transition point. In the tool, the user introduces a value for the efficiency limit for the electric motor to work as generator (initially, value by default = 0.5). In this way, the variable velocity limit to stop regenerative braking is a function of the motor torque.

For the transition between regenerative and friction braking (at low speeds), the software uses a correction factor (Fig. 4), which modifies the regenerative braking torque, taking into account a ratio, obtained from the motor efficiency values, as follows:

$$efficiency\_rate = K = \frac{\eta - \eta_{\min}}{\eta_{\lim x} - \eta_{\min}} \tag{4}$$

where $\eta_{\min}$ is the minimum efficiency of motor and $\eta_{\lim}$ is the limit efficiency introduced by the user.

### 3.3.2 Battery Limitation

In some occasions the motor could give more power then the capacity of the battery. If the previously calculated regenerative power is higher than the value of the maximum battery power, the maximum regenerative power will be the battery limit and the power that is impossible to be recovered will be dissipated by the friction brake (Fig. 5).

## 4 Simulation Results

The principal target of the tool is to obtain a series of look-up-tables that can be used to develop the brake system control of an EV. The look-up-table has to give to the control system the needed braking torques in each motors when the vehicle

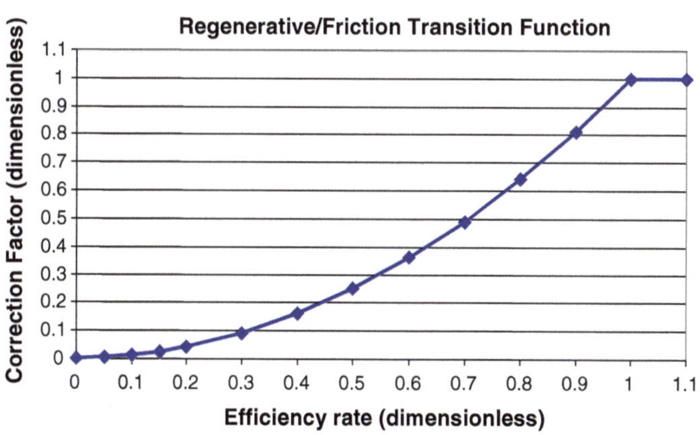


**Fig. 4** Implemented transition function

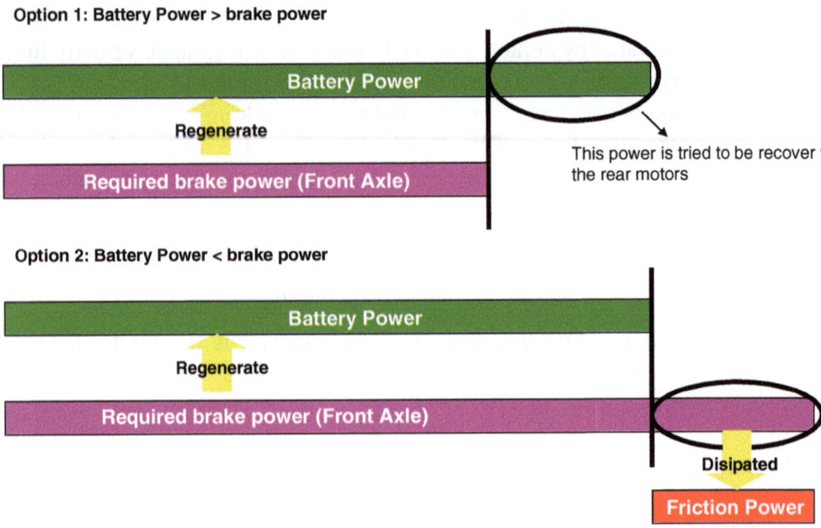


**Fig. 5** Battery limitation

is braking with a determined deceleration and vehicle speed, under the hypothesis of maximum recovery of energy. Also, the tool generates additional look-up-tables with all the calculated parameters.

The look-up-tables are bi-dimensional matrixes, where the inputs are the deceleration and the speed of vehicle and the outputs are the recovered power, motor torques and so on.

To identify the strategy that allows the highest recovering of energy, different driving cycle profiles can be simulated. Following, the result for a specific driving cycle is shown (FTP 72). According to the previous methodology, each working point of the driving cycle is placed over the surface (Fig. 6) to obtain the recovered power. With the integration of the power obtained in each working point, the total recovered energy for each strategy has been obtained. The next table shows the obtain energy:

| Strategy | Recovered energy (KJ) |
|---|---|
| Maximum regenerated power | 3,638 |
| Optimal braking distribution | 3,180 |
| Bilinear regenerative + friction braking | 3,068 |

Using the strategy #1, at the end of the driving cycle, the motors have recovered 3,638 kJ, more than a 14 % over the recovered energy using the Strategy #2—Optimal braking distribution (3,180 kJ) (Fig. 7).

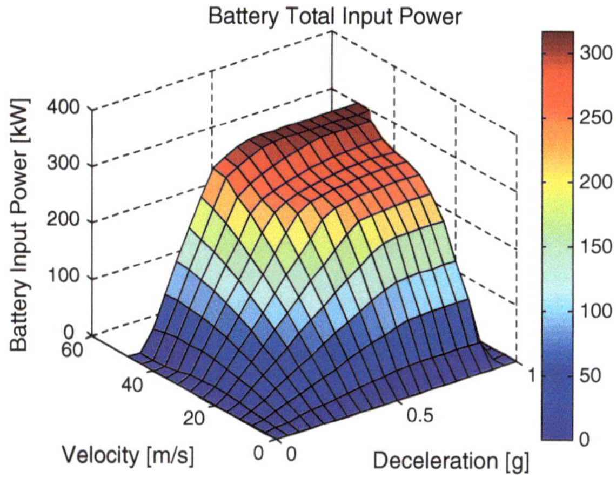


**Fig. 6** Look-up-table: battery total input power

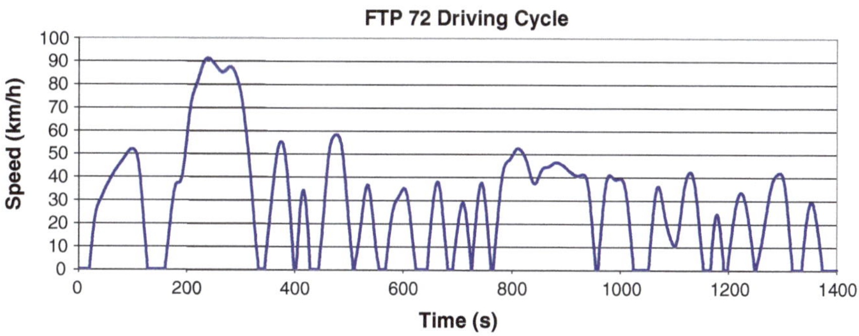


**Fig. 7** FTP 72 driving cycle speed profile

## 5 Conclusion

With the work carried out and the deep analysis of the obtained results, the following conclusions have been achieved:

The development of the braking distribution tool has demonstrated that this type of tool was necessary to reach an adequate distribution of the braking system. This tool has allowed the project partners making a wide number of calculations with different strategies and data, in order to reach the best alternative for the braking distribution between regenerative and friction brakes, from the point of view of energy recovering.

The selected global braking system (electric and friction brake) to be implemented into the demonstrator, decelerates the vehicle under all the use conditions, recovering also a wide amount of energy. The tool has demonstrated it is possible to recover almost 70 % of the kinetic energy if a correct distribution of braking force in all motors is found.

**Acknowledgments** The shown work in this paper was possible thank to the received funding from the European Community Seventh Framework Programme.

# References

1. Ehsani M, Gao Y, Emadi A (2010) Modern electric, hybrid electric and fuel cell vehicles: fundamentals, theory and design, 2nd edn. CRC Press, Boca Raton
2. Limpert R (1999) Brake design and safety, 2nd edn. SAE, New York
3. Gillespie TD (1992) Fundamentals of vehicle dynamics. SAE, New York

# Part V
# Components and Systems

# Innovative MEMS Sensors in Advanced Positioning Systems

**Marco Ferraresi, Gianvito Giuffrida and Nicola Palella**

**Abstract** Micro-mechanical sensors, originally introduced in the car as key components for passive safety systems, are now used also in a number of emerging applications like security, positioning, and telematics. Bundled together with the TeseoII$^{TM}$ chipset, the first-in-the-market positioning receiver able to simultaneously handle satellite signals from GPS/Galileo/Glonass/QZSS constellations, STMicroelectronics' accelerometers and gyroscopes can address and help to solve those issues arising from satellite obscuration and/or degraded reception in various environmental conditions. To combat the complexity of today's driving on multi-level roads and the increasing demand of anti-spoofing and anti-jamming algorithms in eToll applications, other MEMS sensors like pressure and compass may find their way into these emerging applications. This paper addresses the latest sensing solutions offered by ST in this field, with an outlook to future evolutions.

**Keywords** Navigation · Dead reckoning · Gyroscope · Accelerometer · Pressure sensor · Magnetometer · Modules · Data fusion · Sensor fusion · DOP · EKF

M. Ferraresi (✉)
STMicroelectronics, Via Tolomeo, 1 20010 Cornaredo, MI, Italy
e-mail: marco.ferraresi@st.com

G. Giuffrida
STMicroelectronics, Strada Primosole, 50 95121 Catania, CT, Italy
e-mail: gianvito.giuffrida@st.com

N. Palella
STMicroelectronics, Via C. Olivetti, 2 20864 Agrate Brianza, MB, Italy
e-mail: nicola.palella@st.com

J. Fischer-Wolfarth and G. Meyer (eds.), *Advanced Microsystems for Automotive Applications 2013*, Lecture Notes in Mobility, DOI: 10.1007/978-3-319-00476-1_31,

# 1 ST Positioning Solutions

The GPS radio navigation system is increasingly helping land, sea, and airborne users determine their exact location, velocity, and time, 24 h a day, anywhere in the world. The original civilian application that drove the growth of GPS technology was automotive navigation, but today, GPS is present in many other applications such as PNDs, PDAs, digital cameras, and handheld devices for trekking and sailing. ST has been developing and producing silicon solutions for GPS applications for 20 years, contributing significantly to the growth and spread of such applications. This extensive, field-proven experience has allowed ST to develop turnkey solutions for the market, enabling customers to build competitive products. The latest satellite positioning receiver from ST is the TeseoII (see block diagram in Fig. 1). The products belonging to the Teseo family are capable of receiving signals from multiple satellite navigation systems, including GPS (USA), Galileo (Europe), GLONASS (Russia) and QZSS (Japan). One member of this family, the STA8088EXG, is even able to handle the Compass (BeiDou-2) constellation (China).

As a multi-constellation receiver, the TeseoII is able to achieve more accurate positioning compared with any other GPS-only receiver currently available in the market. Position accuracy is a key parameter in the new growing, satellite based, market application. Services associated with telematics and road tolling applications are strongly linked to user position [1, 2]. The accuracy of a satellite positioning receiver is mainly dependent on the number of satellites in view, and where they are located in the sky. The metric that associates the number and geometry of

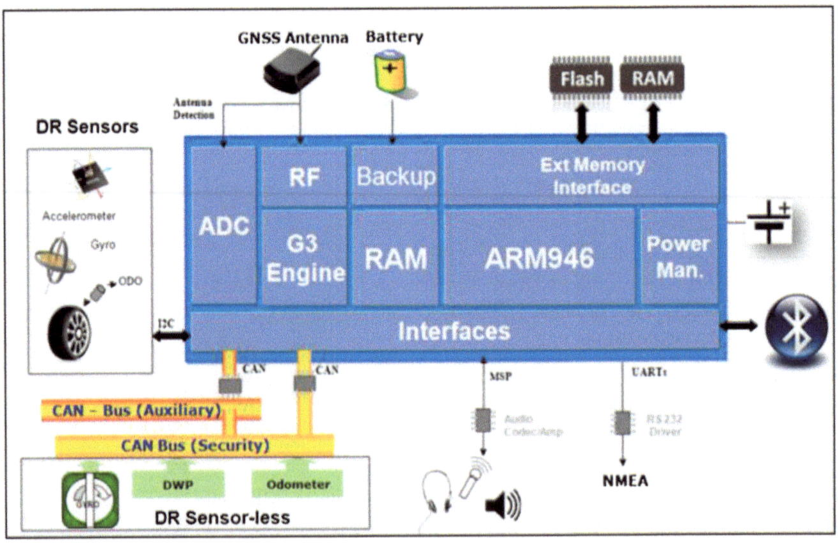


**Fig. 1** Block diagram of the TeseoII chipset

the satellites with accuracy is known as the Dilution of Precision (DOP). When satellites visible to the receiver are close together in the sky, the geometry is said to be weak, and in this case the DOP value is high. When the satellites in view are far apart, the geometry is said to be strong, and the DOP value is low. A greater number of visible satellites normally results in a lower DOP and thus better accuracy.

GPS-only receivers give acceptable position accuracy when they are in open sky or in a moderately obstructed scenario. In severely obstructed environments like the urban canyon, GPS-only receivers are often unable to generate an accurate position, and sometimes face difficulties in even getting a fix. In these scenarios, the receiver's ability to handle signals coming from different satellite systems in parallel allows a significant improvement in the position accuracy. A GPS-only receiver may see 10–12 satellites in an open sky, dropping to 3/4/5 direct signals in the urban canyon, which is marginal for positioning, while the multi-constellation receiver would see 17–20 satellites in an open sky, dropping to 5–8 direct signals in the urban canyon, giving continuous positioning with improved accuracy. Field tests performed in different scenarios confirm the TeseoII receiver achieves better position accuracy versus any other GPS-only receiver.

TeseoII's ability to simultaneously track multi-constellation satellite signals makes it the right solution for all customer applications requesting a high level of position accuracy. To manufacture this complex chipset integrating RF and digital blocks, ST uses an advanced CMOS technology with lithography of 65 nm.

## 2 Sensors

STMicroelectronics is well recognized at worldwide level as a key supplier of MEMS-based sensors.

Accelerometers, gyroscopes, magnetometers, pressure sensors, and microphones are now at the core of many different applications in the automotive, industrial, medical, and consumer domains.

In positioning and navigation systems, it is well know that a degradation of the signal coming from the satellites may lead to a progressive loss of accuracy which eventually prevents the operability of the system.

Two different situations may occur:

- The signal scenario is not suitable for a good fix (for instance, in reflective environments, partially obscured areas, urban canyons, see Fig. 2).
- The GNSS signal is not available (tunnels, indoor, obscured areas, see top section of Fig. 3).

To overcome this problem, modern positioning systems implement the so called Dead Reckoning (aka DR) algorithm, which is the process of calculating one's current position by using a previously determined position, or fix, and advancing that position based upon known or estimated speeds over elapsed time, and course.

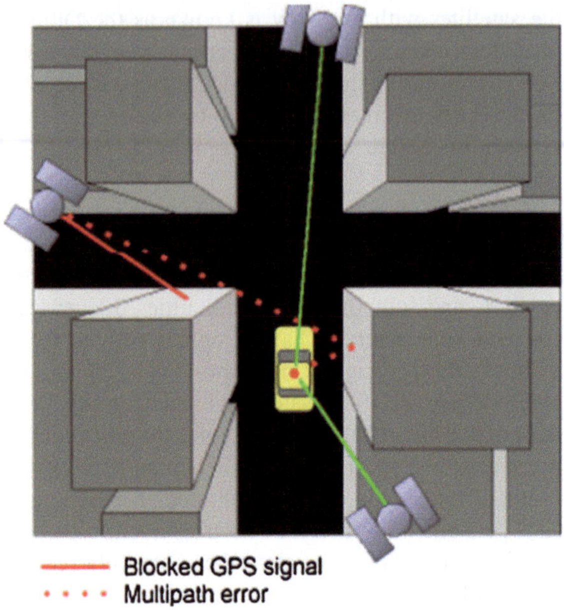


**Fig. 2** An example of urban canyon, critical for a correct GPS fix

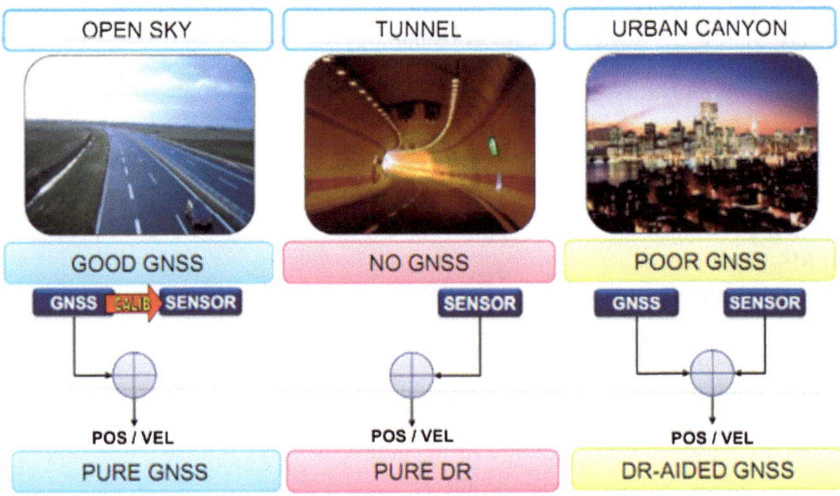


**Fig. 3** Various navigation modes depending on the environment

The algorithm, based on sophisticated mathematical models, is fueled from the information gathered from a set of sensors, namely distance sensors (odometer, wheel speed, ABS wheel tick) and yaw rate sensors (gyroscope, differential wheel pulses, differential wheel speed).

Among all possible solutions, the one using a classical odometer plus a gyroscope (Figs. 3 and 4) is the most accurate in production and all errors (temperature drifts,

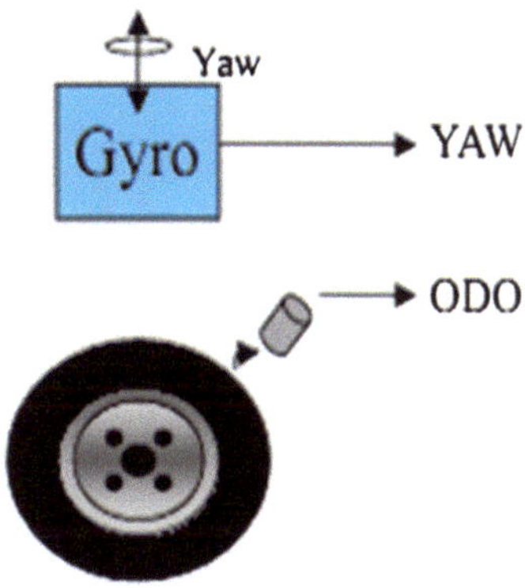


**Fig. 4** The best implementation of a DR algorithm

noise, and improper orientation of the gyro sensor in the vehicle) can be easily predicted and compensated. A possible drawback might be the cost: whenever the ESC (Electronic Stability Control) gyro signal is not available on the CAN bus (majority of the cases) and a dedicated yaw-rate gyroscope has to be used.

STMicroelectronics began mass production of MEMS-based gyroscopes in 2008. Since then, several hundred million units have been produced, shipped, and now are at the core of a number of consumer and industrial applications. Leveraging on this huge manufacturing experience, it was decided in 2011 to bring gyroscopes also into the automotive industry, targeting in-dash navigation and, in general, telematics applications. While in principle a single-axis yaw gyroscope is all that is needed to implement dead reckoning algorithms in vehicle positioning and navigation, a quick technical and marketing analysis concluded that a 3-axis gyroscope does a better job, by solving constraints related to the mounting orientation of the navigation board in the dashboard (Fig. 5).

STMicroelectronics was in fact the first company to introduce the market to a 3-axis gyro based on a single mechanical structure able to respond simultaneously to tri-dimensional stimuli and was the first company to propose this kind of component to the automotive industry (product number A3G4250D). This innovative compact mechanical transducer is named "The Beating Heart" because it

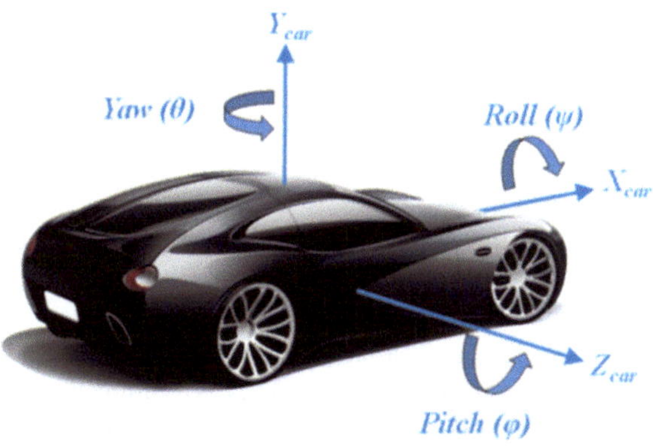


**Fig. 5** Axes definition in vehicles

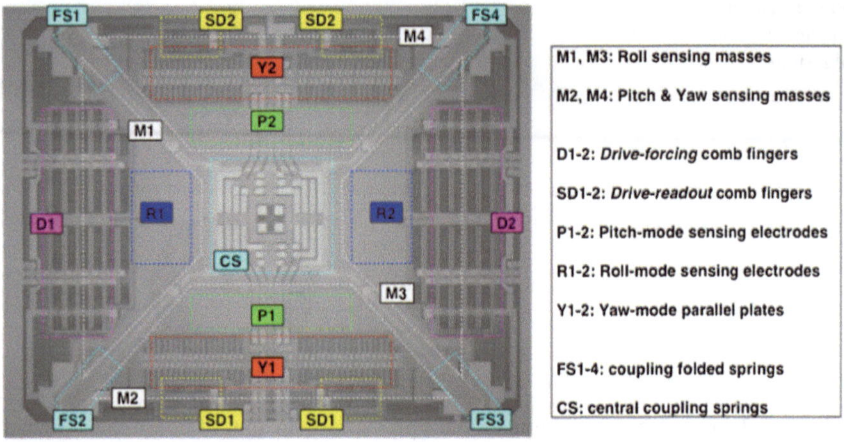


**Fig. 6** The mechanical structure of ST's 3-axis gyroscope

combines a triple tuning-fork structure within a single vibrating mass and because of its working mode, explained below.

The Beating Heart achieves excellent performance in terms of thermal stability, cross-axis error, and acoustic noise immunity by using a small die size. Furthermore, the presence of a single primary vibration mode for the excitation of the three tuning-forks simultaneously, together with the possibility of sensing the pickoff modes in a multiplexing fashion, allow the design of a small area, low power, companion chip. The advantage of having a single driving mass is the main reason for the commercial success of STMicroelectronics' gyroscopes. In fact, the presence of a single vibrating mass instead of multiple vibrating masses avoids the problems of mechanical frequency mixing and interference, extremely deleterious in other manufacturers' gyroscopes (Fig. 6).

For instance, for space constraints, several car makers assemble the navigation board with a pre-determined angle of about 20 mechanical degrees versus the vertical plane. To overcome the loss of sensitivity of a single-axis yaw gyroscope mounted in such a position, sensor manufacturers have to compensate this angle within the package itself. That solution, apart from posing some technical issues at the package level, does not offer the car maker the flexibility to change the mounting position of the board in the dashboard. A three-axis gyroscope instead, coupled with a three-axis low-g accelerometer (optional), provides the maximum degree of flexibility: the accelerometer can calculate the incline angle at which the board is mounted in the dashboard, while the 3D gyroscope can provide all rates necessary to implement the dead reckoning algorithms with precision. The availability of the pitch rate could also be used to extrapolate supplemental information about car elevation. The main problem to overcome is then the fact that $Y_{car}$ may not correspond to $Y_{sensor}$. If a single-axis gyroscope is used, any misalignments between physical stimulus and the gyroscope's rotation axis introduce a non-linear sensitivity reduction that is proportional to the cosine of the mismatch angle: (Fig. 7).

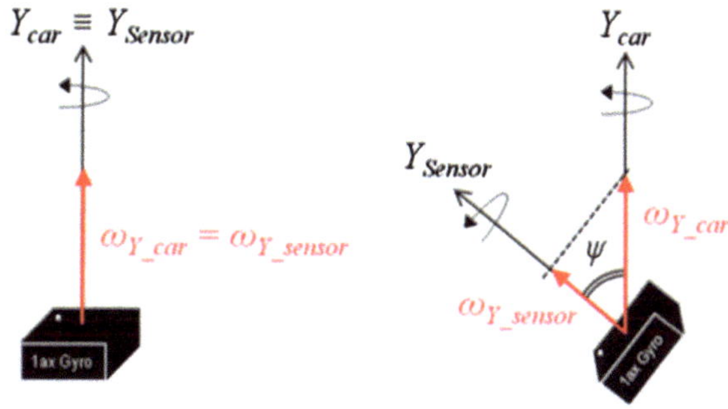


**Fig. 7** Single-axis detection versus orientation

In the worst case, the gyroscope's rotation axis is placed orthogonal with respect to the applied velocity ($\psi = \pm 90°$): in this configuration, a single-axis sensor is not able to detect the desired signal as the cosine factor is equal to zero.

If the board mounting angles are not known a priori, a further enrichment to the solution consists in determining the roll and pitch angles by using an additional three axis linear accelerometer, like the AIS328DQ.

The complete dissertation on this topic is found in the Ref. [3].

STMicroelectronics provides a complete dead-reckoning (DR) solution, as part of the Teseo family SW offer to address the increasing needs of accuracy and precision in modern positioning systems (see Fig. 8).

- The GNSS receiver provides timely measurements about visible SVs (e.g. pseudorange and frequencies).
- The DR Extended Kalman Filter (EKF) tightly integrates GNSS observations with motion sensors, adaptively weighting the contributions and providing optimal data/sensor fusion.
- The wide variety of supported sensors allows ST's DR to be configured in different modes.

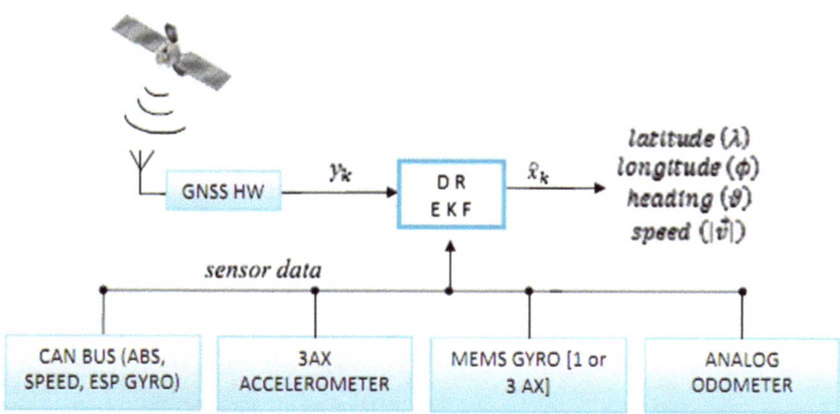


**Fig. 8** Ecosystem of ST's DR solution

Other sensors, like pressure sensors and magnetometers (e-compass) may be used to increase the overall system accuracy. The magnetometer, capable of measuring the earth's magnetic field, gives independent information about the heading angle $\vartheta$ and may help in all those situations when the absence of the GNSS signal for extended periods could lead to a loss of data accuracy from the EKF. The pressure sensor (see Fig. 9), instead, is the best candidate to solve navigation issues in today's complex and dense driving environment, where multi-level roads are often the only way to ease vehicle traffic.

STMicroelectronics is today in mass production with pressure sensors and magnetometers, both in stand-alone versions and also in multi-axis modules (Systems on Board or Systems in Package).

## 3 Field Tests

Extensive field tests were performed by ST using a DR logger box as shown in Fig. 10.

Several dead reckoning configurations were analyzed and compared, but eventually the ones based on MEMS sensors have shown the best results:

Field tests were done in very harsh urban environments, where satellite obscuration, multi-path reflections, and even absence of signal make the

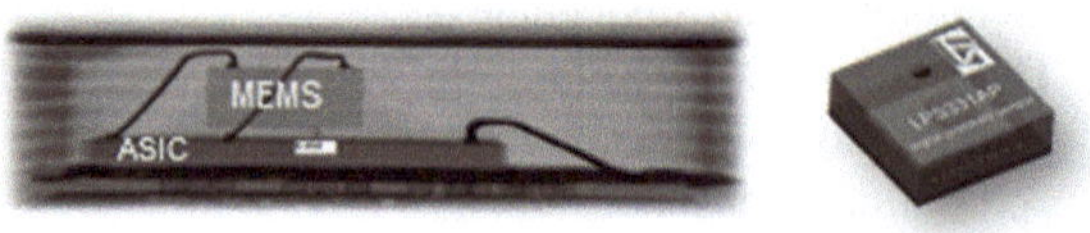


**Fig. 9** LPS331AP, the digital barometric pressure sensor

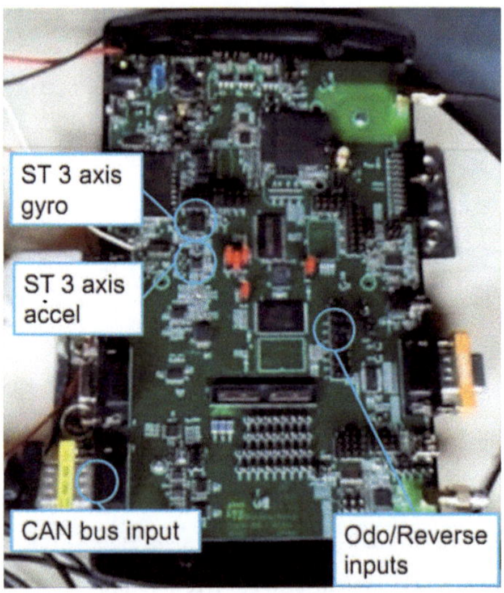


**Fig. 10** ST logger box used in field tests

implementation of dead reckoning using inertial sensors a must to achieve the desired accuracy.

This test made in Chicago clearly shows the improvement in the position performance when a standard navigation system (GPS) is enriched with an effective dead reckoning algorithm based on accurate inertial sensors (GWP, equivalent to the mixed configuration in Fig. 11) (Fig. 12).

Other tests were done while driving through a tunnel where the satellite signals were lost for several minutes: (Fig. 13).

The comparison with a similar test done using CAN differential wheel pulses (DWP) as a yaw sensor shows that gyroscope-based solutions, coupled with an accelerometer to compensate the gyroscope tilt, are more precise, as they result in a smaller percentage error even on a longer distance: (Fig. 14).

| DR CONFIGURATION | YAW RATE SENSOR | DISTANCE SENSORS | OTHER SENSORS |
|---|---|---|---|
| CLASSIC | MEMS gyroscope (1 or 3 axis) | Discrete odometer | Discrete reverse signal<br>3 axis accelerometer option<br>(for 3-axis gyro automatic tilt-compensation) |
| CAN GYRO | CAN gyroscope (ESP) | CAN odometer | CAN reverse signal |
| DWP | CAN differential wheel pulses (ABS) | CAN odometer | CAN reverse signal |
| MIXED | MEMS gyroscope (1 or 3 axis) | CAN odometer | CAN reverse signal<br>3 axes accelerometer option<br>(for 3-axis gyro automatic tilt-compensation) |

Fig. 11 Dead reckoning configuration used during field tests

Fig. 12 ST platform tested in the lower wacker drive of Chicago

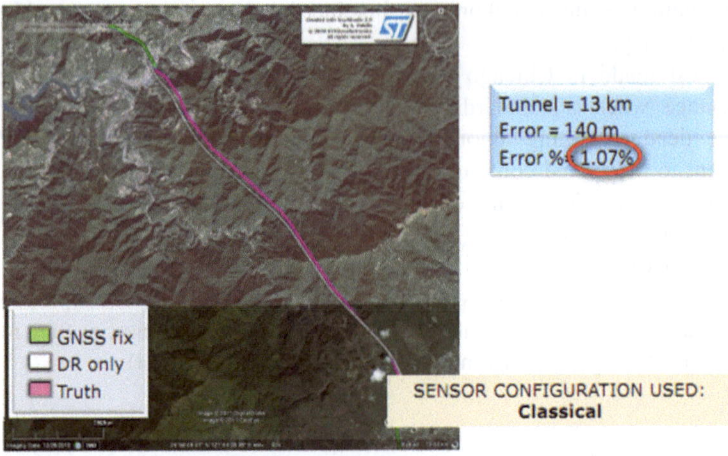


**Fig. 13** Classical dead reckoning across a long tunnel, connecting the cities of Taipei and Yilan, Taiwan

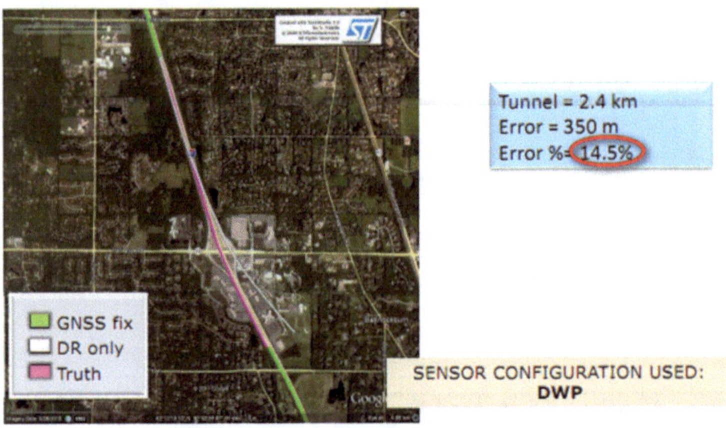


**Fig. 14** Dead reckoning using DWP (test done in Chicago area)

## 4 The Future

Car positioning and navigation systems in the future will require higher accuracies, along with cost competitiveness, in order to fight the battle with the myriad of Personal Navigation Devices (PND) currently available. The Teseo family is already evolving towards a new generation, which will provide customers with a number of benefits, ranging from higher sensitivity to optimized power consumption. A non-exhaustive list of the benefits follows.

- Multi-constellation (GPS + GLONASS + GALILEO + COMPASS + QZSS)
- High-sensitivity engine and innovative power management architecture
- Different package options, with a minimum number of external components
- Extended set of interfaces (UART, USB, SPI, SD Card, CAN, I2C, I2S)
- Support of both TCXO and Xtal configurations.

In the field of sensors, we are already seeing a clear trend towards integration, embedded processing capabilities, and performance optimization. A distinctive feature of ST's accelerometers and gyroscopes is that both sensors are designed and produced using the same technology platform (THELMA). ST has been in mass production with geomagnetic and inertial modules for several months: the LSM330DLC (6X inertial) and the LSM303DLHC (6X geomagnetic) are well known solutions offered to consumer and industrial customers. Future multi-axis modules will see a further reduction in the number of the individual dies due to the possibility of integrating both mechanical elements and electronics onto the same piece of silicon. STMicroelectronics masters already this technology.

Another trend in parallel with the integration of mechanical and electronic elements is the enrichment of the sensors with intelligence. While the 3-axis accelerometer LIS3DSH contains a simple state machine to execute a limited set of instructions, the recently announced LIS331EB has a real brain inside, powered by an ultra-low-power ARM Cortex-M0 core with 64 KB Flash Memory and 128 KB RAM memory, as well as multiple timers and I/O ports (GPIOs/SPI/I2C/UART). ST has combined the microcontroller, operating as a sensor hub that runs sensor-fusion algorithms, with a high-precision 3-axis digital accelerometer into a single package. The device reduces the demand on the host controller and application processor, and decreases power consumption, which is a key feature especially in portable devices.

When higher computational power is necessary, ST proposes System-on-Board solutions, where a powerful STM32 microcontroller (ARM Cortex-M3 core) is coupled with accurate inertial sensors (in the iNEMO-M/Lx family). Eventually, the combination of sensors, actuators, intelligence, wireless connectivity, smart power management, along with energy harvesting capabilities, will enable the creation of an active and smart wireless sensor node, which will enhance further applications in the field of vehicle and people positioning.

And once again, ST rides the crest of the wave.

## References

1. Mattos PG (2011) Accuracy and availability trials of the consumer GPS/GLONASS receiver in highly obstructed environments, IoN GNSS-2011 conference, Portland Oregon, Sept 2011
2. Mattos PG, Pisoni F (2012) Multi-constellation—to receive everything, IoN GNSS-2012 conference, Portland Oregon, Sept 2012
3. Ferraresi M, Marinoni A (2012) Tri-axial MEMS gyroscope for automotive applications. SSI Conference Mar 2012. ISBN 9783-8007-3423-8

# Flexible and Cost-Optimized Platform of Inertial Sensor Systems

**Stefan Günthner, Bernhard Schmid and Helge Graßhoff**

**Abstract** A platform approach for inertial sensor systems is proposed enabling modular and flexible E/E architectures of light vehicles providing acceleration and angular rate signals for usage in multiple safety-relevant applications. The platform portfolio spans surface mounted devices (SMD) and satellite components and makes usage of synergies on each sub-component level like MEMS, ASIC, software and package.

**Keywords** Inertial sensors · Accelerometer · Gyroscope · MEMS · Vehicle E/E architecture

## 1 Motivation

### *1.1 Recent Developments in Vehicle E/E-Architectures*

The proceeding electrification of vehicles during the last decades and the prospect of upcoming vehicle functions and related electronic hardware make it necessary to think about the vehicle E/E architectures. One major problem is the increasing number of electronic control units (ECU) in vehicles. The consequence is an almost non-controllable complexity of vehicle architectures with impact on the manufacturing and development cost while increasing the risks of failures at the same time.

S. Günthner (✉) · B. Schmid · H. Graßhoff
Continental, Chassis and Safety Division, Passive Safety and Sensorics Business Unit, Guerickestraße 7 60489 Frankfurt, Germany
e-mail: stefan.guenthner@continental-corporation.com

B. Schmid
e-mail: bernhard.schmid@continental-corporation.com

H. Graßhoff
e-mail: helge.grasshoff@continental-corporation.com

J. Fischer-Wolfarth and G. Meyer (eds.), *Advanced Microsystems for Automotive Applications 2013*, Lecture Notes in Mobility, DOI: 10.1007/978-3-319-00476-1_32,

Therefore, original equipment manufacturers and their suppliers developed approaches for modular, scalable, flexible and cost-optimized E/E hardware setups [1–3]. Figure 1 shows the approach for the system architecture Continental proposes [4]. The whole vehicle system consists of three major functional blocks:

- Safety Domain—covering active and passive safety functions like crash and rollover protection by airbags, electronic stability control (ESC), vehicle-to-X (V2X) functions etc.
- Comfort and HMI Domain—including navigation, air-condition etc. and the corresponding human–machine-interfaces (HMI)
- Motion and Motion HMI Domain—controlling the movement of the car like accelerating and braking, chassis-control, steering and managing the necessary energy and powertrain incl. the affiliated HMI.

This paper focuses on function blocks related to safety-relevant inertial sensor information, that are pointed out by the red frame in Fig. 1. A possible hardware solution for the functional architecture is depicted in Fig. 2, cf. [5]. Three main ECUs—the Safety Domain Controller, the Motion Domain Controller and the Chassis Domain Options Controller—are hosting the corresponding functions. Having a closer look on the placement of the inertial sensors, there are 4 options:

- Integration into the Safety Domain Controller
- Integration into the Motion Domain Controller
- Integration into the Chassis Domain Options Controller
- Stand-alone ("Motion Information Sensors") as commonly known sensor cluster or as the product M2XPro (Motion Information to X Provider)—a system Continental has developed for lane-level accurate positioning of vehicles

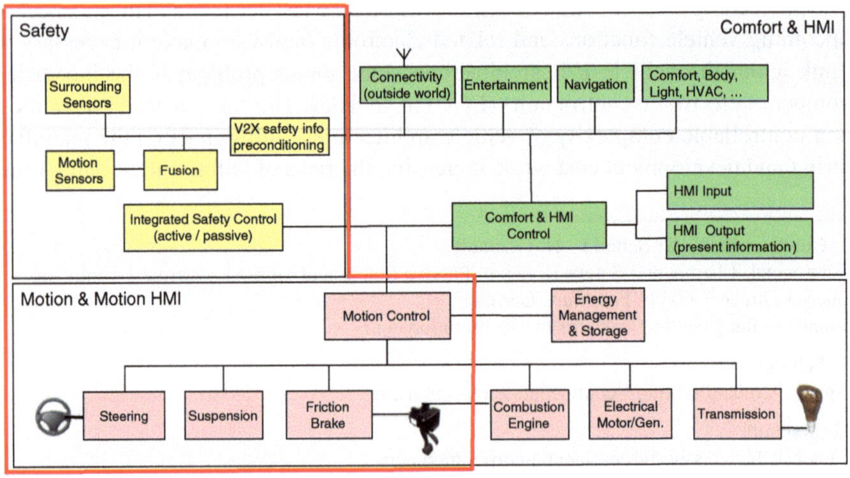


**Fig. 1** Continental's view of the vehicle E/E-architecture on system level

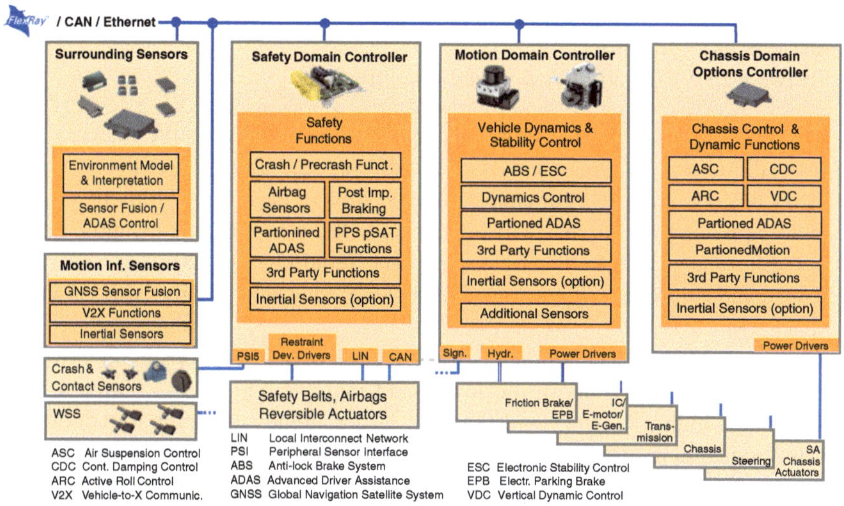


**Fig. 2** Continental's view of the vehicle E/E-architecture on hardware level

for V2X applications that is only based on GNSS[1] information and signals from the wheel-speed sensors, inertial sensors and the steering angle sensor [6].

## *1.2 Vehicle Functions Based on Inertial Information*

Sensor element suppliers have to cope with the requested flexibility for mounting the inertial sensor elements in the car as well as with the fact that the take-rate of certain vehicle functions tends to be almost 100 %, like ESC, Hill Start Assist (HSA) and Front and Side Crash Detection, whereas other functions are only optional or with low take rates, like Vertical Dynamic Control (VDC), Air Suspension Control (ASC), Reversible Restraints Control (RRC), Active Torque Vectoring (ATV), Electronic Power Steering (EPS), Active Body Control (ABC), Active Roll Control (ARC), Rollover Sensing (ROS), Active Front Steering (AFS), Continuous Damping Control (CDC) etc. A more detailed description of some of the listed functions can e.g. be found in [7].

## *1.3 Next Steps in MEMS Technology*

The system architecture of inertial sensor elements consists mainly of 4 blocks:

- Micro-electromechanical systems (MEMS)
- Application specific integrated circuits (ASIC)

[1] GNSS: global navigation satellite system like global positioning system (GPS).

- Package that is housing MEMSs, ASICs and eventually passive components like resistors and capacitors
- Software (SW) for additional signal conditioning.

The inertial MEMS industry was initially driven by automotive demands, e.g. the installation in ESC systems. Indeed during the last years, the consumer industry became the motor for enhancements. Nevertheless, the automotive safety-relevant applications can and will benefit from these advancements, although the major differences will be the tighter lifetime requirements (20 years instead of 2 years), and, of course, the needed safety-level, that is typically ASIL B(D) for ESC, cf. [8].

Currently, MEMS accelerometers for detection of accelerations in one, two and three orthogonal directions in the low-g and mid-g range ($< \sim 1{,}000$ m s$^{-2}$) as well as in the high-g range ($> \sim 1{,}000$ m s$^{-2}$) are already available for application in safety-relevant automotive applications. The same applies to MEMS gyroscopes for measurements of angular velocities about one axis that is either orthogonal to the Silicon substrate plane or parallel to it—commonly called z- and x-gyros, respectively. It is expected that the next development step results in single Silicon chips suitable for safety-relevant system that are sensitive to

- accelerations in two orthogonal directions covering at the same time the low-g and the high-g range,
- angular rates about at least two orthogonal directions.

## 2 The SC300/0 Platform Concept

### *2.1 Basic Sensor Element Architecture*

When creating a best-cost solution for the next-generation sensor elements, four main aspects have to be considered:

First of all, flexibility is required for mounting the inertial sensor elements as stand-alone satellite components on one hand and as surface-mounted devices (SMD) for integration into ECUs on the other hand.

Furthermore, a certain modularity or scalability, respectively, is needed, such that within a vehicle platform, the basic configuration can be equipped with standard elements; vehicles of this platform with an advanced set of features get either additional sensor elements with the missing sensitive axes or dedicated sensor elements providing all necessary axes—depending on the take rates of the corresponding features.

Thirdly, the progress in the development of MEMS-components has to be considered when designing hardware architectures for the next generation inertial sensor elements.

Finally, but most important, the sensor elements have to be competitive. So, the strategy of Continental has been to choose a platform approach. Thus, synergies on all sub-component levels (MEMS, ASIC, package, SW) can be exploited, e.g. by reducing development cost and making usage of quantity effects. Further cost-reduction could be achieved by standardization of certain aspects like the interfaces between ECU and connected sensor element. For SMD components the currently used SPI interface is the optimum approach. Satellite components with the digital PSI5 interface are typically the best-cost solution, not only on sensor element level, but also when taking into account the costs arising on ECU side. At the end, development and manufacturing cost can be decreased by restructuring and standardizing the necessary signal conditioning that is distributed across the sensor-element hardware (ASIC) and software on a Microcontroller Chip (μC) of the corresponding ECU. Pre-conditioning will take place in the ASIC of the sensor-element, whereas post-processing will be performed by a so-called SW-driver that is hosted on the ECU. The SW-driver is designed as a complex device driver in a AUTOSAR[2]-compatible way and consists of modular SW-blocks such that heavy re-use across different sensor-components and hosting ECUs is possible.

Figure 3 depicts high-level system architectures of SMD/SC3000 and satellite/SC300[3] components taking into account the above considerations.

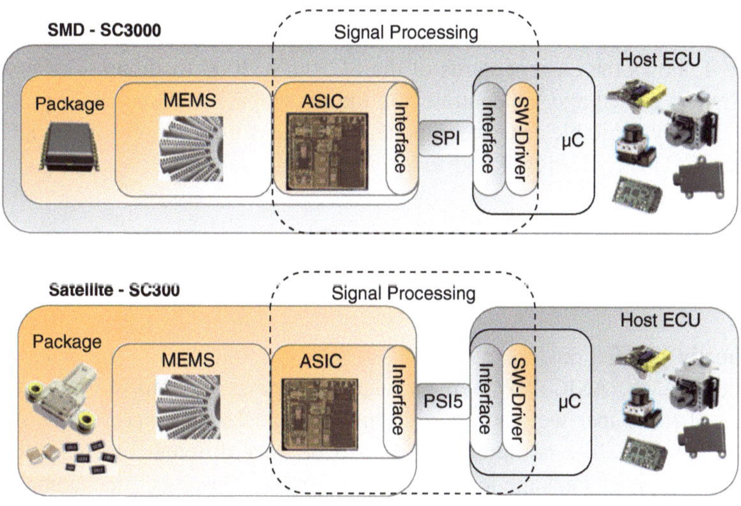


**Fig. 3** High-level system architectures of inertial sensor components

[2] AUTomotive Open System ARchitecture.

[3] Three/four digits denote satellite/SMD components. The last/next-to-last digit counts the number of featured acceleration/angular rate axes.

## 2.2 Portfolio and Roadmap

Based on the derived system architectures a portfolio of inertial sensor components is proposed addressing the functions mentioned in Sect. 1.2. Table 1 gives an overview of the products within this platform ($SC301_{lowG,}$ SC3033) (see footnote 3) and discloses the respective sensitive axes, $a_{x,y,z}$ means linear accelerations in the low-g and mid-g range in x/y/z-direction of the sensor component, $A_{x,y,z}$ means linear accelerations in the high-g range in x/y/z-direction of the sensor component and $\Omega_{x,y,z}$ means angular velocities about the x/y/z-direction of the sensor component. The z-direction of a SMD component is orthogonal to the plane defined by the PCB that carries the sensor; the z-direction of satellite components is given by the symmetry axis of the bushing(s).

Finally, the mounting possibilities of each sensor component are presented: stand-alone installation or integration into an ECU. The main difference between the hosting ECUs is the orientation of the ECUs themselves in the car: whereas the printed circuit boards (PCB) of the Safety Domain Controller (successor of the Airbag ECU) and the Chassis Domain Options Controller are typically parallel to the plane defined by the longitudinal and lateral axis of the vehicle (horizontal installation), the orientation of the PCB of the Motion Domain Controller (the successor of the Brake ECU) is typically vertical to that.

The time-horizon for development and start of series production of the sensor elements takes into account the market needs for the corresponding performance and the availability of the corresponding MEMS and packaging technologies. Figure 4 shows the current status of the SC300/SC3000 roadmap. The first phase has already started and concentrates on the development of successor products of today already available sensor components like sensor-cluster (→SC312), Combo-sensors ($SC3012_{x/z}$), high-g and low-g satellite sensors ($SC301_{highG/lowG}$).

## 2.3 Technological Approaches for the Sensor Platform

As already pointed out, there are 4 main sub-components forming a sensor-component: MEMS, ASIC, SW and package incl. passive components. In the final section of this paper we would like to highlight packaging aspects. The packages of SMD and satellite components seem to be very different on the first glance. Nevertheless, each package level has to perform similar tasks: mechanical fixation of the integrated parts, electrical connection of the sub-components to each other and to the outside world, mechanical fixation to the environment and mechanical as well as electrical protection. Analyzing the conventional technology of inertial satellite sensors, where a SMD component is placed on a PCB in a plastic housing, it becomes obvious that some of the packaging tasks are performed twice, s Fig. 5.

Thus, a platform approach addressing SMD and satellite components could split the packaging tasks in a front-end and back-end part, s. Figure 6. In the front-end

**Table 1** Correlation between sensor component, measurement axes and mounting

| | | SC300/SC3000 platform | | | | | | | | | | | | | |
|---|---|---|---|---|---|---|---|---|---|---|---|---|---|---|---|
| | | $SC301_{lowG}$ | $SC301_{highG}$ | SC312 | $SC3012_z$ | $SC3012_x$ | SC323 | SC3023 | SC314 | SC3014 | SC325 | SC3025 | SC321 | SC333 | SC3033 |
| Sensor-element axes | $a_x$ | x | | x | x | | x | x | x | x | x | x | | x | x |
| | $a_y$ | | | x | x | x | x | x | x | x | x | x | | x | x |
| | $a_z$ | | | | | x | x | x | | | x | x | x | x | x |
| | $A_x$ | | x (x) | | | | | | x | x | x | x | | | |
| | $A_y$ | | | | | | | | x | x | x | x | | | |
| | $A_z$ | | | | | | | | | | | | | | |
| | $\Omega_x$ | | | | | x | x | x | | | x | x | x | x | x |
| | $\Omega_y$ | | | | | | | | | | | | x | x | x |
| | $\Omega_z$ | | | x | x | | x | x | x | x | x | x | | x | x |
| Sensor mounting | Stand-alone satellite | x | x | x | | | x | | x | | x | | x | x | |
| | Safety domain controller | | | | x | | | x | | x | | x | | | x |
| | Motion domain controller | | | | | x | | x | | | | | | | x |
| | Chassis Domain options controller | | | | x | | | x | | | | | | | x |

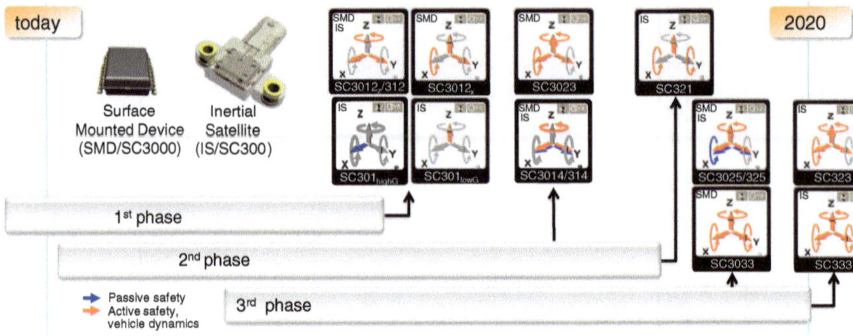


Fig. 4 SC300/SC3000 roadmap

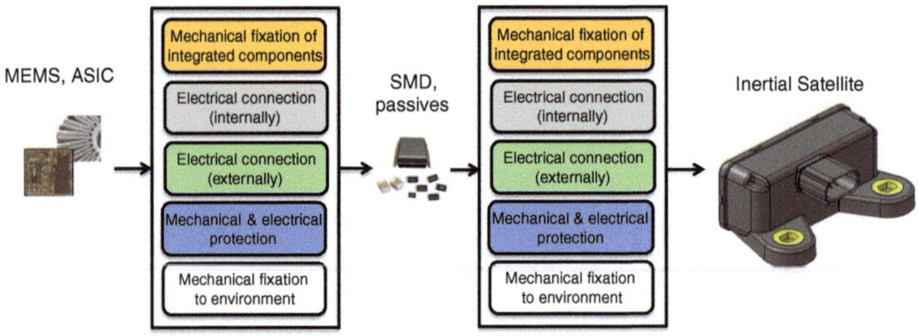


Fig. 5 Conventional packaging approach

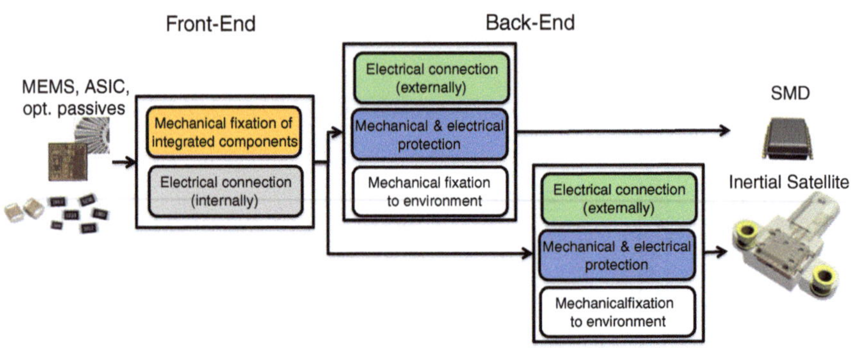


Fig. 6 Possible future packaging approach

stage, the MEMS, ASIC and passive components are mechanically fixed and electrically connected to each other. The solutions of these tasks could be very similar for SMD and satellite components. The back-end part, containing electrical and mechanical connection to the environment as well as electrical and mechanical protection would look different due to varying mission profiles. Apart from consolidating packaging functions and therefore reducing material and processing steps, there would also arise synergies between the production lines of SMD and satellite components.

## 3 Summary and Outlook

Based on the new trends in vehicle E/E architectures, the progress in MEMS technologies and the upcoming developments in vehicle functions, a roadmap for inertial sensors was derived and presented. The next steps and challenges are surely to establish a next-generation packaging technology for inertial satellite sensors and make the MEMS components robust enough for the most promising packaging concept. For a second phase, it is crucial to develop MEMS components with a higher level of integration intended for safety-relevant applications; that means, that one-chip solutions for gyroscopes have to be available in time that provide at least two orthogonal measurement axes for angular rates as well as one-chip solutions for accelerometers with an extended measurement range that fulfill the requirements for HSA on one hand and for crash detection on the other hand. Further steps are pioneered by the consumer industry and will inevitably result in 3D-gyros and accelerometers combined with gyroscopes on one Silicon chip, as soon as the demand by new vehicle features and functions is high enough.

## References


1. Gentner H, Demmeler T, Hohmann S, Wronn O (2009) Architectures for networked active and passive safety. VDI-Berichte Nr 2075:461–475, VDI Verlag, Düsseldorf
2. Knaup V (2006) Integration elektronischer Systeme bei GME, Elektronik-Systeme im Automobil, 7–8 Feb 2006
3. Gallner T (2012) Zukünftige Anforderungen an die E/E Architektur. Elektrik und Elektronik im Nutzfahrzeug, VDI Verlag, Düsseldorf, 10–11 Dec 2012
4. Kelling E (2011) Trends in der Systemvernetzung am Beispiel einer skalierten E/E-Architektur für die Domänen Vehicle Motion und Safety. Elektronik im Kraftfahrzeug, 12–13 Dec 2011
5. Kelling E (2012) Managing complexity system architecture aspects for safe and efficient mobility. Electronica automotive conference, 12–13 Nov 2012
6. Schmid B, Zalewski M, Stählin U, Rink K, Günthner S (2012) Lane accurate position sensing of vehicles for cooperative driver assistance systems. In: Meyer G (ed) Advanced microsystems for automotive applications, AMAA 2012. Springer, Berlin
7. Bosch R (2011) GmbH, Automotive handbook. Wiley, New York
8. Hillenbrand M (2012) Funktionale Sicherheit nach ISO 26262 in der Konzeptphase der Entwicklung von Elektrik/Elektronik Architekturen von Fahrzeugen. KIT Scientific Publishing, 2012

# Smart Soot Sensor for Particulate Filter OBD

**Olivier Brunel, Frederic Duault, Bilal Youssef, Jacques Lavy and Yann Creff**

**Abstract** In the framework of tighter emission requirements and environmental protection, future standards will soon lead to the use of an on board diagnostics (OBD) soot sensor to monitor Diesel particle filter (DPF) leakage. The sensor will first be introduced in the US by MY 2015 and then in Europe in 2017. The resistive ceramic sensing technology has been selected by most OEM as the most appropriate. The smart soot sensor must diagnose filter leakage whatever the driving conditions are, smoother or tighter than the standard NEDC, FTP or WLTC cycles. In order to guarantee a robust detection of failure, Electricfil Automotive and IFPEN have developed strategies capable to accommodate with OEM request and available database. The paper describes the innovative hardware and software solutions and presents the simulated and measured performances.

**Keywords** Diesel particle filter · Electronic control unit · Heavy duty · In use performance ratio · Light duty · Scanning electron microscope · On board diagnostic · Real drive emission · Sensor control module

---

O. Brunel (✉) · F. Duault
Electricfil Automotive, 77, allée des Grandes Combes, Z.I. Ouest Beynost,
F-01708 Miribel cedex, France
e-mail: Olivier.brunel@electricfil.com

F. Duault
e-mail: Frederic.duault@electricfil.com

B. Youssef · J. Lavy · Y. Creff
IFP Énergies nouvelles, Rond point de l'échangeur de solaize,
BP3 69360 Solaize, France
e-mail: Bilal.Youssef@ifpen.fr

J. Lavy
e-mail: Jacques.Lavy@ifpen.fr

Y. Creff
e-mail: Yann.Creff@ifpen.fr

J. Fischer-Wolfarth and G. Meyer (eds.), *Advanced Microsystems for Automotive Applications 2013*, Lecture Notes in Mobility, DOI: 10.1007/978-3-319-00476-1_33,

# 1 Introduction

Despite Diesel Particle Filter has become a standard on diesel engines in most of worldwide countries, there is an increasing concern on human health. Since particles can deposit in the alveoli of human lungs it is in the public interest to have a high level of protection from these particles. The emissions standards are lowering the allowed amount of particles emitted by the vehicle:

- California emission standards, the most stringent, are showing the path with proposal of 3 mg/mi emission limit in 2021 and 1 mg/mi goal in 2028 model year. The OBD threshold limit (OTL) will be 1.7 times the emission limit by 2014 [1].
- European standard for Euro 6 and Euro VI have finally been approved with the respective OTL of 12 mg/km in 2017 and 25 mg/kWh in 2015 [2].

Current DPF OBD is managed by differential pressure sensors, which sense the pressure drop within the DPF. Several studies [3, 4] have shown that this technology will not be able to meet the stricter regulations applicable after 2015–2017. In the frame of tighter emission standards, Electricfil has initiated the development of a soot sensor for OBD purpose 5 years ago [5–8]. The sensor located downstream of the DPF collects the soot in a time cumulative manner so as to monitor any leakage. It contains its own software for signal processing and cleaning strategy and transmits to the engine control unit (ECU) a signal of OTL compliance.

# 2 OBD Requirements

The lowered soot OBD threshold will require the use of a soot sensor after the DPF. This sensor will perform the soot measurement and process the information so as to provide to the ECU "high level information" relative to DPF leakage. For such processing the sensor will also request some information about the engine operating conditions from the ECU. The diagnostic of exceeding threshold limit must be guaranteed during certification phases on a succession of maximum 3 NEDC cycles and also in driving conditions with minimum frequency defined by the IUPR standard (In Use Performance Ratio). The main challenge of the DPF diagnostic strategy is to detect all faulty DPF while minimizing false detections (of good DPF) and to comply with below requirements (Table 1).

# 3 Soot Module Hardware

The Electricfil sensor's is based on the resistive technology: on a insulating substrate, the soot is collected between two platinum interdigitated electrodes. When there is no soot on the sensor, the resistance measured between the electrodes is

**Table 1** Electricfil's PM sensor targets

| Criteria | Requirements |
|---|---|
| Accuracy | 1 ppm false detection<br>100 ppm non detection |
| Response time | Detection on 2nd NEDC Confirmation on the 3rd cycle |
| Max regeneration power | 6A max peak<br>24 W |
| Max regeneration time | 30 s @ whatever the flow conditions are |
| Internal OBD features | Open circuit/short circuit<br>Plausibility<br>Gain, offset, voltage range |
| Lifetime | 200,000 km |

close to the one of the substrate while when the soot is collected on the sensitive window, the resistance is decreasing. A heater is placed on the sensor in order to burn the soot and repeat multiple measures in a row.

## 3.1 Architecture

The soot sensor is composed by a ceramic sensitive element inserted in a body made of different stainless steel alloys and a sensor control module (SCM) that manages the sensor and communicates with the ECU of the vehicle (Fig. 1).

## 3.2 Sensor Design

The ceramic sensitive element of the sensor is protected from breakage thanks to the tip which has a plane area that orientates the sensor in the boss. The ceramic substrate is placed and hold in the tip thanks to different bids. A glass sealing insures the tightness between the exhaust environment and the connection area. The tip is laser welded to the rear tube and a rubber gasket insures the tightness

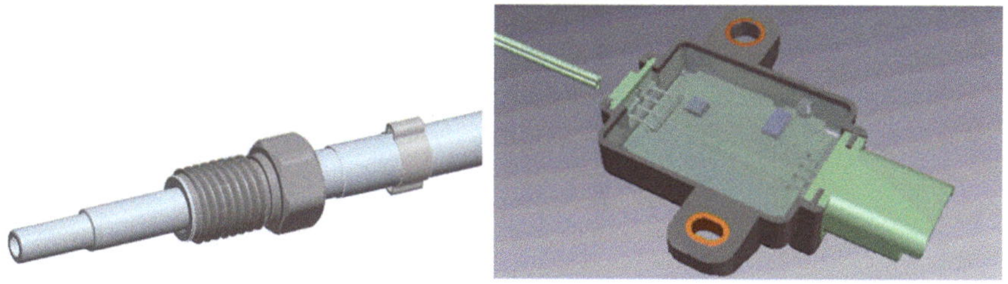

**Fig. 1** Sensor and sensor control module (SCU)

between the ambient environment and the connection area. In the connection area, a ceramic connector and stainless steal terminals apply a mechanical contact on the track pads. The packaging is designed to withstand the severe exhaust environment (three patents deposited and one already accepted WO/2011/004121).

### 3.2.1 Heater

Regeneration is a necessary step in a measuring cycle. This enables the sensor to be cleaned and ready for a new measurement cycle, burning the soot that has been collected on the sensitive element. That is why a platinum track (heater) is ink printed on the sensor substrate. A CAD simulation has been performed in order to achieve the quickest regeneration with highest reliability (Fig. 2).

### 3.2.2 Tip

The tip concept designed is a tip with "reverse" collecting windows. The substrate is protected by a shield without opening upstream of the substrate and the sensitive area is downstream of the substrate. The exhaust gas are sucked up in the sensor tip with a reduced velocity and according to their mass thus their diameter the particles too. The Fig. 3 shows the CAD of the sensor tip and the result of a simulation with the optimized tip in terms of stationary velocity field.

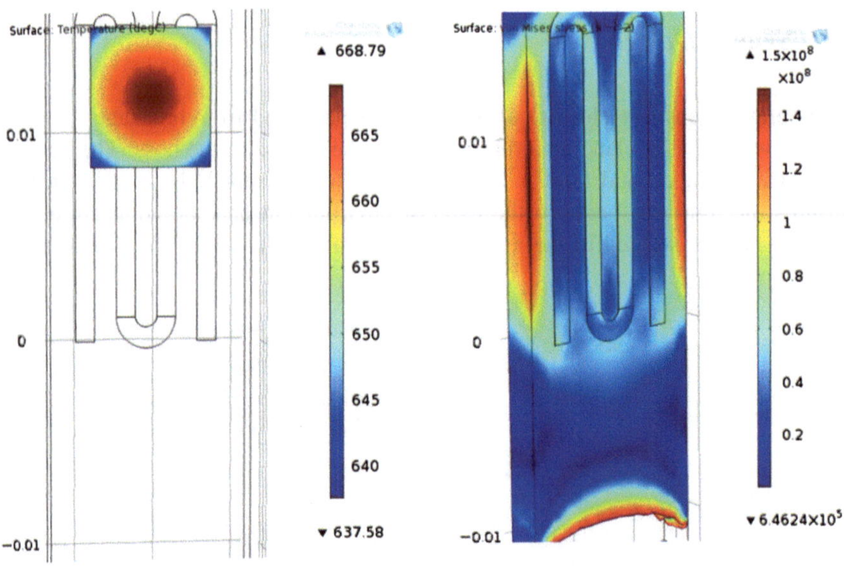


**Fig. 2** Simulations of heated sensitive element and Von Mises stress cartography

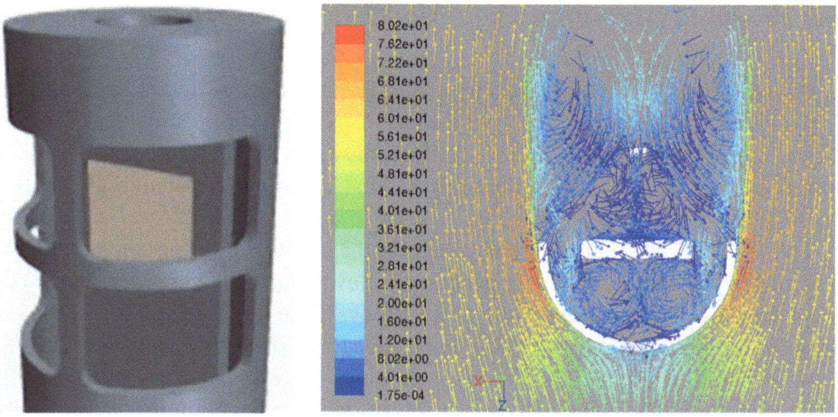


**Fig. 3** CAD of the sensor tip and flow simulation around the optimized tip of the sensor

## 4 OBD Strategies

The sensor accumulates particulate matter (soot) on a ceramic surface covered with interdigitated electrodes. Once the (conductive) soot has bridged the electrodes, the resistance begins to decrease. It keeps on decreasing with soot accumulation, until a predefined threshold is reached. The sensor is then «regenerated» by oxidation of the soot layer (through sensor electrical heating). So, when the sensor is located downstream of a particulate filter, its resistance x(t) is a function of the soot quantity Q(t) getting out of the filter. Once the threshold $x^{\min}$ is reached, a quantity $Q^{\max}$ has got out of the filter. It is possible to compute an average quantity per kilometer $\phi$ (from the $n$ kilometers covered between the last sensor regeneration and the crossing of $x^{\min}$, $\phi = Q^{\max}/n$). However, the standard gives a maximum bound $\phi^{\max}$ *for a normalized cycle*. So, it is not possible to directly compare $\phi$ and $\phi^{\max}$:

- possible false alarm, if real driving PM emissions are larger than emissions on the normalized cycle;
- Possible non detection, if real driving PM emissions are smaller than emissions on the normalized cycle.

This observation regarding LD standards (mass per kilometer) directly transposes to HD standards (mass per unit of energy). The objective is then to find an algorithm that is able to diagnose the PM filtration function, whatever the driving conditions are. The first idea is to compare the signal of the real sensor to a simulated signal based on a sensor model, and to derive alarms from comparisons of these two signals. It appears that implementing such an approach is hard, reaching the required accuracy with difficulty: it turns out to be a clumsy approach. This led EFI and IFPEN to develop alternative approaches, more easily implementable and calibrated, one with models and one without model. In this paper we will focus on the non model approach.

### 4.1 Proposed Algorithm: Non Model Based Approach

In this section, a non model based approach for DPF diagnostic is suggested. This approach has the advantage to avoid the use of models for the particulate filter and for the sensor. Furthermore, no soot estimator is needed in this case, as engine soot emissions are indirectly caught through the analysis of various signals available in the ECU.

The approach described here is based on the direct use of the temporal signal provided by the sensor in order to generate a relevant residual which is related to the leakage failure. This residual, coupled with available variables related to engine operating conditions (IMEP, EGR, engine speed, rail pressure,…), is then used for alarm generation. Hence, the driving conditions are implicitly considered in the diagnostic algorithm, which leads to a robust decision enabling low rates of false alarm and non detection. The resulting diagnostic scheme is simple and easy to implement.

## 5 Performances

In this section, the proposed diagnostic scheme has been evaluated based on simulated and experimental data. As will be seen in the sequel of the paper, the obtained results are good and show the suitability of the soot sensor coupled with the non model based approach to diagnose particulate filter failure with high performance.

### 5.1 Simulation Results

The diagnostic algorithm based on the non model based approach described above has been implemented using Matlab/Simulink simulation. The simulation platform is built using DPF and sensor models and an experimentally calibrated soot estimator. In order to take into account model uncertainty and estimation error, relevant dispersions were added to the sensor model as well as to the soot estimator. Diagnostic results in terms of false alarm and non detection rates are shown in Table 2 for different engine soot dispersions.

**Table 2** Diagnostic results based on data collected from 1,500 simulations for six driving cycles, model dispersion of 3 $\sigma = 45$ %

| Simulation results (%) | False alarms | Non detections (ppm) |
|---|---|---|
| Case 1 engine soot dispersion 3 $\sigma = 25$ | 0 | 60 |
| Case 2 engine soot dispersion 3 $\sigma = 35$ | 0 | 75 |
| Case 3 engine soot dispersion 3 $\sigma = 45$ | 0 | 90 |

## *5.2 Driving Test Results*

A C4 with a 1.6 L diesel engine €4 vehicle has been instrumented in order to test the DPF diagnostic strategy. The C4 battery supplies current for the sensors as well as for the instrumentation. An EXXOtest CAN logger is used to record data from the sensors, the thermocouple signal, vehicle data and the soot concentration. The vehicle can be retrofitted with DPF that have leakages in the range of 0–12 mg/km on a NEDC cycle.

Four soot sensors (PMS1 to PMS4) can be fitted on the exhaust tube knowing that PMS4 has a very long response time due to a bad positioning.

The strategy has been calibrated on the C4 driving the vehicle on a private circuit. The aim is to drive the vehicle around 10 times three different cycles (called type A, B and C) (Fig. 4) in a repeatable manner with three different DPF.

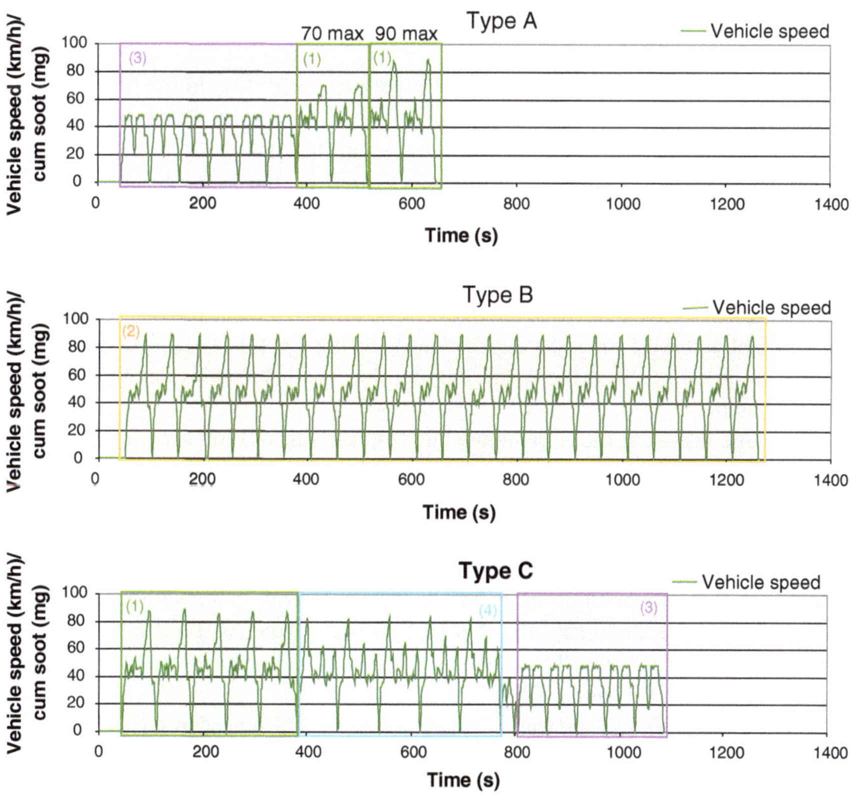


**Fig. 4** Vehicle speed during the 3 kind of driving cycles defined on the circuit

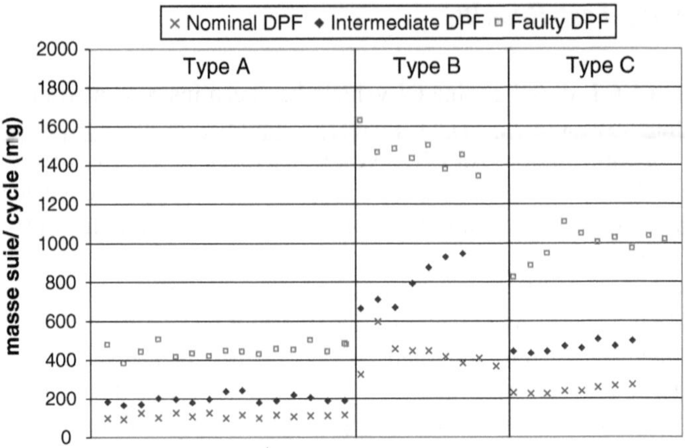


**Fig. 5** DPF leakage in mg/km during the 3 different driving cycles with a DPF characterized at 4.5 mg/km on a hot NEDC mounted on the exhaust line

The three different DPF were chosen as follows: one close to the homologation certification limit (nominal: 4.5 mg/km), one close to the OBD limit (faulty: 12 mg/km) and the last one with an intermediate leakage.

Figure 5 shows the emission results measured by the MSS during the whole campaign. A first analysis shows that all the DPF can be recognized (nominal, intermediate or faulty). That means that with such a measuring tool it is possible to classify DPF leakage, but it is impossible to say what would be the leakage during a homologation cycle. However the repeatability of the measurement seems to be not so good and with some drifts, especially during Type B cycle. This dispersion can be assigned to the drivers, the engine, the weather conditions and the MSS measurement.

The same emission measurements, but with the distance driven correction give data shown in Fig. 6. As it could has been expected, the data show that even with a real time measurement apparatus, one vehicle driven in a different manner or on a different travel with the same DPF, could lead to different emission rates. Furthermore, the emission rates calculated are far away from the one measured with the weighting method on a NEDC cycle. In this case, without any more information, some could say that this DPF is faulty whereas on the NEDC cycle this was the nominal part. That means that a DPF diagnostic strategy is essential for the detection of a faulty DPF.

Finally, Fig. 7 shows the transfer function of the C3 sensor in terms of number of regeneration as function of the cumulated mass of soot measured by the MSS. The curve shows a "step" wave form response because of the resolution of the sensor, with a good correlation coefficient of 0.95.

It is possible to classify the three DPF as well as we did with the MSS even if it is impossible without DPF diagnostic strategy to say if the nominal DPF is nominal neither if the faulty DPF is faulty on a NEDC cycle.

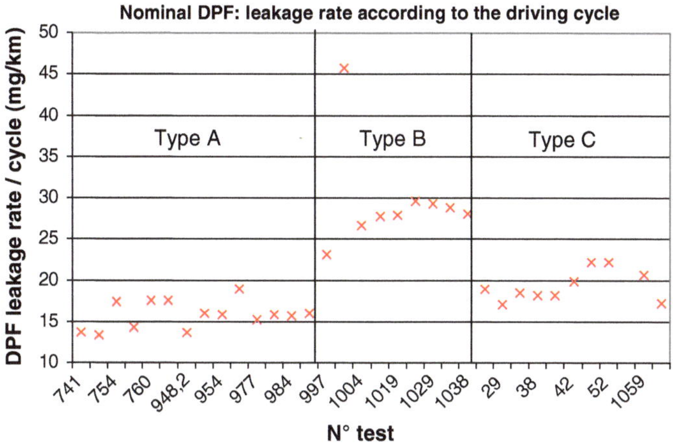


**Fig. 6** C4 DPF leakage rate with the nominal filter during the 3 type of cycle (mg/km)

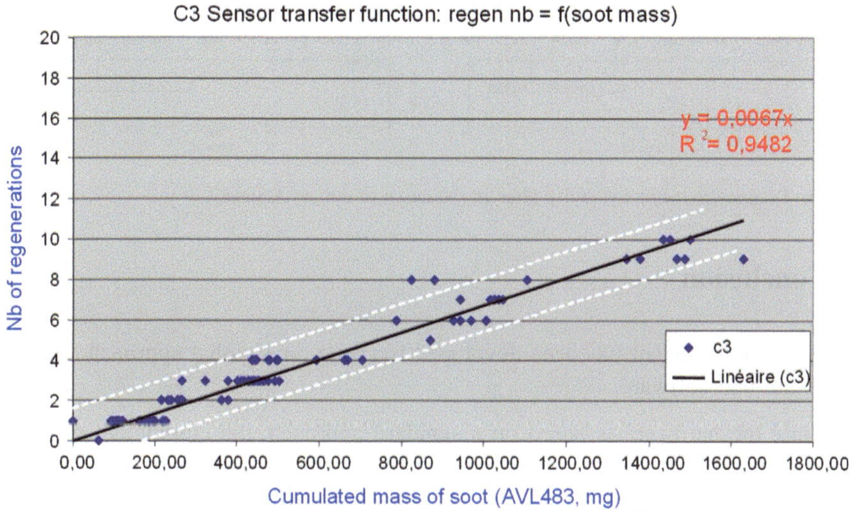


**Fig. 7** C3 sensor's transfer function (number of regeneration in function of the cumulated mass of soot)

The proposed non model based diagnostic algorithm has been evaluated using the experimental data described above. In all cases, the diagnostic procedure successfully noted the DPF failure when the faulty filter has been used while no alarm is generated in the case of nominal filter. Figure 8 show typical diagnostic results for nominal filter. Examples of the diagnostic results in the case of faulty filter are shown in Fig. 9.

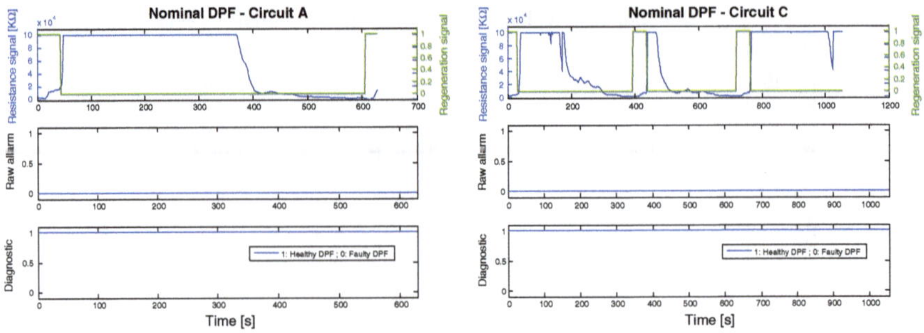


**Fig. 8** Diagnostic result for nominal filter in the cases of circuit *A* and *C*

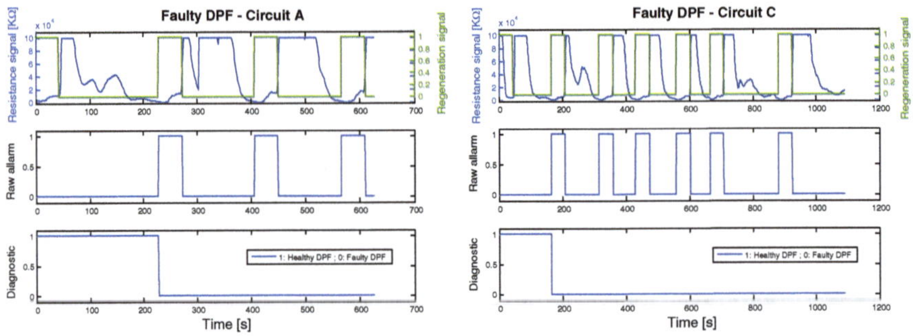


**Fig. 9** Diagnostic result for faulty filter in the cases of circuit *A* and *C*

# 6 Conclusion

A new particle sensor has been developed focusing on a global approach on both hardware and software:

- While the hardware is based on a standard approach, the cumulative resistive technology, our work led to unique features aimed at providing an optimal sensitivity/accuracy performance. This has been achieved thanks to innovative sensing area, collecting tip and processing electronics.
- The software developed with IFPEN enables high diagnostic capability with robustness against driving conditions and allows to detect DPF leakage without engine soot cartography which offers to OEM development time and cost benefits.

At that stage of the project, sensor hardware is enduring qualification and software has been validated through intensive simulations.

Tests in real running conditions are going on through the technical feasibility project conducted by the EC which should prove in early 2013 the compliance of our solution with OEM and emission standard requirements.

# References

1. California Code of Regulations, Title 13, LEV III final regulation order—amended 22 March 2012
2. Commission Regulation (EU) , TCMV 19 Dec (2012)—technical committee—December 2011, No 459/2012, 29 May 2012
3. Van Nieuwstadt M, Trudell D (2004) Diagnostics for diesel particles filters. SAE technical paper 2004-01-1422. doi:10.4271/2004-01-1422
4. Uchiyama T (2010) DPF diagnostic; new approach to meet challenging future requirement in 201X, SAE OBD symposium
5. Zein H (2011) Soot sensing for DPF OBD: bottlenecks and challenges. CTI 3rd international conference, "vehicle emission reduction technologies—criteria pollutants and $CO_2$", May 2011
6. Duault F (2011) Resistive soot sensors. 4th international CTI conference: "emission relevant sensors", Nuremberg, 12–13 July 2011
7. Lavy J, Millet CN, Serrano D, Breuil P, Viricelle JP (2010) "On-board soot sensor study and development", diesel engines facing competitiveness challenges. SIA Congress, Rouen, May 2010
8. Lavy J, Millet CN, Reveillé B, Abassi H, Landry L, Creff Y, Duault F, Raquin S (2012) PM sensor development for on-board diagnosis of diesel particulate filter failures. 16th ETH-conference on combustion generated nanoparticles, Zurich, June 2012

# Air Flow Meter 8. Generation: A Modular Approach for New Diesel System Challenges

**Michael Rittmann, Rainer Moritz, Stefan Bauer and Uwe Konzelmann**

**Abstract** Increasing precision and robustness are the main drivers for the next generation of air flow meters for automotive applications while cost reduction is the key factor for a successful market entry. This contradiction can be resolved by increasing the sensor intelligence by efficient digital signal processing and by evaluation of all information available within the sensor system. Compared to a stand-alone sensor, further cost reduction is realized by integration of additional pressure and/or humidity sensors in the air flow meter. The information of pressure and/or humidity near the air flow meter provides additional relevant parameters to the engine controller. A digital SENT interface is used to submit the information to the engine controller with a minimum number of wires.

**Keywords** Hot-film air flow meter · HFM8 · Integration pressure and humidity sensor · SENT

## 1 Introduction

Strict emission regulations in the United States and Europe are a major challenge for engine management systems. Besides the optimization of injection and combustion processes, emission-related components have to meet increased accuracy requirements.

M. Rittmann (✉) · R. Moritz · S. Bauer · U. Konzelmann
Robert Bosch GmbH, Wernerstraße 51 70469 Stuttgart, Germany
e-mail: michael.rittmann@de.bosch.com

R. Moritz
e-mail: rainer.moritz@de.bosch.com

S. Bauer
e-mail: stefan.bauer5@de.bosch.com

U. Konzelmann
e-mail: uwe.konzelmann@de.bosch.com

J. Fischer-Wolfarth and G. Meyer (eds.), *Advanced Microsystems for Automotive Applications 2013*, Lecture Notes in Mobility, DOI: 10.1007/978-3-319-00476-1_34,

Especially in diesel systems the air flow meter plays an important role, as it provides key data by which the combustion strategy and exhaust gas recirculation is determined. A more accurate determination of the air mass can reduce the effort for exhaust gas after treatment and eventually the total system cost. A high basic accuracy of the sensor is only one aspect to be considered. In order to comply with emission regulations, a reduced tolerance of the sensor during the entire lifetime is required. Here, the robustness against contamination plays an important role together with the robustness against environmental influences, such as climatic stress.

For new more complex engine management systems the information of pressure and/or humidity near the air flow meter increases the precision of the air flow metering and, even more important, provides additional relevant parameters to the engine controller.

In order to cover the wide spectrum of customer requirements, a highly flexible mounting position and a set of electrical interfaces to the ECU are required by an air flow meter.

The new air flow meter HFM8 of Robert Bosch GmbH meets these requirements through a combination of:

- flexible electrical interface to the engine control unit,
- integration of additional temperature, humidity and/or pressure sensors,
- digital signal processing with the air flow meter,
- innovative, aerodynamic design to protect the sensor against contamination, and
- an integrated heater to protect against oil contamination.

## 2 HFM8 Design Concept

For an air flow meter the so-called package plays a decisive role with respect to the measurement task. The package of an air flow meter is determined, on one hand, by the need of forming a part of the intake tract of the engine and, on the other hand, by the need of a carrier of sensor element, electronics circuit, flow passage and connector. For the Bosch micro silicon air flow meter a modular design concept was chosen. The air flow meter consists of the metering duct, which has flanges to fit to the adjacent parts of the intake tract. Flow straighteners, which can become necessary to provide a stable air flow, are also part of that metering duct. Sensor element, electronics circuit, flow passage and connector are integrated in a plug-in module (Fig. 1). This module can be inserted in metering ducts of various sizes or directly into an air cleaner.

### *2.1 Measuring Principle*

The HFM8 air flow meter is a hot-film air mass sensor and therefore a thermal air flow meter. From the intake air flow, a portion of the total mass air flow will pass across the micromechanical sensor element in the bypass duct [1, 2].

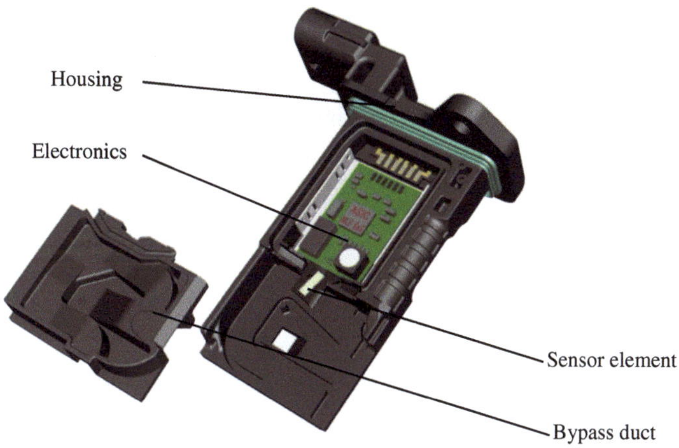


**Fig. 1** Design of HFM8 air flow meter plug-in sensor

A diaphragm is located on the sensor element. It is generated by an etching process from the back side of the silicon material wafer. In the centre of the diaphragm, a heating zone is implemented with symmetrically mounted temperature sensors (Fig. 2).

The heating zone is controlled to keep a certain temperature, which depends on the temperature of the intake air. Without air flow, the temperature from the heating zone to the edges of the diaphragm decreases linearly (Fig. 2: red curve), and the temperature sensors up- and downstream (Fig. 2: marked $T_1$ and $T_2$) of the heating zone indicate the same value. With air flow, the sensor diaphragm area

**Fig. 2** Measuring principle of Bosch air flow meters

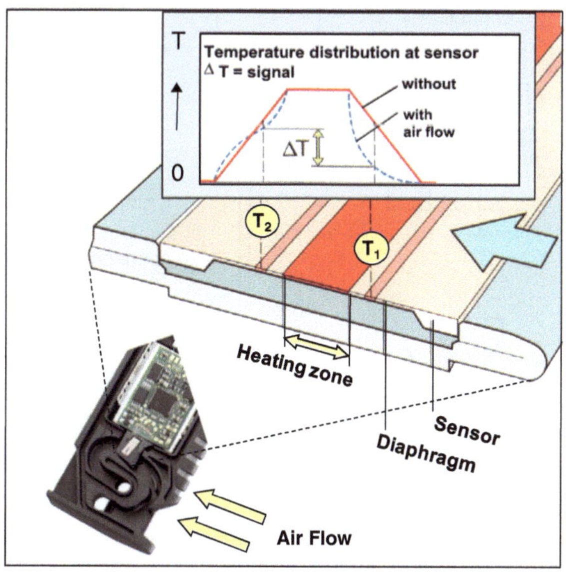

upstream will be cooled by the heat transfer in the boundary layer. The downstream temperature sensor will keep its temperature because the air is heated as it passes over the heating zone.

The temperature sensors show a temperature difference which depends on amount and direction of the air flow. The difference between the signals of the two temperature sensors is evaluated in a bridge circuit configuration.

## 2.2 *Signal Formation and Calibration*

The necessary function blocks for an air flow meter are as following [3]:

- heater control,
- signal amplifier,
- offset correction, and
- decoupling of electric disturbances (overvoltage, EMC).

The heating zone of the diaphragm consists of the resistors Rh (heater, Fig. 3) and Rhf (sensor of heater temperature; Fig. 3). Outside of the diaphragm the resistor Rlf (sensor for ambient temperature, Fig. 3) is located on the silicon sensor element. The heater is controlled with the temperature information of Rlf and Rhf. Due to the temperature dependency of Rlf, the heater temperature changes depending on the ambient temperature. Hereby, the sensor can compensate the temperature dependency of the air flow signal.

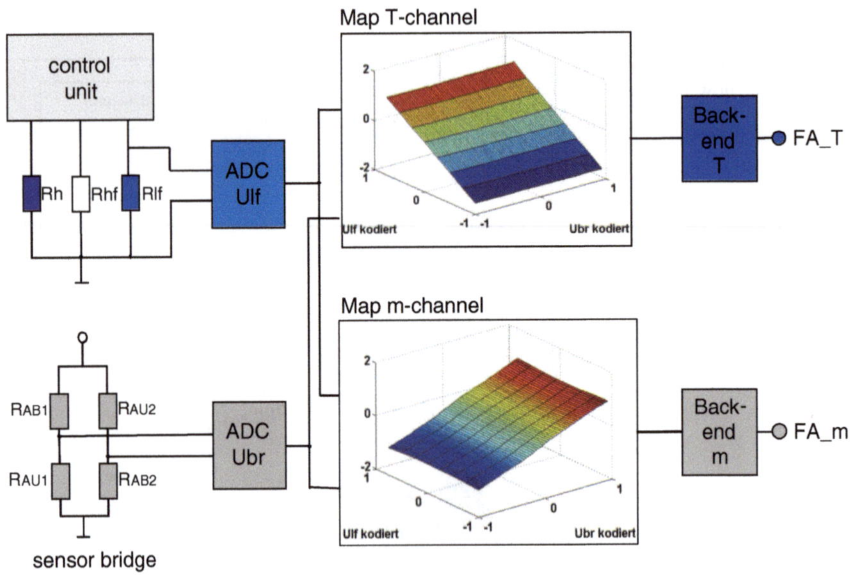


**Fig. 3** Digital signal processing of HFM8

The resistors Rab1, Rau1, Rab2, Rau2, located upstream and downstream of the heating zone on the diaphragm of the sensor element, are arranged in a Wheatstone bridge (sensor bridge, Fig. 3). The output of the sensor bridge Ubr corresponds to the air flow rate.

The voltage at Rlf (Ulf) and the output of the sensor bridge Ubr are converted to a digital signal Slf and Sbr in an AD-converter (Fig. 3). The digital signal is used as input for 2 correction maps; one for temperature, one for air mass. The map stores correction values, which are added to Slf and Sbr respectively. The corrected values are eventually forwarded to the output drivers (back end) as the air mass signal and the temperature signal. A frequency signal is generated to transmit the two signals.

Beside the bridge voltage Ubr, also the voltage at the heater (Uh) depends on the air flow rate. This additional information can be used to improve the accuracy of the air mass signal.

Air mass and temperature signal are calibrated at the end of the production process. For this purpose a measurement is performed at zero air flow rate at 2 temperatures. The output signal of air mass is measured at different air flow rates at room temperature. Then a correction map is calculated optimizing the target characteristic. The digital concept using correction maps enables a high precision air flow measurement. The precision largely depends on the measuring effort during the calibration procedure.

## 3 Integration of Additional Sensors

Beside the air mass, the intake air temperature provides important input data for today's engine management systems. The HFM8 can measure the intake air temperature via an optional temperature sensor (NTC) mounted into the housing.

Based on the temperature sensor on the silicon sensor element an alternative temperature signal can be provided. The temperature sensor on the silicon sensor element is thermally coupled with the housing of the air flow meter and the housing temperature affects the measurement at Rlf. This effect can be compensated using the temperature information measured on the electronics of the air flow meter. With this information a precise ambient temperature signal can be calculated without the necessity of an additional NTC temperature sensor.

The knowledge of other physical quantities such as pressure and relative humidity at the installation position of the air flow meter provides new and enhanced features for engine control systems as well as exhaust gas treatment systems. For example, the air mass signal can increase the precision of the exhaust gas modeling using the information of the optionally integrated relative humidity sensor. On the other hand, future emission regulation requirements can be fulfilled using new diagnostic functions based on this physical information. Using the additional sensor information of relative humidity and/or pressure, it is also possible to reduce system costs. For example, it is possible to avoid an additional NOx sensor realizing an increased precision of the NOx model within the exhaust gas modeling.

System cost reduction and new diagnostic functions are enabled with an intelligent integration of the additional sensors into the air flow meter and the use of digital interfaces for transmission of the set of measured values. In comparison to the use of individual sensors for each physical information, the number of components and wiring can be significantly reduced by the integration of additional sensors in the air flow meter. On top of this, the cost for the wiring harness can additionally reduced using a digital single wire interface. HFM8 provides a SENT interface. Including the power supply line and ground connection line, a three wire harness realizes the transfer of air mass, temperature, pressure and relative humidity information to the engine control system.

In the Bosch HFM8 air flow meter the additional sensors pressure and/or humidity can optionally be integrated directly on the printed circuit board (PCB, see Fig. 4) using surface mounted design (SMD) components. The integration on the PCB is flexible and cost-efficient as a standard PCB equipment can be used.

Both the pressure and the humidity sensor are stand-alone sensors, which operate independently from each other and independently from the air flow meter. Data acquisition and signal pre-processing are performed in the respective sensor components. The PCB connects the pressure and the humidity sensors to the application specific integrated circuit (ASIC) of the HFM8 (see Fig. 5). The connected sensors communicate with the HFM8 ASIC via internal digital data-bus where the measured values are sent to the ASIC. Measured data (pressure, humidity, temperature, air mass) are collected by the HFM8 ASIC and with the SENT interface submitted to the engine controller.

# 4 Electrical Interface

For the current series production air flow meters the frequency interface has become an established interface to the engine control unit. The air mass information is coded in the cycle time of a periodic signal. Figure 6 shows an idealized

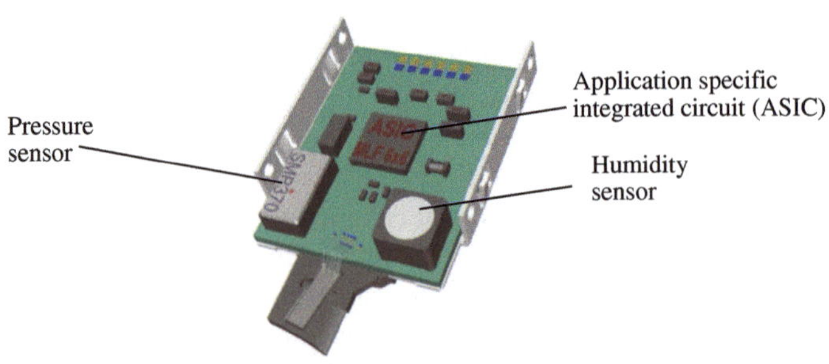


**Fig. 4** Integration of pressure and humidity sensor on HFM8 electronics

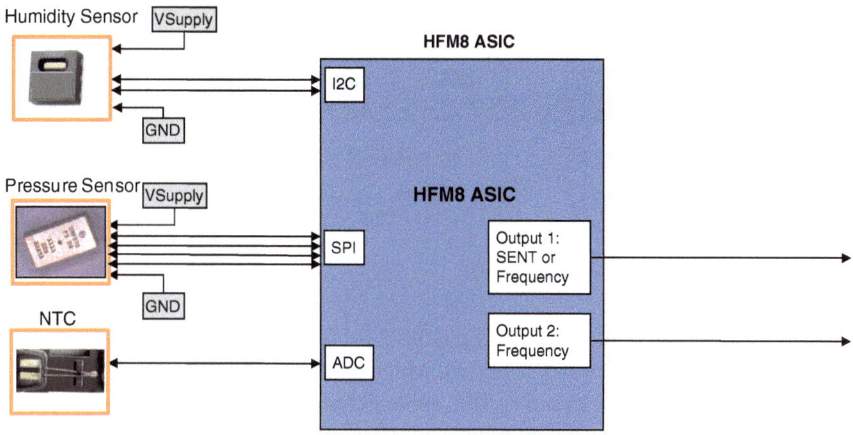


**Fig. 5** Link from optionally integrated sensors to *HFM8 ASIC*

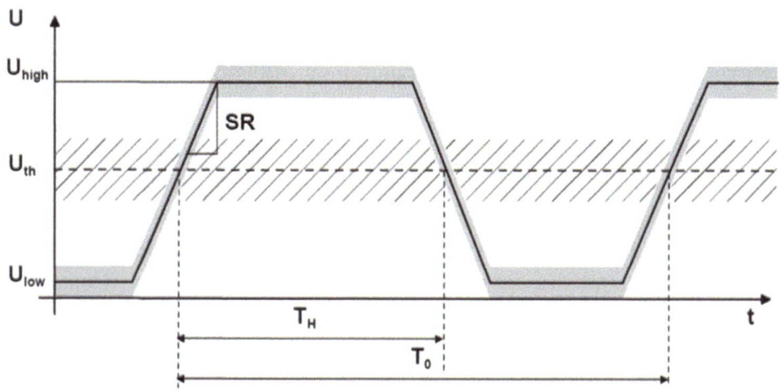


**Fig. 6** Frequency output signal principle

signal. The frequency output interface meters the cycle time $T_0$ between two falling edges. Due to the fact, that only falling edges are evaluated, the part-to-part variation of the slew-rate (SR) does not influence the cycle time metering.

With a frequency interface only one single physical information can be transmitted using one wire. Today in many applications also the ambient temperature is measured within the air flow meter. Hence, the wire for submission of the air mass signal has to be added to an additional wire necessary for the transmission of the ambient temperature.

The integration of additional sensors in the air flow meter will only be successful, if significant cost improvements can be achieved [4]. A major contribution to cost improvement is the re-use of existing wiring harnesses transmitting additional sensor information. This can be realized by the use of a digital interface such as the SENT interface.

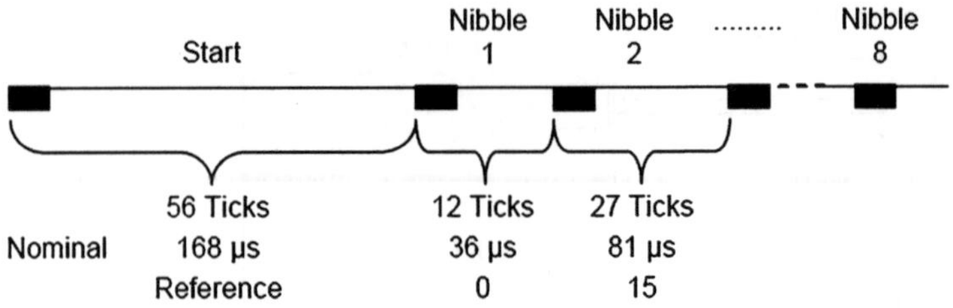


**Fig. 7** Definition of SENT interface signal

SENT is an acronym for Single Edge Nibble Transmission. The SENT specification is an open SAE industry standard, which can be used license free by all OEM and sensor suppliers. The current valid status is specified in SAE J2716 Ref. [3, 5].

The SENT interface is a single wire interface. Power supply and ground signal are additionally provided by the receiving device using separate wires. The actual valid status of SENT interface signal definition is demonstrated in Fig. 7.

The signal consists of a sequence of so called nibbles. The nominal cycle time of these nibbles depends on the data, which are transferred. The physical cycle time is a multiple of a basic time unit called ticks. Ticks have a specified cycle time of $3 \pm 0.6$ µs, which must be guaranteed by the sender. This tolerance can be covered easily by a cost effective RC oscillator.

The start sequence has a nominal cycle time of 168 µs. This nominal cycle time is longer than all other possible cycle-times in the SENT specification. The start sequence has two functions: First, it indicates the start of the information transfer. Second, it allows the calibration of sender and receiver clocks. The serial communication nibble 1 allows device specific information transfer. The serial communication nibble is a 4 bit signal. Via the serial communication nibble low-rate (i.e. medium temperature and humidity information) and low-priority diagnostic data (i.e. sensor characteristic or error codes) are transmitted to the receiver over several SENT frames. Nibbles 2–7 represent the so called fast channel with a 24 bit signal. Via the fast channel i.e. air mass and temperature information are transmitted within one SENT frame. Checksum Nibble 8 allows checking signal of the fast channel regarding transmission errors.

With these features the SENT interface offers a high resolution and precision without the use of a quartz oscillator or of a separate reference signals.

Due to various customer requirements and applications the new air flow sensor HFM8 will be available with frequency or SENT interface.

## References

1. Strohrmann M, Lembke M, Hüftle G, Konzelmann U. et al, Sensoren und Messsysteme, VDI-Berichte 1829, VDI Verlag, Düsseldorf, 535–542
2. Rittmann M, Konzelmann U, Brückner J, Gleisberg T (2006) HFM7—Neuer thermischer Luftmassenmesser mit hoher Flexibilität in den Anwendungsmöglichkeiten. Haus der Technik, Essen

3. Marek J et al (2003) Sensors for automotive applications. Wiley-VCH, Weinheim
4. Grossmann A, Gleisberg T, Strohrmann M (2005) SENT interface—innovative and cost effective improvement of sensor system tolerances: Sensors & Actuators for Advanced Automotive Application
5. SAE (2010) International SENT—single edge nibble transmission for automotive applications, SAE J2716 Rev. 3. www.sae.org

# A Methodology for Design, Validation and Performance Analysis of Vehicle Electronic Control Systems

**Alexander Hanzlik and Erwin Kristen**

**Abstract** The Data Time Flow Simulator (DTF) has been developed at the AIT Austrian Institute of Technology in the course of the EU ARTEMIS Project POLLUX which is related to the design of electronic control systems for the next generation of electric vehicles. In previous work, we have introduced a generic approach for performance analysis of vehicle electronic control systems using the DTF. The main issue of this paper is to demonstrate the practicability of this design methodology by means of a specific use case, namely the performance analysis of a vehicle control system for an electric city car, designed using inputs from different POLLUX project partners from the automotive industry.

**Keywords** Simulation · Model based design · Automotive control systems · Network performance analysis · Electric vehicles

## 1 Introduction

Overcoming the problem of steadily increasing system complexity is an engineering challenge today and it can be expected that it will remain to be a challenge in the future. This is especially true for vehicle electronic control systems. The trend in the development of such systems shows that mechanical and hydraulic components are more and more replaced by electronic components. Representative evidence for this evolution are the emerging brake-by-wire and steer-by-wire technologies that aim for the mechanical de-coupling of driver actions and the

A. Hanzlik (✉) · E. Kristen
AIT Austrian Institute of Technology GmbH, Donau-City-Straße 1, Vienna, Austria
e-mail: alexander.hanzlik.fl@ait.ac.at; ahanzlik@gmx.at

E. Kristen
e-mail: erwin.kristen@ait.ac.at

J. Fischer-Wolfarth and G. Meyer (eds.), *Advanced Microsystems for Automotive Applications 2013*, Lecture Notes in Mobility, DOI: 10.1007/978-3-319-00476-1_35, 

physical reaction of the vehicle. Introducing these technologies in vehicle control systems will not only increase system complexity, but will also increase the demands on functional safety for vehicles.[1]

On the one hand, steadily increasing system complexity and safety requirements raise the need for regulations and standards for the development process of safety-relevant systems. On the other hand, development time is costly and the automotive industry operates in a very competitive market. Moreover, it is a market of mass production. Products are released in large quantities, and car manufacturers and suppliers are very ambitious to keep the costs per piece as low as possible. In this area of tension between system complexity, quality considerations, safety requirements and time constraints it may become hard to maintain product quality and safety at affordable costs unless there are no appropriate methods deployed that support system design processes.

A model based development approach for the design of vehicle electronic control systems is one possibility to overcome this situation. In this paper, we present a design methodology that consists of two steps. Step one is a model based design phase where a simulation model of the system under consideration is built. Step two is the validation of pre-defined requirements in the course of simulation experiments, where the requirements are focusing on the communication load observed in the control system and the latencies of time-critical control signals.

The described methodology is used for the design and performance analysis of an electronic control system for an electric vehicle. Using this architecture, we validate the requirements of two time-critical action paths, where an action path denotes the flow of a control signal from a sensor to an actuator.

## 2 Network Performance Analysis: Principle of Operation

Modern vehicle control systems are distributed systems, built from spatially separated electronic control units (ECUs) interconnected via shared communication resources, like bus systems [1]. ECUs are embedded systems that control one or more of the electrical or mechanical systems in a vehicle. For the following considerations, we assume that a vehicle electronic control system consists of different *segments*, interconnected via *gateways*. Each segment contains ECUs that communicate via a shared *communication resource* (a CAN [2] or a FlexRay [3] bus). *Sensors* and *actuators* are attached to the ECUs and form the interface between the control system and the controlled object. A more detailed description of this system model is given in [4].

---

[1] The recently released norm ISO 26262 "Road vehicles—Functional safety" defines, among others, functional safety aspects of the entire development process and provides requirements for validation and confirmation measures.

To analyze the performance of the control system, signals are issued to sensor elements and signal propagation is observed, both in the domains of value and time, from the sensors through the communication network to the actuators. From the signal propagation time through the communication network important information can be gained with regard to the dynamics and responsiveness of the control system, especially for safety-relevant signals that are usually subjected to real-time constraints, like the accelerator pedal position control signal.

Figure 1 shows the workflow of a control system performance analysis using the DTF[2] based on a drive cycle. The drive cycle consists of a set of events that are issued to the sensor elements of the control system and that contain typical driving scenarios like the accelerator pedal positions or the positions of the steering wheel during the drive cycle. Using this set of events, the system is put under stress and the system behavior is observed. The analysis of the system performance consists of two phases:

- In the **System Assessment** phase, the maximum loads of different network segments during the drive cycle are determined. If a given maximum network load is exceeded, the network model is modified.
- In the **Signal Assessment** phase, single control signal latencies are examined. If a given maximum signal latency is exceeded, the network communication schedule is modified.

This process is iterated until all requirements are fulfilled.

## 3 City Car Control System Assessment

We now apply the network performance analysis methodology presented in the last section on the electronic control system for an electric city car, using the DTF simulator. This electronic control system has been developed in the course of the POLLUX project and it consists of by the following components:

- SEG_BODY Body network segment (grey)
- SEG_BATT Battery network segment (red)
- SEG_CHASSIS_1 Chassis network segment 1 (green)
- SEG_CHASSIS_2 Chassis network segment 2 (green)
- SEG_PT_FRONT Front power train network segment (blue)
- SEG_PT_REAR Rear power train network segment (blue)
- GW_BODY Body/Battery gateway (light grey)
- GW_PT_CH Power train/Chassis gateway (light grey)

---

[2] The DTF is a discrete-event simulation [5] environment for design, evaluation and performance analysis of electronic control systems. More information about the DTF is given in [6].

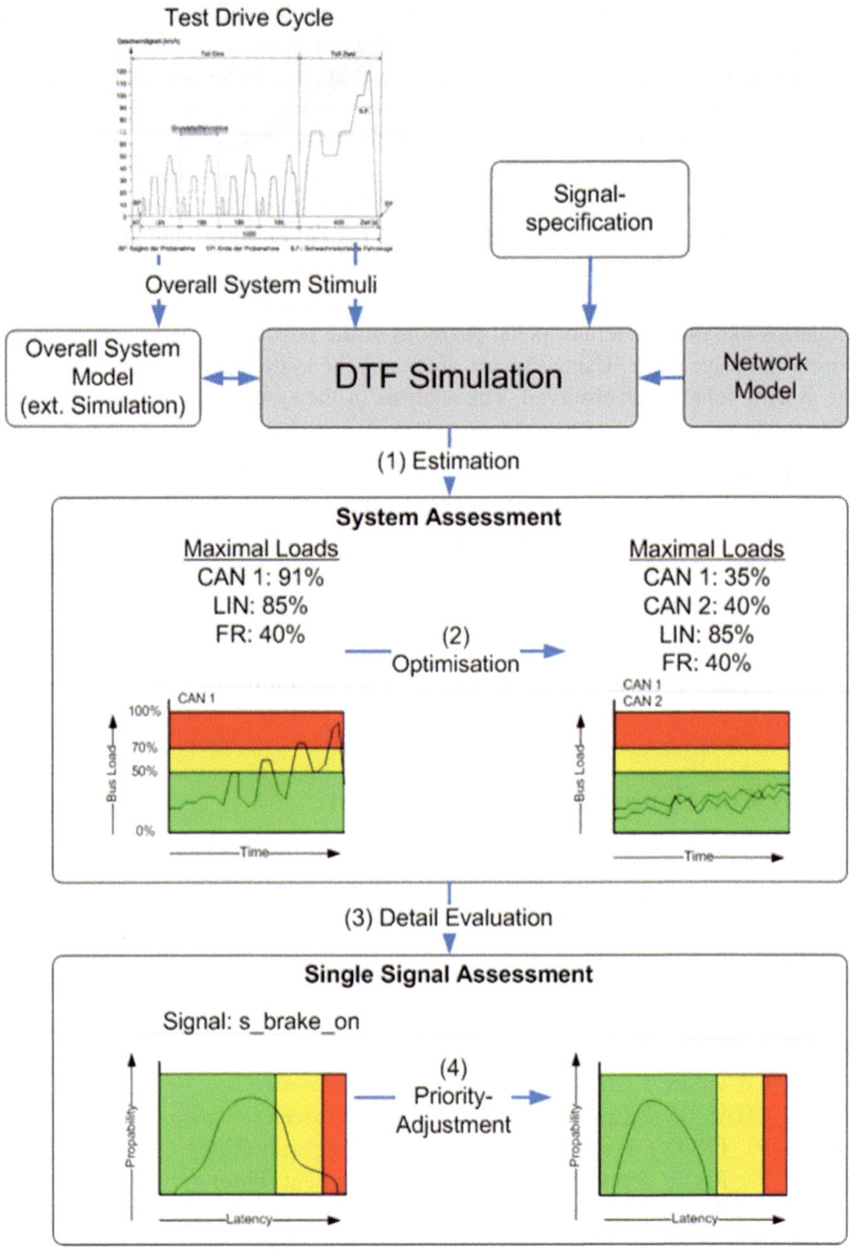


**Fig. 1** Control system performance analysis—principle of operation

Figure 2 shows the DTF model of the control system. Each segment contains 10 inputs, each connected to one *TX ECU* (TX ECUs send frames on the bus), a communication bus (based on FlexRay for the powertrain segments and on CAN

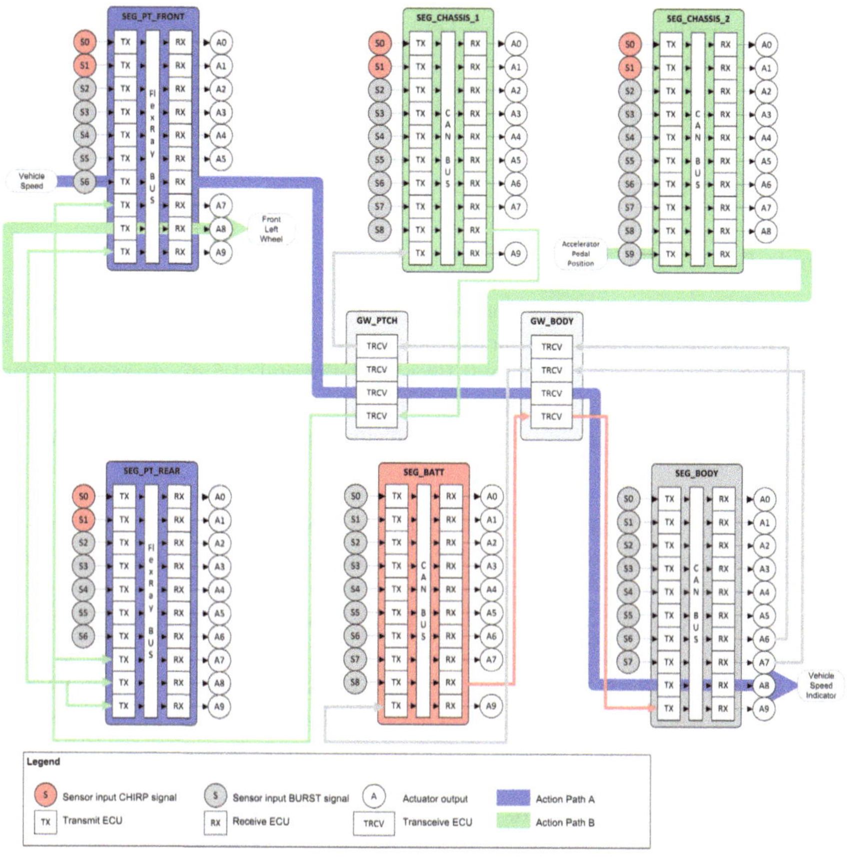


**Fig. 2** City car control system with action path A (*blue*) and action path B (*green*)

for all other segments), and 10 *RX ECUs* (RX ECUs receive frames from the bus), each connected to one output. A segment input is either connected to a sensor or to a gateway output, a segment output is either connected to an actuator or to a gateway input.

For the assessment, we consider two time-critical action paths:

- **Action Path A**. Over this action path, the vehicle speed is communicated, from the speed sensor *S6* in *SEG_PT_FRONT* to the vehicle speed indicator *A8* in *SEG_BODY*.
  The maximum latency for Action Path A shall be 40 ms. This requirement is critical because the signal is used for the cruise control regulator circuit.
- **Action Path B**. Over this action path, the accelerator pedal position is communicated, from the accelerator pedal sensor *S9* in *SEG_CHASSIS_2* to the front left wheel actuator *A8* in *SEG_PT_FRONT*.

The maximum latency for Action Path B shall be 30 ms. This requirement is critical because the signal is used for the motor speed control regulator circuit.

## 3.1 Model Parameters

**Stimuli**. For the stimulation of the sensors, two types of signals are used:

- **Burst** The burst signal has a constant period of 20 ms and a constant value 10.
- **Chirp** The chirp signal has a varying period. Starting with a period of 10 ms, the period is iteratively shortened by 0.1 ms until a period of 5 ms is reached. From this point in time, the chirp signal behaves like a burst signal with a constant period of 5 ms. That means that the chirp signal periods are as follows: 10, 9.9, 9.8,…, 5 ms.

Figure 2 shows which stimuli are assigned to which sensors in the six segments.

**Message allocation**. Figure 3 (CAN message priorities/FlexRay static slots) shows the assignment of message priorities to segment inputs. Data from segment inputs are transformed into messages (CAN or FlexRay frame) by the TX ECUs. In CAN segments, the messages are sent on the CAN bus with the assigned message priority (which is e.g. 10 for Segment input 0 in segment *SEG_BODY*). In FlexRay segments, the messages are sent on the FlexRay bus in the given slot in the static segment (which is e.g. 1 for segment input 1 in segment *SEG_PT_FRONT*).

**Bandwidth assignments**. For the different segments, Fig. 3 (Bus protocols and bandwidth) shows the settings for the bus protocols and the bandwidths.

## 3.2 Requirements and Conditions

For the simulation experiments, we define the following requirements and conditions:

**Requirements**

- In all segments, the maximum bus load shall be **below 70** %.
- The maximum control signal latency for Action Path A shall be **40** ms.
- The maximum control signal latency for Action Path B shall be **30** ms.

**Conditions**

- The duration of a simulation experiment is one second real-time ($10^6$ simulation ticks).
- The payload length of all messages is 8 bytes.
- Each message transports one control signal.

First Assessment

Bus protocols and bandwidth

| Segment | Protocol | Bandwidth |
|---|---|---|
| SEG_BODY | CAN | 250kbit/s |
| SEG_BATT | CAN | 125kBit/s |
| SEG_CHASSIS_1 | CAN | 250kbit/s |
| SEG_CHASSIS_2 | CAN | 250kbit/s |
| SEG_PT_FRONT | FlexRay | 1Mbit/s |
| SEG_PT_REAR | FlexRay | 1Mbit/s |

CAN message priorities / FlexRay static slots

| Sensor | BODY | BATT | CHASSIS_1 | CHASSIS_2 | PT_FRONT | PT_REAR |
|---|---|---|---|---|---|---|
| S0 | 10 | 10 | 10 | 10 | 0 | 0 |
| S1 | 20 | 20 | 20 | 20 | 1 | 1 |
| S2 | 30 | 30 | 30 | 30 | 2 | 2 |
| S3 | 40 | 40 | 40 | 40 | 3 | 3 |
| S4 | 50 | 50 | 50 | 50 | 4 | 4 |
| S5 | 60 | 60 | 60 | 60 | 5 | 5 |
| S6 | 70 | 70 | 70 | 70 | 6 | 6 |
| S7 | 80 | 80 | 80 | 80 | 7 | 7 |
| S8 | 90 | 90 | 90 | 90 | 8 | 8 |
| S9 | 100 | 100 | 100 | 100 | 9 | 9 |

Second Assessment

Bus protocols and bandwidth

| Segment | Protocol | Bandwidth |
|---|---|---|
| SEG_BODY | CAN | 125kbit/s |
| SEG_BATT | CAN | 125kBit/s |
| SEG_CHASSIS_1 | CAN | 250kbit/s |
| SEG_CHASSIS_2 | CAN | 250kbit/s |
| SEG_PT_FRONT | FlexRay | 1Mbit/s |
| SEG_PT_REAR | FlexRay | 1Mbit/s |

CAN message priorities / FlexRay static slots

| Sensor | BODY | BATT | CHASSIS_1 | CHASSIS_2 | PT_FRONT | PT_REAR |
|---|---|---|---|---|---|---|
| S0 | 10 | 10 | 10 | 10 | 0 | 0 |
| S1 | 20 | 20 | 20 | 90 | 1 | 1 |
| S2 | 30 | 30 | 30 | 30 | 2 | 2 |
| S3 | 40 | 40 | 40 | 40 | 3 | 3 |
| S4 | 50 | 50 | 50 | 50 | 4 | 4 |
| S5 | 60 | 60 | 60 | 60 | 5 | 5 |
| S6 | 70 | 70 | 70 | 70 | 6 | 6 |
| S7 | 80 | 80 | 80 | 80 | 7 | 7 |
| S8 | 90 | 90 | 90 | 20 | 8 | 8 |
| S9 | 100 | 100 | 100 | 100 | 9 | 9 |

**Fig. 3** Simulation results before and after optimization cycle

- The FlexRay communication cycle duration is 5 ms.
- In FlexRay segments, frames are only sent in the static segment. The static segment has a duration of 2.5 ms. Each slot in the static segment has a duration of 0.25 ms.
- In CAN segments, only data frames in base frame format are sent.

## 3.3 *First Assessment*

A simulation experiment using the settings shown in Fig. 3 (First assessment) delivers the following results:

### 3.3.1 System Assessment

The following bus loads can be observed in the segments that are involved in the action paths:

- **SEG_PT_FRONT Segment (FlexRay)**. Despite of the low bandwidth of 1Mbit/s, the bus load of the *SEG_PT_FRONT* segment is well below 20 %. This is due to the fact that the FlexRay communication cycle is 5 ms, while the duration of the static segment is 2.5 ms. Frames are only sent in the static segment, the dynamic segment is not used. Also, the payload length of 8 bytes is rather small for FlexRay frames.
- **SEG_BODY Segment (CAN)**. The bus load of the *SEG_BODY* segment is about 25 %, using a bandwidth of 250 kBits/s. This is about a third of the maximum allowed bus load of 70 % and gives room for possible reduction of the bandwidth.
- **SEG_CHASSIS_2 Segment (CAN)**.The bus load of the *SEG_CHASSIS_2* segment is about 50 % and clearly below the maximum allowed bus load of 70 %.

A first assessment of the bus load shows that all requirements are fulfilled. Further, the bus load in the *SEG_BODY* segment is so low that a reduction of the bandwidth can be taken into consideration.

### 3.3.2 Signal Assessment

Figure 3 shows the latency over simulation time and the latency distribution for the vehicle speed control signal (Action Path A) and the accelerator pedal position control signal (Action Path B).

- **Action Path A**. The signal latency is well below 40 ms, thus fulfilling the requirement.
- **Action Path B**. The signal latency is exceeding 30 ms, thus violating the requirement.

### 3.3.3 Summary

The first assessment delivers the following results:

- All bus load requirements are fulfilled. The bus load of the *SEG_BODY* segment is particularly low.
- The control signal latency requirement for Action Path A is fulfilled.
- The control signal latency requirement for Action Path B is not fulfilled.

## 3.4 *Optimization Cycle*

From the observations made in the first assessment, the following optimization actions are derived:

- To improve the bus resource utilization in the *SEG_BODY* segment, the bandwidth is reduced from 250 to 125 kBit/s. This will also reduce the electromagnetic interference (EMI) induced by the CAN bus. However, this action will have an impact on the vehicle speed control signal latency due to the longer transmission time over the CAN bus of the *SEG_BODY* segment.
- The accelerator pedal position control signal latency has to be reduced. This will be done by changing the priority of the message for the accelerator pedal position in the *SEG_CHASSIS_2* segment. The original message priority is 90, which is the second lowest message priority in the *SEG_CHASSIS_2* segment. By raising the message priority up to 20 (which is the second highest one in the *SEG_CHASSIS_2* segment), the message transmission time is reduced, resulting in a reduction of the overall accelerator pedal position control signal latency.

## 3.5 *Second Assessment*

A simulation experiment using the settings shown in Fig. 3 (Second assessment) delivers the following results:

### 3.5.1 System Assessment

Figure 3 (Bus protocols and bandwidth) shows the used bus protocols and the bandwidths for the six segments of the system. The bandwidth of the *SEG_BODY* segment CAN bus has been reduced from 250 to 125 kBit/s.

This optimization measure raises the bus load from 25 % to about 50 %. The bus load requirement can be fulfilled while reducing the original bandwidth. The bus loads in the *SEG_PT_FRONT* and the *SEG_CHASSIS_2* segments are not affected by the optimization actions and remain unchanged.

### 3.5.2 Signal Assessment

Now we have to consider the effects of the optimization measures to the control signal latencies. As argued in Sect. 3.4, a reduction of the bandwidth of the *SEG_BODY* segment CAN bus results in extended latency for the vehicle speed control signal as shown in Fig. 3.

- **Action Path A.** Compared to the first assessment, the signal latency has increased from 35 to 39 ms. However, this latency is below 40 ms, thus the requirement is still fulfilled.
- **Action Path B.** Compared to the first assessment, the signal latency has remarkably improved from 32 to 27 ms. After the message priority adaptation described in Sect. 3.4, the accelerator pedal position control signal now fulfills the requirement.

### 3.5.3 Summary

The second assessment after the optimization measures described in Sect. 3.4 delivers the following results:

- All bus load requirements are fulfilled. A reduction of the *SEG_BODY* segment CAN bus bandwidth leads to a better utilization of the communication bus, while preserving the bus load requirement.
- The requirement for the vehicle speed control signal latency (Action Path A) is fulfilled.
- The requirement for the accelerator pedal position control signal latency (Action Path B) is fulfilled.

## 4 Conclusion

This paper presents a simulation based control network assessment and optimization workflow based on a practical example, namely the analysis of an electronic control system for an electric city car. In the simulation experiments, we have validated two time-critical control signals. The assessment of the proposed POLLUX vehicle network architecture shows that the architecture has sufficient reserves for extension. The following arguments support that claim:

- **Network bandwidth**. The bandwidths chosen for the six segments are at the lower limit. This is especially true for the FlexRay segments with a bandwidth of 1 Mbit/s. Under the given assumptions concerning the segment input stimuli, even with these low bandwidths, the bus load in all segments is below 50 %. While in the FlexRay segments the bus load is constant (as long as only the static segment is used), the bus load observed in the CAN segments provides reserves to fulfill the bus load requirements in case of frame losses and the need for re-transmission of control signals, without the need to increase the network bandwidth.
- **Network utilization**. The utilization of the available bandwidth is not "optimal". In the proposed configuration, each CAN and each FlexRay frame transports only one control signal. Packing more control signals into one frame

will lead to a better utilization of the network resources. Additionally, the chosen payload data length in the FlexRay segments is considerably low (8 bytes), compared to the maximum FlexRay frame payload length of 254 bytes. The dynamic segment is not used at all. So from the point of view of network utilization there is also considerable room for optimization, without the need to increase the network bandwidth.
- **Network load**. The observed network load also provides reserves with regard to the adding of ECUs to the different segments. This is meaningful for the expandability of the control system. While adding components to the control system will increase the bandwidth requirements and increase the network load, the observed reserves show that there is also room for adding ECUs to the segments, providing new functionalities, without the need to re-design the control network architecture.

This application example shall illustrate the principle of operation and shall deliver a proof of concept. In a "real" control system, there exist more than two critical control signal paths whose requirements have to be validated. Thus, the validation effort will be higher, presumably requiring lots of network modifications and subsequent simulation runs. However, the simulation based validation approach allows the detection of requirement violations in early development phases, and shall help to avoid costly re-design activities in later development phases. Considering only three parameters (network bandwidth, control signal latencies and segment input stimuli frequencies) essential properties of the control system can be assessed without the need to consider functional aspects of the system. Therefore, this method is also a step towards a generic approach for vehicle control network design. Providing the methodology for description of vehicle network architectures and the possibility for modification of essential network parameters, this approach shall enable manufacturers and suppliers to develop control systems without the need to share corporate know-how with third parties, like tool developers.

**Acknowledgments** This work has been supported by the European Union ARTEMIS Project POLLUX (JU Grant Agreement number 100205).

## References

1. Zimmermann W, Schmidgall R (2010) Bussysteme in der Fahrzeugtechnik—Protokolle,Standards und Softwarearchitektur. Vieweg + Teubner, 4. Auflage
2. Etschberger K (ed) (2001) CAN controller area network—Grundlagen, Protokolle, Bausteine. Anwendungen. Hanser Fachbuch, Deutschland
3. FlexRay Communications System Protocol Specification Version 3.0.1. Available from http://www.flexray.com
4. Ortmeier F, Daniel P (eds) Computer safety, reliability, and security. LNCS 7613. Springer, Berlin, pp 373–381

5. Banks J, Carson J, Nelson B (2000) Discrete-event system simulation. Prentice Hall, Englewood Cliffs
6. Hanzlik A, Kristen E (2011) DTF—A simulation environment for communication network architecture design for the next generation of electric cars. In: Meyer G, Valldorf J (eds) Advanced microsystems for automotive applications 2011, pp 259–267

# Smart Sensor Networks for Structural Health Monitoring

**Thilo Bein and Dirk Mayer**

**Abstract** Condition based maintenance (CBM) concepts are widely used in the industry and are also applied to modern vehicles. Since structural components are becoming increasingly complex, e.g. due to lightweight composite materials, it is useful to apply CBM also to those parts. This requires the application of sensors to the components. The permanent instrumentation with systems for structural health monitoring is yet implemented for high performance structures, e.g. in military aircraft. Those rather expensive systems are not directly applicable to a road vehicle. This paper discusses ways to adapt the idea of structural health monitoring to vehicles using smart sensor networks, including signal processing, low-cost hardware platforms and energy supply of wireless sensor nodes.

**Keywords** Structural health monitoring · Damage detection · Lightweight structures · Smart sensor networks · Energy harvesting

## 1 Introduction

Modern vehicles are designed with respect to loads derived from selective measurements on prototypes following pre-defined usage profiles [1]. This should enable a cost-effective and lightweight design. Of course, also safety aspects are regarded. Therefore, some misuse and worst case scenarios are included in the design, and structural parts are always oversized by a certain safety margin.

---

T. Bein (✉) · D. Mayer
Structural Durability and System Reliability, Fraunhofer Institute, Darmstadt, Germany
e-mail: thilo.bein@lbf.fraunhofer.de

D. Mayer
e-mail: dirk.mayer@lbf.fraunhofer.de

J. Fischer-Wolfarth and G. Meyer (eds.), *Advanced Microsystems for Automotive Applications 2013*, Lecture Notes in Mobility, DOI: 10.1007/978-3-319-00476-1_36,

Normally, this margin is never exploited, and at the end-of-life of the vehicle, parts are recycled which could be used for a longer period.

From many industrial sectors, the benefits of condition based maintenance have been proven, which aim at optimizing maintenance intervals and using less spare parts [2]. Those are most effective, if the actual condition of the part under consideration is monitored automatically by a sensor system. For example, in today's passenger cars oil quality is monitored automatically by a sensor and the driver is advised to exchange the oil when necessary and not at pre-defined intervals.

Regarding lightweight design, structural parts will become more complex, e.g. by the use of hybrid components from fiber-reinforced plastic materials and metals. By monitoring the condition of those parts for every particular vehicle, maintenance intervals could be adjusted to the individual usage profile.

For structural components, permanent *structural health monitoring* (SHM) or *health and usage monitoring* (HUMS) has been considered mainly for infrastructure objects [3], aircrafts [4], and ship hulls [5]. However, the monitoring systems for a military airplane are not supposed to be directly applicable for instrumentation of a passenger car in series production. Solutions for a road vehicle have to be more compact and low cost in order to generate an economic benefit from the instrumentation.

Meanwhile, embedded systems have been improved towards small and cheap units with a number of possibilities for communication and signal processing, which enabled the development of *smart sensor networks*. Those can successfully be applied to the tasks of structural health monitoring, e.g. for bridges [6]. However, several challenges have to be met when developing a smart sensor network for SHM in a mobile application.

In this paper, several relevant topics will be touched and examples for investigations on solutions will be discussed. This includes the specific requirements for the system design of a smart sensor network for SHM, the utilization of cost efficient hardware and the energy supply for wireless sensor nodes. Finally, a prototype application of a smart sensor network for a compact mobile system is described [7].

## 2 System Design with Smart Sensor Networks

An important point is the type of data that has to be acquired and processed for SHM. Structural loads change quite quickly with time compared to the status of oil quality, i.e. monitoring of vibrational loads is attended by high data rates of several hundreds of Hertz. Further, data has to be acquired at a number of structural parts in order to cover all critical spots of a larger structure. Thus, decentralized signal processing and analysis have to be implemented on the nodes of the smart sensor network in order to avoid the use of an expensive broad band communication network. However, this raises the challenge to realize a signal analysis on small microcontroller platforms. For example, many vibration analyses require the correlation of the sensor signals. The autocorrelation of a segment of $N$ samples of the signal $x$ is calculated by:

$$R_{XX}(\tau) = \frac{1}{N} \sum_{n=1}^{N-1-n} x(t_n + \tau)x(t_n).$$

For larger values of $N$, this procedure is computationally quite complex. Alternatively, an algorithm using Fourier transforms (FFT) and inverse transforms can be used, but this still seems to be not a suitable solution for the implementation on a sensor node. In contrast, the Random Decrement (RD) method represents a simple, yet effective method for the estimation of correlation functions. Originally being invented for the damage detection of aerospace structures in wind tunnel tests [8], it was also used in SHM applications [9]. The RD technique estimates a scaled auto-correlation function of a system under random input loads by averaging time series which are taken when a given trigger condition is fulfilled. Cross-correlation functions between two system outputs can be estimated by generating averaged time series by still defining a trigger condition for the first output $x$, but synchronously taking time series of a second output $y$. If a simple level crossing with level $a$ of the signal $x$ is assumed, the mathematical expression of the RD technique reads:

$$D_{XX}(\tau) = \frac{1}{N}\sum_{n=1}^{N} x(t_n + \tau)\bigg|\, x(t_n) = a \quad D_{YX}(\tau) = \frac{1}{N}\sum_{n=1}^{N} y(t_n + \tau)\bigg|\, x(t_n) = a.$$

The scaling factor is proportional to the trigger level and the variance $\sigma_X^2$ of the signal:

$$R_{XX}(\tau) = D_{XX}(\tau)\frac{\sigma_X^2}{a}.$$

Due to the averaging process, also a significant rate of data compression is gained. A system layout for a networked RD estimation was derived in Friedmann et al. [10], see Fig. 1. The only data transmitted in the sensor network in real-time is the trigger event, which saves communication bandwidth.

As a general consideration, when processing mechanical vibrations in a smart sensor network, the transmission of real-time data seems not favorable. As much signal processing as possible should be implemented on the network nodes.

## 3 Hardware for Prototyping

Smart sensor nodes consist of a front-end for data acquisition from a sensor, e.g. an accelerometer, a signal processing unit, and a communication unit. Usually, the processing unit is a microcontroller suited for low-energy applications. Though

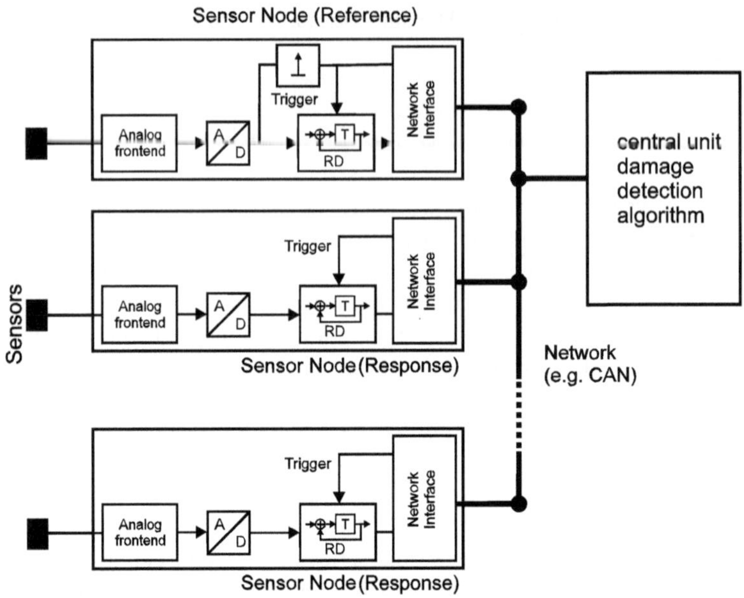


**Fig. 1** System layout for the distributed vibration signal processing [10]

computational power might be small, such a platform can still implement a vibration analysis, e.g. based on the RD technique described above [10]. Those controllers can be used for the development of a low-cost monitoring system which can be produced in high volumes. However, also functional prototyping and test is possible with frameworks like *Arduino* [11] or *Waspmote* [12], using microcontrollers like ATMega328P or a 32-bit ARM Cortex-M3. *Shields* to enable network communication, data acquisition and storage are readily available. Due to the open-source framework, also the development of tailored boards and sensor frontends is easily possible.

For vibration monitoring, mostly acceleration signals are relevant. A potential low-cost solution for this task is the technology of micro-electro-mechanical systems (MEMS). These are used in huge volumes in vehicles and consumer products. For the prototyping of monitoring systems, a small circuit board with *ADW22035Z* for mono-axial and *ADXL325* for tri-axial measurements (both from *Analog Devices*) and housing have been developed for the application of structural parts with a minimum of needed space and weight (Fig. 2). The housing is realized by selective laser melting (SLM), which allows for the fast and cost-efficient production of small lots of parts. To prevent water and soil from the sensor inside, the housing is completely filled with epoxy. A back-to-back calibration with commercially available quartz sensor from *PCB Piezotronics* shows a sufficient precision over the relevant frequency range (Figs. 3, 4).

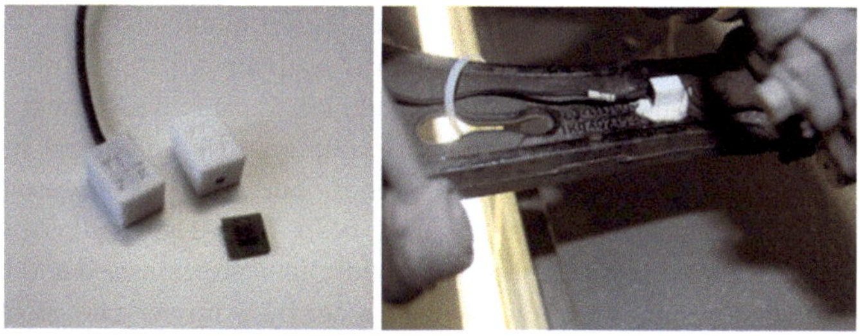

**Fig. 2** MEMS accelerometer on the circuit board (*left*) and with housing applied to a suspension part (*right*)

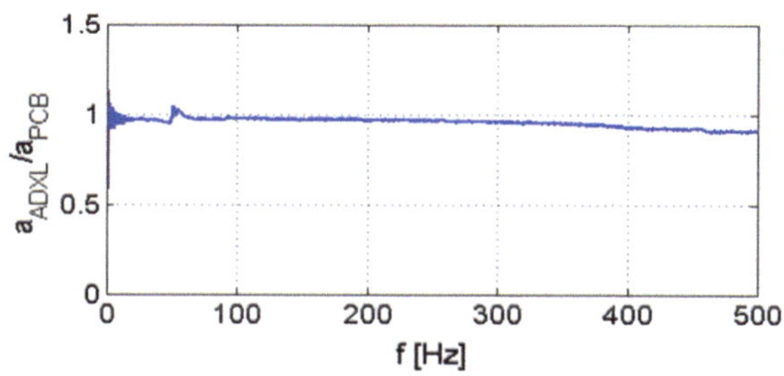


**Fig. 3** Back-to-back test of the designed MEMS based low-cost sensor against a quartz accelerometer

**Fig. 4** Smart sensor node with frontend for accelerometer and Bluetooth wireless communication (*left*) and cased into a SLM housing (*right*)

# 4 Energy Supply of Wireless Sensors

The wireless implementation of a smart sensor enables the installation e.g. in suspension parts or rotating shafts in the powertrain. Mostly, power supply is implemented by batteries. However, for a fully autonomous operation, energy has to be generated directly at the smart sensor. This *energy harvesting* can use different sources of environmental energy. For a vibrating structure, obviously the mechanical oscillations are an interesting source of energy [13].

A widely used design comprises a mechanical oscillator with an integrated electromechanical transducer. Besides electrodynamic transducers [14], mainly piezoelectric elements are used [15]. The cantilevered beam with piezoceramic layers which is applied to a structure has become a common design [13]. In Fig. 5, a sketch of such a harvester and the mechanical equivalent system is shown for the case of a piezoelectric transducer with capacitance $C_p$ and an electrical load $R_L$. It is worth notice, that the electro-mechanical coupling $\alpha$ is influenced by the coupling coefficient of the piezoelectric material, but also by the geometry and positioning of the transducer on the vibrating beam [16]. The oscillator is coupled to a point of the structure, where significant vibrations are expected and is tuned to a suitable frequency, e.g. a structural resonance, by a proper choice of stiffness $k$ and mass $m$.

A prototype was tested in a self-powered sensor system for hotbox detection on a freight car [12]. The harvested energy is sufficient to load a capacitor of 1.5 F to a voltage level of 3.6 V and enable a complete cycle of temperature measurement and wireless communication every 15 min.

For the supply of a vibration monitoring system, which would have to collect and process far more data, the system still has to be improved. It is known, that the amount of harvested power is proportional to the mass $m$ [14]. If the ratio of $m$ to the effective vibrating mass of the support structure becomes significant, the energy harvester will also work as a tuned vibration absorber (TVA). Since a number of tuned absorbers are commonly used in vehicles, they might be also used to supply sensor nodes.

As an example, an oscillator attached to a single-degree-of-freedom (SDOF) system with a mass $M = 1$ kg and a resonance of 50 Hz is studied [17]. First, the oscillator is tuned to the resonance and the mass ratio is chosen to 0.01 to simulate an energy harvester. The potentially harvested power is estimated by the dissipated power in the viscous damper (damping ratio $\theta = 0.01$). The vibrations of the host structure are nearly unaffected (Fig. 6). If the mass ratio $\mu$ = m/M is chosen to 0.2, and resonance and damping are chosen following the well-known tuning rules for TVAs [18], a clear effect is observed. However, in both cases a similar amount of energy is dissipated in the damper, which demonstrates the potential for the integration of harvesting and vibration absorber function.

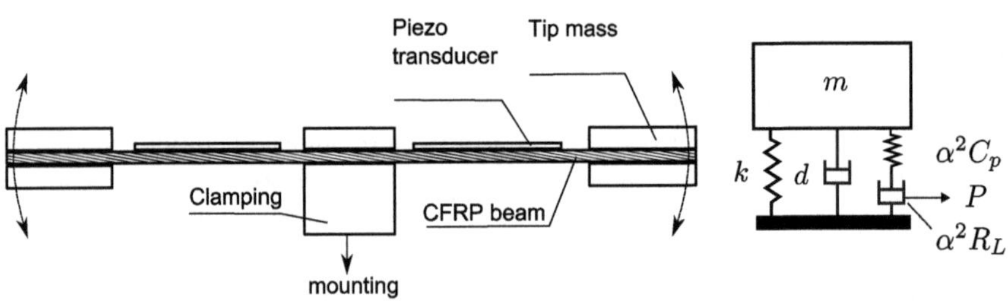


**Fig. 5** Energy harvesting system with cantilever beam and piezo transducer (*left*), electromechanical equivalent (*right*)

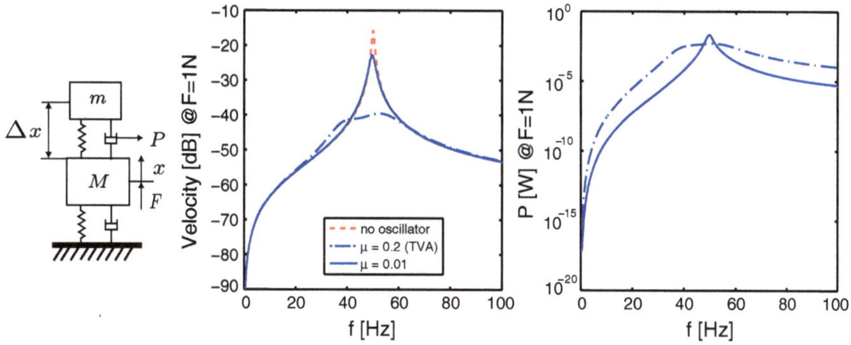


**Fig. 6** Energy harvesting oscillator attached to a SDOF system (*left*), effect on the vibration level (*middle*) and dissipated power in the damper (*right*) [17]

# 5 Application Example

The sensor system described in "A Flexible Environment Perception Framework for Advanced Driver Assistance Systems" was used to develop a smart sensor network for damage detection in automotive suspensions [7]. The system uses accelerometers distributed among the car (Fig. 7). One at each wheel carrier and one at the center of gravity (COG). A local damage, e.g. at a damper of the suspension will alter the characteristics of the vibrations at the sensor nearby, while the vibrations at the other points will remain unaffected. Changing road conditions or loading of the car are supposed to influence the accelerations at all wheels. By calculating normalized cross-correlations, between the sensor signals from the wheel carriers and from the COG, it is possible to decide whether damage is present at a specific wheel. The correlation coefficient is derived from the autocorrelation $S_{pp}$ measured at the COG and the cross-correlation $S_{pw}$ between the signals taken at the COG and the wheel:

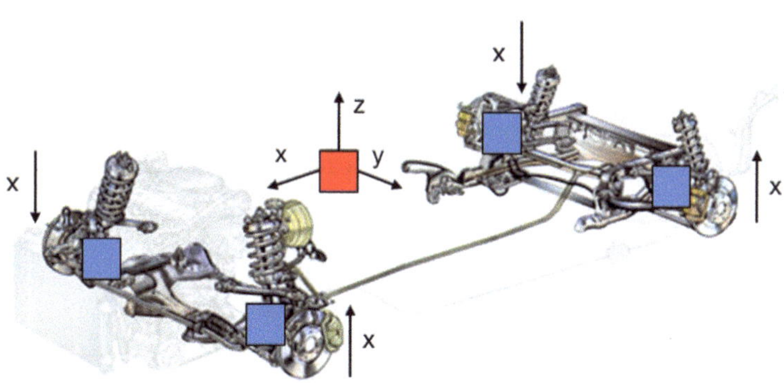


**Fig. 7** Sensor positions for the implementation of damage detection in the suspension, *blue* wheel mountings, *red* COG [7]

**Fig. 8** RC car for first tests and instrumentation [7]

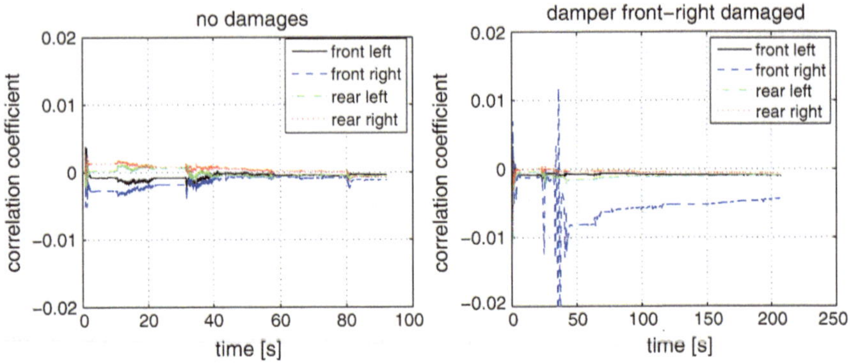


**Fig. 9** Calculated correlation coefficients for each sensor, undamaged state (*left*), damaged damper (*right*) [7]

$$Cc_{pw}(t) = \frac{S_{pp}}{S_{pw}}.$$

As mentioned above, such an algorithm can be efficiently implemented with a wireless network of smart sensors.

To evaluate the potential of the system, a scaled radio controlled (RC) car, which still has a realistic suspension, was instrumented with the sensor network (Fig. 8). To simulate damage, oil was removed from one damper. For both states, a test track was passed to generate operational vibrations which were acquired and analyzed by the sensor network. When comparing the calculated correlation coefficients from both the healthy and the damaged state the damaged damper can be clearly identified (Fig. 9).

## 6 Summary and Conclusions

As modern materials and components of vehicles are more complex and lightweight, SHM becomes a relevant part of condition based maintenance concepts. In contrast to SHM for infrastructure or aerospace, the use of low cost smart sensors is essential to enable cost efficient solutions. The utilization of rather weak hardware and low communication bandwidth requires distribution of damage detection algorithms in the network and the application of efficient signal processing. For a fully autonomous operation, energy harvesting, e.g. from vibrations, is an interesting solution. To generate the needed amount of energy for a vibration monitoring system, an idea is to integrate the harvesting into a tuned vibration absorber. Future work will have to include the application of energy efficient sensors, e.g. low power accelerometers, the optimization of the energy harvesting systems, and the integration of the systems into long-term tests.

**Acknowledgments** Parts of the research presented was conducted within the FP7 project *MoDe - Maintenance on Demand*, (grant agreement no. 233890), funded by the European Commission, *LOEWE AdRIA*, funded by the German federal state of Hesse (grant no. III L 4 518/ 14.0004) and *EA-TSM*, funded by the German Ministry of Education and Research (contract no. W40027).

This financial support is gratefully acknowledged. The authors are also indebted to the project teams for their contributions to the work and to the LBF scientists Dipl.-Ing. Thorsten Koch and Dipl.-Ing. Chip Sabirin for the realization of the MEMS based accelerometers.

# References

1. Haibach, E. Betriebsfestigkeit, 3rd ed., Springer, 2006, ISBN: 3540293639
2. Khazraei, K (2011) Design, organization and implementation of a methods pool and an application systematics for condition based maintenance, PhD Thesis, Technische Universität Dortmund, 2011
3. Carden EP, Fanning P (2004) Vibration based condition monitoring: a review. Struct Health Monit 3:355–377
4. Gunther G (2001) Loads monitoring and HUMS, in RTO Lecture Series 218. Structural and Other Subsystem Aspects, Aging Aircraft Fleets
5. Slaughter S et al. (1997) State of the art in hull response monitoring systems, report SSC 401. Ship Structure Committee, Washington D.C
6. Lynch JP (2007) An overview of wireless structural health monitoring for civil structures. Philos Trans R Soc. Math, Phys Eng Sci 365(1851):345–372
7. Buff H, Kloepfer J, Hansmann J (2013) Automatic damage detection in vehicle's shock absorbers. ICEDYN, Sesimbra
8. Cole HA (1973) On-line failure detection and damping measurement of aerospace structures by random decrement signatures, NASA Report NASA-CR-2205
9. Friedmann A, Mayer D, Koch M, Siebel T (2011) Monitoring and damage detection in structural parts of wind turbines, in advanced topics of wind power, InTech, 207–232
10. Friedmann A, Mayer D, Kauba M (2010) An approach for decentralized mode estimation based on the Random Decrement method. Shock Vibr 17:579–588
11. Russell L, Steele AL, Goubran R (2012) Low-cost, rapid prototyping of IMU and pressure monitoring system using an open source hardware design. In: IEEE international instrumentation and measurement technology conference (I2MTC), 2695–2699
12. Koch M, Kurch M, Mayer D (2012) On a methodical design approach for train self-powered hot box detectors. In: First international conference on railway technology: research, development and maintenance
13. Elvin NG, Lajnef N, Elvin AA (2006) Feasibility of structural monitoring with vibration powered sensors. Smart Mater Struct 15:977–998

14. Elliott S, Saba R, Baumann O (2008) Vibration power harvesting from head motion. Eurodyn, Southampton
15. Anton S, Sodano H (2007) A review of power harvesting using piezoelectric materials (2003–2006). Smart Mater Struct 16(3):1–21
16. Kaal W, Herold S, Kurch M (2009) Design strategies of energy harvesting devices with piezoelectric transducers. In: 4th ECCOMAS thematic conference on smart structures and materials, Porto
17. Kurch M, Mayer D (2011) On modeling of energy self-sufficient vibration absorber sensor modules. In: 5th ECCOMAS thematic conference on smart structures and materials, Saarbrücken
18. Den Hartog J (1985) Mechanical vibrations. Dover Publications, New York

# Author Index



J. Fischer-Wolfarth and G. Meyer (eds.), *Advanced Microsystems for Automotive Applications 2013*, Lecture Notes in Mobility, DOI: 10.1007/978-3-319-00476-1,

# Subject Index




J. Fischer-Wolfarth and G. Meyer (eds.), *Advanced Microsystems for Automotive Applications 2013*, Lecture Notes in Mobility, DOI: 10.1007/978-3-319-00476-1,

MIX
Papier aus verantwortungsvollen Quellen
Paper from responsible sources
FSC® C105338

If you have any concerns about our products,
you can contact us on
**ProductSafety@springernature.com**

In case Publisher is established outside the EU,
the EU authorized representative is:
**Springer Nature Customer Service Center GmbH**
**Europaplatz 3, 69115 Heidelberg, Germany**

Printed by Libri Plureos GmbH
in Hamburg, Germany